Bio-IT and AI

As the fields of Information Technology (IT) and Artificial Intelligence (AI) rapidly evolve, their potential to revolutionize biomedice, personalized healthcare, and drug discovery becomes increasingly evident. This series will provide comprehensive insights into how modern computational and IT methods, AI, generative AI, and machine learning (ML) techniques are being integrated into biomedicine and healthcare to advance research, diagnostics, and treatments.

The scope of the series includes, but is not limited to, the following:

- Fundamentals of Bio-IT and AI: Basic concepts, tools, theoretical and mathematical principles, methods, technical foundations, and software technologies and solutions;
- Latest developments in bioinformatics and systems biology research, including topics such as multi-omics data integration, applications of large language models, and network biology;
- AI in genomics and precision medicine: Exploring the use of AI and generative AI in analyzing genomic data, understanding genetic diseases, and the development of personalized treatment strategies;
- Machine learning, AI, and generative AI for drug discovery, focusing on how ML algorithms accelerate the process of drug discovery, from target identification to predicting drug responses;
- Biomedical and health informatics and data analytics, covering the use of big data and analytics, including patient data management, electronic health records (EHR) analysis, and predictive modeling for patient care.

Aim

The series is targeted at a broad audience that is keen on exploring both the fundamental theories of IT, AI and their practical uses in healthcare and biotechnology. It caters to various groups including academic researchers and students in the fields of biomedicine and computer science, specialists in IT and data science, AI and generative AI, HPC and clouds, healthcare professionals, and industry experts.

Arli Aditya Parikesit · Arif Nur Muhammad Ansori
Editors

Recent Computational Techniques in De Novo Drug Design

Springer

Editors
Arli Aditya Parikesit
Department of Biotechnology, School of Health and Life Sciences
i3L University
Jakarta, Indonesia

Arif Nur Muhammad Ansori
Postgraduate School
Universitas Airlangga
Surabaya, Indonesia

ISSN 3059-2895 ISSN 3059-2909 (electronic)
Bio-IT and AI
ISBN 978-3-032-20427-1 ISBN 978-3-032-20428-8 (eBook)
https://doi.org/10.1007/978-3-032-20428-8

This Springer imprint is published by the registered company Springer Nature Switzerland AG
The registered company address is: Gewerbestrasse 11, 6330 Cham, Switzerland

Preface

In the past decade, computer-based methods have significantly advanced rational drug design. De novo drug design now allows researchers to use computational tools to create compounds with specific properties. Progress in algorithms and models, including machine learning and molecular modeling, has supported these developments. The book *Recent Computational Techniques in De Novo Drug Design* gives a broad overview of the field. It highlights new algorithms, predictive modeling, and advanced computer simulations. Experts in theoretical chemistry, cheminformatics, structural biology, and artificial intelligence have contributed to this volume. The book also covers the latest progress in computer-assisted molecular discovery.

The chapters discuss different methods, such as generative models for molecular design, fragment and reaction design, enhanced sampling, and quantum optimization. The book also explains how deep learning, reinforcement learning, and cloud computing are making de novo design better. The chapters are elaborated as following:

Chapter 1: Fundamentals of Generative AI in Drug Development

Chapter 1 covers generative models like variational autoencoders, generative adversarial networks, and diffusion models and explains how they are used in molecular design. It shows how these models handle chemical data and create new structures for drug development.

Chapter 2: Computational Design of mRNA-Based Cancer Immunotherapies

Chapter 2 explains computational methods for designing mRNA-based cancer immunotherapies, focusing on sequence optimization, epitope selection, and structural stability. It shows how AI predicts antigens, optimizes codons, and checks immunogenicity to make treatments more effective and safer.

Chapter 3: Protein and Peptide Design with Generative AI

Chapter 3 looks at how generative AI helps design proteins and peptides, showing the links between sequence, structure, and function. It explains how deep learning and reinforcement learning are used to create new proteins and peptides with specific binding, stability, and bioactivity.

Chapter 4: Exploring Chemical Space with Generative Models

Chapter 4 explains how generative models help explore and expand chemical space. It covers how these models search for new options and focus on promising candidates, suggesting molecules that are diverse, easy to make, and biologically important.

Chapter 5: Prediction of Drug Likeness and Synthetic Accessibility Using AI: A Case Study on Potential Therapeutic Compounds from Plants for SARS-CoV-2 Treatment

The chapter shows how cheminformatics descriptors, predictive models, and filters are used to choose plant-based compounds for preclinical testing against SARS-CoV-2.

Chapter 6: Challenges in AI-Generated Drug Candidates

Chapter 6 looks at the main challenges of AI-generated drug candidates, including data quality, model bias, and the difficulty of understanding how models work. It also covers issues like off-target effects, drugs that are hard to make, and related regulatory and ethical concerns.

Chapter 7: AI-Driven Network Pharmacology in Natural Product-Based Drug Design: A Systems Bioinformatics Perspective

Chapter 7 looks at AI-driven network pharmacology for designing drugs from natural products, using a systems bioinformatics approach. It explains how multi-omics data, interaction networks, and machine learning work together to clarify polypharmacology and how complex phytochemicals work.

Chapter 8: Applications of Generative AI in Small Molecule Drug Discovery

Chapter 8 reviews how generative AI is used in small molecule drug discovery, from finding early hits to refining leads. It gives examples of how generative models work with docking, molecular dynamics, and ADMET prediction to improve drug candidates.

Chapter 9: Generative AI for Antimicrobial and Antiviral Drug Design

Chapter 9 looks at generative AI methods for making antimicrobial and antiviral drugs, explaining how these models consider resistance, conserved targets, and features of each pathogen to create strong anti-infective candidates.

Chapter 10: Integration of AI-Driven Drug Design in Industry and Academia

Chapter 10 reviews molecular modeling and AI techniques for designing new drugs from scratch, focusing on optimizing across different scales and parameters. It highlights how physics-based methods, data-driven models, and multi-objective algorithms work together to create strong drug candidates.

Chapter 11: Trends of Open-Source Tools and Frameworks for AI Drug Design

The chapter covers the growing ecosystem of open-source tools, libraries, and frameworks for AI-driven drug design, evaluating their features, compatibility, and limits, and giving advice for building reliable and transparent computational pipelines.

Chapter 12: Future Horizons: Collaboration of Advanced Pharmaceutical Labs and Computational Sciences

Chapter 12 explores how advanced pharmaceutical labs and computational sciences can work together, using digital twins, automated synthesis, and closed-loop cycles for designing, making, testing, and analyzing drugs. It emphasizes the need for teamwork across disciplines, shared data systems, and ethical oversight to guide the future of AI-driven drug discovery.

This book is intended for researchers and professionals in computational chemistry, bioinformatics, pharmacoinformatic, and structural biology. Readers will find practical methods and new approaches for computer-aided drug discovery that they can use in their own work. We hope it serves as a helpful reference for both current and future developments in the field. We thank the faculties and institutions of i3L University and the Department of Research and Community Service (RCS i3L University) for their support and collaboration. Herewith, we declare that the review and editing of this book were improved with the Grammarly Premium software. However, all editors and authors took full responsibility of the contents. We appreciate the dedication of everyone involved, which made this volume possible. We hope this book encourages more research in de novo design.

Jakarta, Indonesia — Arli Aditya Parikesit
Surabaya, Indonesia — Arif Nur Muhammad Ansori

Contents

Chapter 1
Fundamentals of Generative AI in Drug Development

Arli Aditya Parikesit, Arif Nur Muhammad Ansori, Viol Dhea Kharisma, and Muhammad Hermawan Widyananda

Abstract Drug discovery and development are one of the most challenging, costly, and time-consuming endeavours in modern science. 12–15 years and more than $2.5 billion may be required to progress a single lead compound from bench to clinic through traditional approaches. A very different paradigm, generative artificial intelligence (AI) is revolutionizing this process through accelerated navigation of chemical space, improved lead optimization speed, and new molecular design. Generative AI techniques, such as transformer-based models, generative adversarial networks (GANs), variational autoencoders (VAEs), and recurrent neural networks (RNNs), have previously unheard-of potential for de novo drug discovery, producing new chemical entities with desired therapeutic qualities. These approaches are complemented by knowledge-enhanced frameworks that integrate structured biological information to guide molecular generation. Alongside further maturation, data quality, interpretability of the model, and experimental evidence issues need to be

A. A. Parikesit (✉)
Bioinformatics Streaming, Department of Biotechnology, School of Health and Life Sciences, i3L University, Jakarta, Indonesia
e-mail: arli.parikesit@i3l.ac.id

A. N. M. Ansori · V. D. Kharisma · M. H. Widyananda
Drug and Vaccine Innovation Research Group, Virtual Research Center for Bioinformatics and Biotechnology, Surabaya, Indonesia
e-mail: arif.nma-17@fkh.unair.ac.id

V. D. Kharisma
e-mail: viol.dhea.kharisma-2022@fst.unair.ac.id

M. H. Widyananda
e-mail: mh.widyananda@ub.ac.id

A. N. M. Ansori
Postgraduate School, Universitas Airlangga, Surabaya, Indonesia

V. D. Kharisma
Department of Biology, Faculty of Science and Technology, Universitas Airlangga, Surabaya, Indonesia

M. H. Widyananda
Biosystem Study Center, Universitas Brawijaya, Malang, Indonesia

A. A. Parikesit and A. N. M. Ansori (eds.), *Recent Computational Techniques in De Novo Drug Design*, Bio-IT and AI, https://doi.org/10.1007/978-3-032-20428-8_1

resolved to realize generative AI's potential for revolutionary drug development and, through quicker release of better, more efficacious therapy, improve patient outcomes.

Keywords Generative AI · Drug discovery · GAN · Transformer-based models · Chemical space

1.1 Background and Motivation

1.1.1 The Crisis in Drug Discovery and Development

Even with the amazing advancements in genomics, molecular biology, and high-throughput screening technology, the pharmaceutical industry still faces significant challenges in bringing novel drugs to market. With an estimated 10^{60} possible drug-like molecules, the vast chemical space poses an impossible obstacle for conventional exploration techniques [1]. The overwhelming majority of these molecules remain unexplored, constituting what researchers refer to as the "dark chemical space" or "dark chemistry" [2].

These errors significantly impact the increased costs and delayed release of innovative treatments to the market. Adding to these difficulties is the growing difficulty in discovering new chemical scaffolds and therapeutic targets because many of the "low-hanging fruit" have already been picked, leaving more difficult and complicated targets for subsequent drug development initiatives [3, 4].

1.1.2 Cost and Time Constraints in Traditional Approaches

The conventional drug development method necessitates significant time and cost commitments. It usually takes 12–15 years and costs more than $1 billion to develop a new medicine from concept to commercial launch [5]. This amount is significantly larger, according to more recent estimates, at almost $2.5 billion for each successful medication [6]. The financial consequences of high failure rates and opportunity costs are also reflected in these startling costs, in addition to the direct expenditures of research and development.

Long development schedules present extra difficulties, especially when responding to new diseases or urgent public health requirements. The conventional method of drug development, which mostly depends on iterative experimental testing and optimization, is time-consuming by nature and frequently unable to keep up with changing medical needs. Every phase of the procedure, from lead optimization and clinical trials to target identification and validation, has the potential to be a bottleneck that delays the delivery of treatments that could save patients' lives [7,

8]. These inherent delays also have broader implications, as they influence which diseases receive priority for research investment.

The cost of drug development also influences research priorities. Rare diseases or ailments that are common in lower-income areas may be overlooked as resources are frequently allocated to diseases with larger markets and greater profit potential. More effective discovery techniques could help alleviate some of the inequalities in healthcare innovation caused by this economic reality [9, 10]. Addressing these limitations requires a paradigm shift toward approaches that reduce both the time and cost of drug discovery.

1.1.3 The Emergence of Computational Methods in Drug Design

Computational techniques have become useful instruments to improve the efficacy and efficiency of drug discovery in response to these issues. Pharmaceutical discovery pipelines in both academic and industrial contexts have increasingly incorporated computer-aided drug design (CADD) techniques [11]. By using these computational methods, scientists can more methodically investigate chemical space and choose which molecules to manufacture and test experimentally.

The application of computational approaches in drug discovery has been consolidated in recent decades. These techniques now constitute one of the pillars of pharmaceutical discovery pipelines in many academic and industrial environments. Their implementation has been demonstrated to tremendously improve the speed of the early discovery steps, allowing for the proficient and rational choice of proper compounds for desired therapeutic needs from the vast drug-like chemical space [11]. Within this computational landscape, QSAR modeling emerged as one of the earliest and most impactful approaches to guide compound selection.

Among the first and most significant computer methods in drug development are quantitative structure–activity relationship (QSAR) models. QSAR models facilitate the selection of viable candidates for synthesis and investigation by predicting attributes for untested compounds and establishing a correlation between molecular structures and biological activities [2]. More complex techniques like 3D-QSAR, molecular docking, and virtual screening have significantly improved the capabilities of computational drug design as techniques and computer power have grown [12]. These advancements laid the foundation for increasingly sophisticated computational strategies that continue to transform modern drug discovery.

1.1.4 The Promise of Artificial Intelligence in Drug Development

The next step in computational drug discovery is artificial intelligence (AI). AI has the potential to transform drug research by improving precision, cutting costs and timeframes, and enabling AI-driven computer-aided drug design [13] (Fig. 1.1) by utilizing sophisticated algorithms and large datasets. The use of AI in drug discovery includes several phases of the development pipeline, such as target identification, hit discovery, lead optimization, and even clinical trial design [14].

Generative AI, in particular, has emerged as a powerful tool for de novo drug design, which is the creation of novel chemical entities tailored to specific therapeutic targets. This approach has been considered the "holy grail" of drug discovery [15]. By learning the patterns and properties of known bioactive molecules, generative AI models can propose new chemical structures that possess desired pharmacological characteristics while maintaining drug-likeness and synthetic feasibility. This represents a paradigm shift from traditional screening-based methods toward a more directed and efficient exploration of chemical space.

Recent advances in generative models have established state-of-the-art criteria for the production of compounds and novel treatment candidates [16]. These models may speed up the early phases of drug discovery and lessen the need for massive physical chemical libraries by promptly suggesting promising drug candidates that satisfy particular efficacy and safety requirements. Accelerating the discovery process,

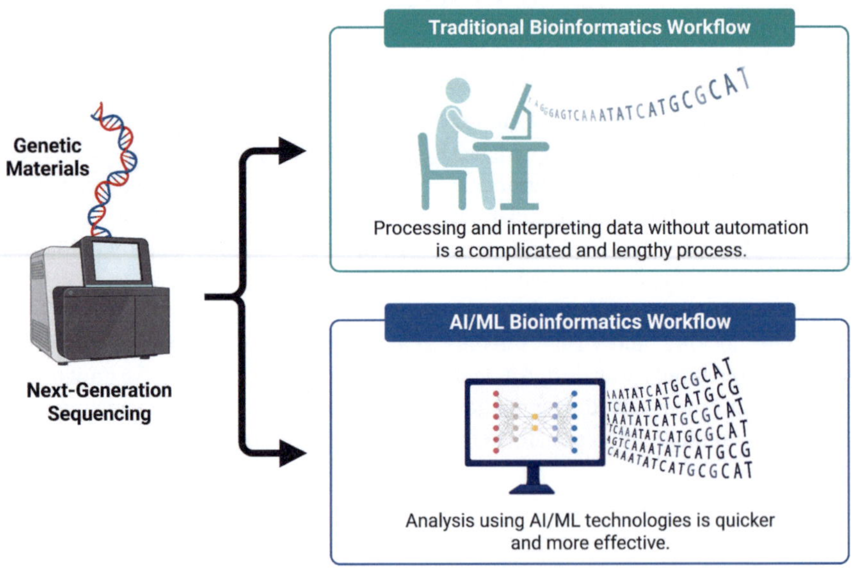

Fig. 1.1 Comparison of traditional and AI/ML-based bioinformatics workflows (created with Biorender.com)

finding new molecular scaffolds, exploring uncharted chemical space, and creating multi-target medications with optimal pharmacological properties are all possible with generative AI.

1.2 The Foundation of Generative AI in Drug Discovery: From Traditional QSAR Models to Modern Deep Learning Techniques

1.2.1 Historical Perspective of QSAR Models

One of the oldest and most significant computational methods in drug development is the quantitative structure–activity relationship (QSAR) methodology. QSAR was established several decades ago on the core tenet that a compound's chemical makeup is inextricably connected to its biological activity, chemical structure, and properties [17]. The traditional QSAR approach involves correlating the biological activities of compounds with their physicochemical properties using statistical methods [18].

In classical QSAR studies, researchers established correlations between the affinities of ligands to their binding sites, inhibition constants, and other biological endpoints with atomic, group, or molecular properties such as lipophilicity, polarizability, and electronic and steric characteristics (Hansch analysis), or with specific structural features (Free-Wilson analysis) [18]. This approach allowed scientists to develop predictive models that could estimate the biological activities of untested compounds based on their structural properties.

QSAR has evolved into a practical forecasting tool for pharmaceutical and agricultural design. Even while trial and error is an unavoidable aspect of drug research, QSAR significantly reduces the number of compounds that need to be synthesized by making it easier to find the most promising candidates 18]. Because of its efficacy, QSAR has become a cornerstone of rational drug design, encouraging medicinal chemists to examine the relationships between biological activities and structural features in greater detail.

1.2.2 Evolution to 3D-QSAR and Computer-Aided Drug Design

Despite their value, traditional QSAR methods are not very useful for creating novel molecules because they do not take into account the three-dimensional structure of the compounds. Due to this constraint, 3D-QSAR was developed as a logical progression of traditional methods. Using reliable chemometric methods such as artificial neural networks (ANN), genetic/PLS (G/PLS), and partial least squares

(PLS), 3D-QSAR uses the three-dimensional characteristics of ligands to forecast their biological actions [18, 19].

The process of creating a practical 3D-QSAR model consists of gathering information from databases or the literature, calculating suitable descriptors from molecular representations, establishing connections between biological activity and particular descriptors, and using the QSAR model to forecast biological characteristics for novel compounds [12, 18]. More complex computational approaches to drug discovery have been built upon this methodological framework.

The evolution of QSAR methodologies coincided with broader advancements in computer-aided drug design (CADD). CADD approaches have become integral to pharmaceutical discovery pipelines, tremendously improving the speed of early discovery steps and allowing for the rational selection of compounds from the vast drug-like chemical space [11, 20, 21]. These computational techniques have enabled researchers to rationalize biochemical and interactive processes of pharmaceutical interest at the molecular level, facilitating the rational 3D design and optimization of chemical entities based on structural information about their targets.

1.2.3 Machine Learning Approaches in Drug Discovery

The introduction of machine learning has greatly increased the powers of computational drug discovery. Virtual screening (VS) approaches, crucial computational tools in the drug development process, have made substantial use of machine learning techniques, such as QSAR analysis [2, 22]. QSAR's high-throughput screening capabilities and good hit rates make it especially valuable.

Machine learning approaches in drug discovery encompass a wide range of techniques, from traditional statistical methods to more sophisticated algorithms capable of learning complex patterns from large datasets. These approaches have been applied to various aspects of drug discovery, including target identification, bioactivity prediction, and the design of compounds with specific properties [23]. Building on these foundations, specialized machine learning applications such as multi-target QSAR have emerged to address the complexity of modern pharmacology.

Multi-target QSAR (mt-QSAR), which allows the computation of a compound's probability of activity against many biological or pharmacological targets, is a noteworthy use of machine learning in drug discovery [24]. By allowing he formulation of a single equation for numerous outputs, mt-QSAR improves the effectiveness and applicability of prediction models in contrast to standard QSAR techniques, which usually concentrate on a single target or feature. In polypharmacology, where drugs are made to interact with several targets implicated in complex disorders, this feature is especially useful.

Machine learning techniques have grown more complicated, utilizing larger and more varied datasets and combining increasingly intricate algorithms as techniques and processing capacity have improved. Deep learning techniques have been developed as a result of this evolution, significantly changing the field of computational drug discovery [25]. These deep learning approaches represent the current frontier, offering unprecedented predictive power and adaptability across diverse stages of the drug discovery pipeline.

1.2.4 Deep Learning Architectures for Molecular Generation

Deep learning, a branch of machine learning that uses multi-layered neural networks, has become a potent method for drug discovery and molecular creation. De novo drug design has been approached using a variety of deep learning architectures, each with special advantages and uses [1, 26]. Below are a few examples of deep learning architectures.

1.2.4.1 Recurrent Neural Networks (RNNs)

Recurrent Neural Networks (RNNs), particularly those incorporating Long Short-Term Memory (LSTM) cells, have shown remarkable success in molecular de novo design. These models capture the syntax of molecular representations, such as SMILES (Simplified Molecular-Input Line-Entry System) strings, with high accuracy. The learned pattern probabilities can then be used for de novo SMILES generation, eliminating the need for virtual compound library enumeration [27].

RNN-based approaches offer several advantages for molecular generation. They can be fine-tuned through transfer learning for specific molecular targets, enabling virtual compound design without requiring secondary or external activity prediction that could introduce error or unwanted bias. This capability makes RNN-LSTM systems particularly valuable for high-impact use cases such as low-data drug discovery, fragment-based molecular design, and hit-to-lead optimization for diverse drug targets [27, 28]. Beyond these general advantages, RNNs have also been adapted for creative strategies that integrate user-defined structural information.

One particularly inventive application of RNNs in drug development is the "fragment growing procedure," where the model grows a user-provided chemical fragment instead of starting from scratch with molecules. This technique ensures that specific structural motifs are preserved in the molecules that are created, while allowing for creative exploration of possible expansions [27]. This capability is very helpful in medicinal chemistry campaigns when specific substructures are known to be essential for activity or binding to a target.

1.2.4.2 Variational Autoencoders (VAEs)

Another family of deep learning architectures that has been effectively used for molecule creation is Variational Autoencoders (VAEs). VAEs are made up of a decoder network that reconstructs molecules from points in a continuous latent space and an encoder network that maps input molecules to this space. By sampling from the latent space and decoding the generated vectors, this architecture enables the creation of new molecules [6].

For molecular design, the latent space representation in VAEs provides several benefits. VAEs facilitate the systematic exploration of chemical space and smooth interpolation between existing compounds by mapping molecules to a continuous space where similar structures are positioned near each other. Furthermore, certain properties can be optimized by manipulating the latent space, which enables the directed generation of molecules with the desired traits [29]. As a result, VAEs bridge the gap between generative modeling and rational design, allowing chemists to fine-tune molecular structures with greater precision.

VAEs have demonstrated effectiveness in generating diverse yet valid molecular structures, performing continuous optimization in the latent space, and incorporating property prediction and optimization into the generation process. These capabilities make VAEs valuable tools for exploring and expanding chemical space in a directed and efficient manner for drug discovery applications [30]. Together, these features position VAEs as a cornerstone method in generative chemistry, complementing other architectures such as GANs and transformers that are also being explored for molecular design.

1.2.4.3 Generative Adversarial Networks (GANs)

Generative Adversarial Networks (GANs) ability to produce new molecules with desirable features has made them popular in drug development. GANs are made up of a discriminator and two neural network generators that have been adversarially trained. As the discriminator compares the new molecular structures produced by the generator to existing instances, it provides input that aids in the generator's improvement [31].

GAN-based frameworks have been applied to various aspects of drug discovery, including molecular de novo design, dimension reduction of single-cell data in preclinical development, and de novo peptide and protein design [32]. Studies have shown that GANs can generate novel molecular compounds that could be considered potential therapeutic agents, such as anticancer drugs [33].

Hybrid techniques that integrate reinforcement learning (RL) with GANs have shown particular promise. When compared to recurrent neural networks or GANs alone, models like ORGAN and ORGANIC have shown better performance, especially when it comes to producing molecules with desired characteristics like drug-likeness [34]. Since GANs n be used with different methods to solve particular drug discovery difficulties, these hybrid approaches demonstrate the versatility and flexibility of GANs in the context of molecular design.

1.2.4.4 Transformer Models

Transformer models were first created for applications involving natural language processing, but they have recently shown promise when used for molecule production. Transformer-based chemical language models (CLMs) present novel molecular design opportunities in drug development [35]. Transformers make use of attention mechanisms to better learn complicated patterns in chemical data by capturing the links between various components of a molecular representation. When dealing with the complex syntax and semantics of chemical representations, like SMILES strings, this functionality is especially helpful. Transformer-based CLMs' predictive capabilities for molecule production heavily rely on attention processes [36].

Any compound or target data that can be tokenized and presented in a sequential style can be used with these models, which enable the learning of mappings between various sequence types. Protein sequence motifs, for instance, can be used to predict active compounds, and CLMs can be used to extend medicinal chemistry analog series with novel compounds and describe them as chemical sequences [37]. Because of their adaptability, transformer models are useful instruments for a range of drug discovery applications, especially those that entail creating molecules with certain characteristics or functions.

1.2.4.5 Reinforcement Learning in Molecular Design

A potent method for maximizing molecular generation models toward particular goals is reinforcement learning, or RL. In RL-based molecular design, a generative model (such as an RNN, VAE, or GAN) is trained to maximize a reward function that captures important molecular attributes like drug-likeness, synthetic accessibility, or anticipated biological activity [15].

The Models like ORGAN and ORGANIC integrate RL with GANs to generate molecules with specific properties. In these apps, the generator's parameters are optimized to produce strings that score highly according to an external scoring function, which can be designed to prioritize low sulfur content, high predicted target activity, or other desirable characteristics [8]. ORGAN has demonstrated better performance than recurrent neural networks or GANs alone on various molecular design tasks [38].

Reinforcement learning offers a versatile framework for guiding molecular generation toward several, possibly conflicting goals. This ability is especially helpful in the drug discovery process, where it can be difficult to design molecules that simultaneously meet multiple safety, efficacy, and developability requirements. RL-based methods can direct the exploration of chemical space toward areas that are most likely to produce promising drug candidates by carefully crafting reward functions that strike a balance between these factors [39].

1.2.5 Knowledge-Enabled Generative Models

Recent developments in generative AI in drug discovery have emphasized that structured biological knowledge and molecular generation must be combined. A knowledge graph, where entities, like molecules, proteins, and diseases, are nodes represented by edges that represent relations between two nodes, is a rich and informative resource, yet constraining, for molecular generation [16].

Knowledge-embedded generative models bridge generative models with large biomedical knowledge represented in knowledge graphs. These methods integrate knowledge graphs' contextual knowledge with governing generative models of novel drug candidates with preferred properties of interest in a rational and synthesizable way [16].

Such a solution was seen to facilitate knowledge-aware generation by building a scalable process that increases knowledge graph functionality without losing semantic quality. Integrating knowledge graph embeddings and diffusion-based generative models offers a robust mechanism for producing novel drug candidates with desired properties. This approach has outperformed state-of-the-art generative models on both unconditional and targeted generation tasks [40].

One promising approach to overcoming some of the drawbacks of existing generative approaches—specifically, their propensity to ignore structured biological knowledge—is to use knowledge-enhanced generative models. These models can impose biologically relevant constraints during the generation process by integrating relational information from knowledge graphs, which could result in more significant and potentially more effective drug candidates. The use of such models in drug discovery has advanced significantly with this combination of generative capabilities and domain knowledge [41].

1.3 Difference from Classical Drug Development Approach

1.3.1 Traditional Drug Discovery Pipeline

The conventional drug development pipeline is a long, multi-stage activity that starts with target identification and validation and culminates in the selection of a candidate molecule for clinical development. This process typically involves several key stages, including assay development, high-throughput screening, hit identification, lead optimization, and preclinical testing [42–44]. Target selection and validation are the initial two steps of drug development, when scientists choose a protein or a pathway already linked with a disease state and define evidence for it as a drug target. This evidence can be obtained from a number of sources, including academic and clinical research as well as commercial investigations. Building a convincing body of supporting evidence for a target can take many years before it is selected for a costly drug discovery program [5, 45, 46].

Once a target has been validated, researchers develop assays to screen compounds for activity against the target. High-throughput screening (HTS) is then employed to test large libraries of compounds, often numbering in the hundreds of thousands or millions, to identify those with desired activity (hits). These hits undergo further characterization and optimization to improve their potency, selectivity, and drug-like properties, eventually leading to the selection of lead compounds [47–49].

Lead optimization is the most costly phase of conventional drug discovery. Drug chemist intentionally diversify lead compounds to optimize their properties based on structure–activity relationships and iterative screening. It can take synthesis and screening of hundreds to thousands of analogs before the designation of an analog showing potential for preclinical development, substantially adding to the time and cost of getting new drugs to the clinic and significantly to the time and cost of bringing new drugs to market [50, 51].

1.3.2 Integration of Generative AI in Modern Drug Discovery

The integration of generative AI into the drug discovery pipeline represents a paradigm shift from the traditional approach described above. Rather than relying on the physical screening of existing compound libraries, generative AI enables the computational design of novel molecules tailored to specific targets or with desired properties (Fig. 1.2) [52].

There are several phases in the drug discovery process where generative AI can be applied. Examining biological data and proposing proteins or pathways linked to disease causes can aid in the early discovery of possible targets. The chemical space can be expanded beyond current libraries by using generative models to suggest new scaffolds with anticipated activity against the target during hit identification. AI can direct the methodical alteration of hit compounds to improve their characteristics in lead optimization, which could minimize the quantity of analogs that require synthesis and testing [14].

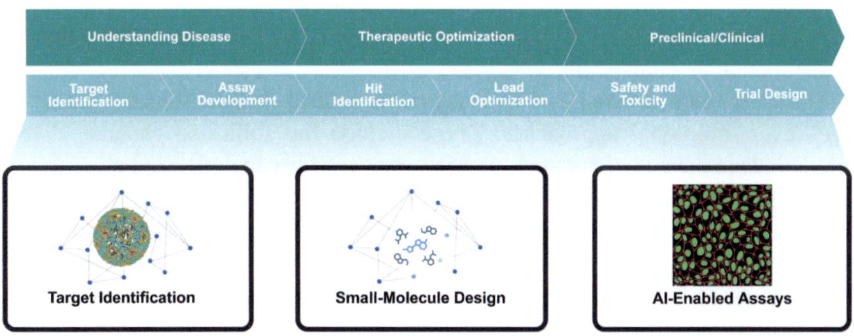

Fig. 1.2 Generative AI in drug discovery and development (created with Biorender.com)

Generative models can rapidly suggest promising drug candidates that meet particular efficacy and safety requirements by automating the molecular design process [16]. Thanks to this capability, researchers can concentrate their experimental resources on the most promising compounds, which also speeds up the exploration of chemical space. Furthermore, compounds for difficult targets that have resisted conventional methods can be designed more easily with the help of generative AI.

The influence of generative AI on drug development is further increased when it is combined with other computational and experimental methods. For instance, integrating generative models with structure-based design techniques can make the targeted creation of compounds expected to interact with particular protein binding sites possible. Similarly, combining generative AI with high-throughput experimental methods produces a potent feedback loop that can speed up lead compound optimization and raise the likelihood of success in subsequent phases of development [13, 53].

1.3.3 De Novo Drug Design Using Generative Models

Creating new chemical entities with desired biological activity is known as de novo drug design, and it is regarded as the "holy grail" of drug discovery. This method has been transformed by generative AI, which makes it possible to computationally create new molecules with particular characteristics [13, 15, 41]. Various generative model techniques have been applied to de novo drug design, including recurrent neural networks, autoencoders, generative adversarial networks, transformers, and hybrid models with reinforcement learning. These approaches offer different strengths and capabilities for molecular generation, and their selection depends on the specific requirements of the drug discovery project [54, 55].

Regulatory organizations are still developing frameworks for evaluating drugs with significant AI contributions. Because some AI models are "black box" in nature, regulatory review—which often requires extensive documentation of the reasons behind design choices—may become more difficult. It is necessary to create standards and procedures for the validation and documentation of AI-generated compounds in order for them to successfully navigate regulatory routes and eventually be licensed as therapeutic medicines [56, 57].

The application of generative models to de novo drug design offers several advantages over traditional approaches. These include the ability to generate diverse compounds to expand compound libraries, the capability to design compounds with specific properties, and the potential to identify novel scaffolds that may not be present in existing chemical libraries. Several publicly available molecular design tools based on generative models have been developed, allowing researchers to generate molecules directly without requiring extensive computational expertise [58, 59].

Recent developments in de novo drug design employing generative models have focused on key challenges like guaranteeing synthetic accessibility, optimizing multiple properties at once, and producing molecules with particular binding modes

[60, 61]. These advancements are making generative models more useful in drug discovery and making it possible to incorporate them into actual drug development initiatives [62, 63].

1.3.4 Target-Specific Drug Generation

Generative AI's capacity to create compounds especially for a target is one of its advantages in drug discovery. Compounds that are expected to interact well with the target can be suggested using generative models by integrating data on the target structure and binding preferences. [6, 37]. Transfer learning has emerged as a valuable approach for target-specific drug generation. By fine-tuning pre-trained generative models on data related to a specific target, researchers can bias the generation process toward molecules that are likely to bind to the target of interest. This approach has proven particularly valuable in "low-data" situations every day in early-phase drug design, where limited information is available about ligands for a specific target [27].

Knowledge-enhanced generative models, which integrate structured data regarding target proteins and their interactions with small molecules, further aid in the development of target-specific drugs. By employing embedded NGS from knowledge graphs to capture complex interactions between molecular shapes and biological aims, these models facilitate more informed molecular design [16, 64]. Compared to previous methods, which typically rely on screening vast libraries of compounds with little regard for target-specific interactions, the capacity to develop target-specific compounds is a considerable gain. Generative AI has the potential to improve the efficiency of the hit identification stage and discover novel chemotypes with distinct binding modes or better properties than known ligands by concentrating the generation process on molecules that are likely to interact with a particular target [65, 66].

1.3.5 Accelerating Hit-to-Lead Ptimization

Hit-to-lead optimization, the process of transforming initial active compounds into drug-like molecules with improved properties, has traditionally been one of the most time-consuming and resource-intensive stages of drug discovery. Generative AI offers the potential to significantly accelerate this process by suggesting modifications to hit compounds that are likely to enhance their properties [67]. Hence, reinforcement learning-based approaches are particularly well-suited for hit-to-lead optimization. By defining reward functions that encapsulate desired molecular properties such as potency, selectivity, and ADME (absorption, distribution, metabolism, excretion) characteristics, generative models can be trained to propose modifications that optimize multiple objectives simultaneously. This capability reduces the need for extensive trial-and-error experimentation and allows medicinal chemists to focus on the most promising directions for optimization [68, 69].

The "fragment growing procedure" implemented in some RNN-based generative models exemplifies how AI can assist in hit-to-lead optimization. This approach allows researchers to specify a molecular fragment (such as a core scaffold or a binding motif) and use the model to suggest potential extensions, ensuring that the essential structural elements are preserved while exploring possibilities for improvement. This capability is particularly valuable for medicinal chemistry campaigns where certain structural features are known to be essential for activity or binding [55, 70].

In multi-parameter optimization, when gains in one feature (like potency) may come at the expense of another (like solubility), generative AI can also assist in navigating the intricate trade-offs. AI can suggest compounds that represent ideal compromises between conflicting features by integrating several objectives into the generation or optimization process. This could help find solutions that conventional methods would overlook [71, 72].

1.3.6 Case Studies and Success Stories

The application of generative AI in drug discovery has led to several notable successes, demonstrating the practical utility of these approaches in real-world drug development projects. While the search results do not provide specific case studies, they mention that generative AI has been used successfully in drug discovery and that there are documented examples of AI-generated compounds progressing to clinical evaluation, a significant milestone in the validation of these approaches [13, 73].

A number of clinical-level assets have been developed through commercial partnerships as a result of integrating generative AI in drug discovery. These examples demonstrate how AI is becoming more widely accepted and used in drug discovery and show how these methods can have real advantages in terms of quicker development times and higher success rates [74, 75].

More case studies and success stories will probably surface as the area develops, further confirming the value of generative AI in drug discovery and revealing best practices for incorporating these methods into drug development processes. These illustrations will be helpful in directing the future advancement and improvement of generative AI techniques and in fostering practitioners' trust in their use in drug discovery [76].

1.4 Challenges and Limitations of the Methods

1.4.1 Data Quality and Availability Issue

The caliber and volume of training data have a fundamental impact on how well generative AI models perform in drug discovery. There are still a number of issues in this field, even with the increasing amount of chemical and biological data available [77]. Given that experimental measurements might be erratic or prone to errors, data quality is undoubtedly a major concern. The settings in which tests are carried out might affect the outcomes, and biological assays frequently show variability. These elements add to the training data's noise, which may affect the generative models' dependability [78].

Another difficulty is finding data, especially for novel targets with few known ligands or rare diseases. The generative models may be biased due to the lack of high-quality data for particular targets or chemical classes, which could restrict their use in particular drug discovery domains. In early-phase drug discovery, this "low-data" scenario is typical and poses a major challenge for generative AI techniques [41, 79].

Moreover, there are other difficulties in representing molecular structures in machine learning-compatible formats. Different molecular representations (such as fingerprints, molecular graphs, and SMILES strings capture different facets of chemical structures, and the representation selection can impact the performance capabilities of generative models. Research is still ongoing to create comprehensive and standardized representations that capture the pertinent biological and chemical data [80, 81].

1.4.2 Model Interpretability and Explainability

The interpretability of generative AI models represents a significant challenge in drug discovery, where understanding the rationale behind predictions and generations is crucial for building confidence in the results. Many deep learning architectures, particularly those with complex structures and numerous parameters, operate as "black boxes," making it difficult to decipher their decision-making processes [82]. Thus, another difficulty is finding data, especially for novel targets with few known ligands or rare diseases. The generative models may be biased as a result of the lack of high-quality data for particular targets or chemical classes, which could restrict their use in particular drug discovery domains. In early-phase drug discovery, this "low-data" scenario is typical and poses a major challenge for generative AI techniques [83].

Moreover, there are other difficulties in representing molecular structures in machine learning-compatible formats. Different molecular representations (such as fingerprints, molecular graphs, and SMILES strings capture different facets of chemical structures, and the representation selection can impact the performance capabilities of generative models. Research is still ongoing to create comprehensive and standardized representations that capture the pertinent biological and chemical data [84, 85].

1.4.3 Validation of Generated Molecules

Validating compounds suggested by generative AI models is an important difficulty. However, many new structures can be produced by computational methods; experimental validation s necessary to demonstrate their validity, synthesizability, and biological activity [86]. Usually, computational validation entails evaluating characteristics including projected biological activity, synthetic accessibility, and drug-likeness. Measures of chemical validity, variety, novelty, and resemblance to known active chemicals are among the benchmarks and metrics that have been created to assess generative models. These computer evaluations, nevertheless, have drawbacks and might not adequately represent th intricacy of actual drug-development standards [87].

Experimental validation remains the gold standard for evaluating generated molecules, but it is resource-intensive and can only be applied to a small subset of the generated compounds. The discrepancy between experimental reality and computational forecasts emphasizes how crucial it is to create trustworthy computational filters and prioritization techniques in order to find the most promising candidates for testing and synthesis. Given that generative models can generate considerably more molecules than can be validated experimentally, this challenge is especially important [88, 89]. The creation of integrated procedures that combine efficient experimental validation techniques with computer prediction is a significant direction for future research in this area. In order to assess and enhance generative models, these techniques may include automated synthesis platforms, high-throughput experimental procedures, and iterative feedback loops, ultimately increasing their applicability in drug discovery [90].

1.4.4 Computational Requirements and Scalability

For training and inference, generative AI models—especially those built on deep learning architectures—frequently require significant amounts of processing power. Research teams with low computational resources may find them unusable because of their intricacy and the substantial datasets needed for hands-on training [14, 91]. Depending on the particular architecture and application, generative models have different computing needs. For example, transformer-based models may perform better for specific tasks but generally demand more computing power than RNN-based methods. In the development and use of generative AI for drug discovery, striking a balance between cost and model performance is a persistent difficulty [92].

When trying to produce and assess a large number of molecules, as may be necessary in a thorough investigation of chemical space for a certain target, scalability problems also occur. Effective screening and prioritization techniques must be developed to concentrate resources on the most promising candidates because the computational cost of predicting attributes for thousands or millions of produced molecules can be prohibitive [93, 94].

To meet these computational difficulties, ongoing improvements in hardware, software, and algorithmic efficiency will be necessary. Cloud computing and distributed computing methodologies present viable remedies by providing scalable computational resources. Algorithmic optimizations and more efficient model structures can simultaneously reduce the processing demands of generative AI systems. For a greater range of drug discovery applications, these developments are essential for improving the usability and accessibility of generative AI systems [95, 96].

1.4.5 Bridging the Gap Between In-Silico Predictions and Experimental Reality

Bridging the gap between computational predictions and experimental reality is one of the biggest obstacles to using generative AI in drug discovery. Although generative models can suggest compounds with predicted features, these predictions might not always result in favorable pharmacokinetic qualities or biological activity [73, 97].

The intricacy of biological systems, the assumptions and simplifications present in computational models, and the ambiguities surrounding the connection between molecular structure and function are some of the variables that contribute to this disparity. Depending on the property or behavior being predicted, computational models' predictive accuracy varies greatly; certain traits are easier to forecast accurately than others [98, 99]. Furthermore, target identification and validation are similarly impacted by the reproducibility issue of generative AI, which may have an impact on the dependability of generative models trained on such data. The accuracy of structure-based generative systems can also be impacted by limitations in protein structure prediction techniques, such as AlphaFold's incapacity to capture numerous protein states and dynamic behaviors [100].

Iterative feedback between virtual and physical screening is necessary to address this difficulty, which calls for the close integration of computational and experimental methodologies. The models can be improved by using experimental data from synthetic substances, increasing th predicted accuracy for later generations. In drug development, this iterative process emphasizes how complementary computational and experimental methods are, and how crucial it is to create integrated workflows that capitalize on each approach's advantages [101, 102].

1.4.6 Ethical and Regulatory Considerations

As these technologies develop further, a number of ethical and regulatory issues are brought up by the use of generative AI in drug discovery. These include concerns about data privacy, intellectual property, and the legality of chemicals produced by artificial intelligence [103]. AI-generated molecular intellectual property rights are a complicated legal topic. There is still disagreement over whether AI-generated inventions may be patented and who should be credited with an idea when AI is heavily involved in the design process. The financial incentives for creating and implementing generative AI in drug discovery may be pacted by these concerns [104].

Training models using sensitive or proprietary biological data raises data privacy issues, especially in the context of customized medicine. Maintaining ethical standards and regulatory compliance requires making sure generative models don't unintentionally expose protected information that is included in their training data. This challenge is particularly relevant given the increasing integration of genomic and other personal health data in drug discovery efforts [105]. Regulatory agencies are still developing frameworks for evaluating drugs designed with significant AI contributions. Regulatory scrutiny, which normally necessitates thorough documentation of the reasoning behind design choices, may be complicated by the "black box" aspect of some AI models. For AI-generated compounds to successfully move through regulatory processes and eventually be approved as medicinal agents, criteria and norms for their validation and verification must be established [106].

1.5 Conclusion and Outlook for the Transformative Impact of Generative AI in Drug Discovery

1.5.1 Current State of Generative AI in Drug Discovery

Generative AI has become a potent instrument in drug discovery, providing fresh methods for target-specific chemical creation, molecular optimization, and de novo drug design. The creation of several generative structures, such as transformers, RNNs, GANs, and VAEs, has increased the potential of computational drug design and made it possible to explore hitherto unreachable areas of the chemical space [107]. Therefore, the current state of the subject is characterized by significant methodological advancements and increased recognition of AI-generated compounds as feasible starting points for drug discovery efforts. Clinical-level assets have been developed as a result of commercial relationships between AI businesses and pharmaceutical manufacturers, illustrating the usefulness of generative techniques. These partnerships demonstrate how generative AI can speed up drug discovery and improve the development process efficiency [108].

Despite these advances, challenges remain in areas such as data quality, model interpretability, validation of generated molecules, and bridging the gap between computational predictions and experimental reality. Addressing these challenges will be crucial for realizing the full potential of generative AI in drug discovery and establishing these approaches as standard tools in the drug development toolkit [109]. Drug discovery professionals now have easier access to generative AI technologies because of an expanding ecosystem of tools and platforms. These include freely accessible generative model-based molecular design tools that let scientists create molecules directly without needing a lot of computational know-how [2]. It is anticipated that this democratization of access to generative AI technologies would hasten their uptake and spur additional advancements in the area [110].

1.5.2 Future Directions and Emerging Trends

A number of new themes are shaping the future of generative AI in drug development. These include applying generative techniques to increasingly complicated molecular design tasks, integrating multimodal data sources, and creating models that are more visible and interpretable [111]. Since they make it possible to integrate structured biological knowledge into the generation process, knowledge-enhanced generative models offer a promising avenue for further investigation. These methods can produce molecules that are both chemically and physiologically valid by utilizing the abundance of information contained in knowledge graphs and other structured data sources. The application of AI to drug discovery has advanced significantly with this combination of generative skills and domain knowledge [112].

Chemical language models (CLMs) based on transformer architectures are gaining prominence in molecular design. These models, which treat molecules as sequences analogous to natural language, can capture complex patterns in chemical data and generate diverse yet valid structures. More complex and powerful generative models will probably result from the further application of developments in natural language processing to molecular production [113]. Combining generative AI with various experimental and computational methods is another significant trend. Targeting particular binding sites more precisely by combining generative mode with structure-based design techniques. Likewise, combining high-throughput experimental methods with generative methodologies may enable quick validation and improvement of produced compounds, generating potent feedback loops that quicken the drug development process [114].

1.5.3 Integration with Other Technologies

The potential of generative AI in drug discovery is amplified when integrated with complementary technologies. The synchronization of genomics, proteomics, and in silico simulation has the potential to revolutionize the drug discovery process, and generative AI can play a central role in this integration [115]. One promising area of integration is with protein structure prediction methods such as AlphaFold [116, 117]. By combining accurate protein structure predictions with generative models for ligand design, researchers can develop more effective approaches for structure-based drug design, potentially leading to molecules with higher binding affinities and selectivity profiles [73]. Targeting proteins with little to no experimental structural information may benefit this integration, increasing the pool of possible therapeutic targets.

Another technology that can be combined with generative AI is digital twins, which are computer models that mimic intricate biological systems. By integrating generative models into digital twin frameworks, researchers can create compounds precisely suited to alter the behavior of intricate biological networks. This advances precision medicine techniques. Better translation of preclinical results to clinical outcomes and more individualized drug development efforts may be made possible by this integration [118].

Large language models (LLMs) like ChatGPT are also being integrated with generative AI for drug discovery. These models can process and synthesize vast amounts of scientific literature, potentially identifying patterns and connections that inform molecular design. The combination of LLMs for knowledge extraction and generative models for molecular design represents a powerful approach to knowledge-driven drug discovery, enabling the incorporation of insights from disparate sources into the design process [119, 120].

1.5.4 Areas Requiring Further Research

Even though generative AI for drug development has advanced significantly, there are still a number of areas that need more study to overcome current constraints and open up new possibilities. These include developing more efficient validation techniques, resolving computational scaling concerns, improving model interpretability, and improving data quality and representation [121]. The efficiency and potential of generative models may be improved by investigating new molecular representations and architectures created especially for chemical data. Current methods frequently modify structures created for other fields (like computer vision or natural language processing) for molecular production. However, the distinctive features of chemical structures and their connections to biological activity are better captured by domain-specific architectures [122].

The creation of benchmark datasets and uniform assessment criteria for generative models in drug discovery would speed up advancements in the area and enable a more thorough comparison of various strategies. It is challenging to evaluate the relative benefits and drawbacks of various approaches and to monitor advancement over time because current studies frequently include a variety of sometimes conflicting measures [123]. Examining multi-objective optimization techniques for generative models may help with the prob m of creating compounds that meet several safety, developability, and efficacy requirements at the same time. More advanced optimization techniques can effectively negotiate the many trade-offs involved in drug discovery, producing more balanced and developable candidates, even if present methods frequently employ weighted combinations of individual property scores [124].

1.5.5 Long-Term Impact on the Pharmaceutical Industry

From the economics of drug development to the kinds of diseases that might be successfully targeted, generative AI is expected to have a profound long-term impact on the pharmaceutical sector [52]. Generative AI has the potential to drastically cut down on the expenses and timeframes involved in introducing new medications to the market by speeding up the early phases of drug discovery and improving the success rates of compounds entering clinical development.

Additionally, generative AI may democratize drug research by allowing smaller businesses and university institutions to make significant contributions to the area. Computational methods may reduce the entry barriers for drug development by displacing the need for substantial experimental infrastructure and physical compound libraries, which could result in more creative and varied treatment options. A more open and cooperative innovation ecosystem in pharmaceutical research may result from the democratization [125, 126].

The druggable proteome may grow, and new therapeutic opportunities may arise from the use of generative AI in conventionally difficult fields like protein–protein interactions, allosteric modulators, and multi-target medications. AI could aid in the treatment of illnesses that have proven resistant to conventional methods, producing compounds especially made for these intricate targets, which could result in advances in fields with a high unmet medical demand [127].

Generative AI has the potential to drastically alter the pharmaceutical industry as it develops and combines with other technologies. Although there are still issues with data quality, model interpretability, and experimental validation, the field's trajectory indicates that generative AI will become more and more important in the search for and creation of the next generation of medicinal agents [52, 73]. This change has the potential to alleviate some of the long-standing issues that have hindered he effectiveness and productivity of pharmaceutical research while also hastening the delivery of novel medications to patients.

Acknowledgements The authors would like to thank the Department of Research and Community Service of Indonesia International Institute for Life Sciences (LPPM i3L University) for their heartfelt support. Moreover, this manuscript was improved and proofread with Grammarly Premium software. As the authors, we assume full responsibility for the content of the published work. There is no competing interest to declare, and all authors agree on the content.

References

1. Lin E, Lin C-H, Lane H-Y. Relevant applications of generative adversarial networks in drug design and discovery: molecular De Novo design, dimensionality reduction, and De Novo peptide and protein design. Molecules. 2020;25:3250. https://doi.org/10.3390/molecules25143250.
2. Achary PGR. Applications of quantitative structure-activity relationships (QSAR) based virtual screening in drug design: a review. Mini Rev Med Chem. 2020;20:1375–88. https://doi.org/10.2174/1389557520666200429102334.
3. Ogel DB. Factors associated with clinical trials that fail and opportunities for improving the likelihood of success: a review. Contemp Clin Trials Commun. 2018;11:156–64. https://doi.org/10.1016/j.conctc.2018.08.001.
4. Sun D, Gao W, Hu H, Zhou S. Why 90% of clinical drug development fails and how to improve it? Acta Pharm Sin B. 2022;12:3049–62. https://doi.org/10.1016/j.apsb.2022.02.002.
5. Hughes JP, Rees S, Kalindjian SB, Philpott KL. Principles of early drug discovery. Br J Pharmacol. 2011;162:1239–49. https://doi.org/10.1111/j.1476-5381.2010.01127.x.
6. Gangwal A, Ansari A, Ahmad I, Zad AK, Kumarasamy V, Subramaniyan V, Wong LS. Generative artificial intelligence in drug discovery: basic framework, recent advances, challenges, and opportunities. Front Pharmacol. 2024;15:1331062. https://doi.org/10.3389/fphar.2024.1331062.
7. Pham P, Nuyen VTD, Cho KH, Hy T-S. rugPipe: generative artificial intelligence-assisted virtual screening pipeline for generalizable and efficient drug repurposing. Biol Methods Protocols. 2025;10:bpaf038. https://doi.org/10.1093/biomethods/bpaf038.
8. Zhao AP, Li S, Cao Z, Hu PJ-H, Wang J, Xiang Y, Xie D, Lu X. AI for science: predicting infectious diseases. J Saf Sci Resil. 2024;5:130–46. https://doi.org/10.1016/j.jnlssr.224.02.002.
9. Sertkaya A, Beleche T, Jessup A, Sommers BD. Costs of drug development and research and development intensity in the US, 2000–2018. JAMA Netw Open. 2024;7:e2415445. https://doi.org/10.1001/jamanetworkopen.2024.15445.
10. Yates N, Hinkel J. The economics of moonshots: value in rare disease drug development. Clin Transl Sci. 2022;15:809–812. https://doi.org/10.1111/cts.13270.
11. Bassani D, Moro S. Past, present, and future perspectives on computer-aided drug design methodologies. Molecules. 2023;28:3906. https://doi.org/10.3390/molecules28093906.
12. Bosc N, Atkinson F, Felix E, Gaulton A, Ersey A, Leach AR. Large-scale comparison of QSAR and conformal prediction methods and their applications in drug discovery. J Cheminf. 2019;11:1–16. https://doi.org/10.1186/S13321-018-0325-4/TABLES/6.
13. Gangwal A, Lavecchia A. Unleashing the power of generative AI in drug discovery. Drug Discov Today. 2024;29:103992. https://doi.org/10.1016/j.drudis.2024.103992.
14. Abbas KG, Rassam A, Karamshahi F, Abnora R, Abouseada M. The role of AI in drug discovery. ChemBioChem. 2024;25:e202300816. https://doi.org/10.1002/cbic.202300816.
15. Tong X, Liu X, Tan X, Li X, Jiang J, Xiong Z, Xu T, Jiang H, Qiao N, Zheng M. Generative models for De Novo drug design. J Med Chem. 2021;64:14011–27. https://doi.org/10.1021/acs.jmedchem.1c00927.

16. Malusare A, Aggarwal V. Improving molecule generation and drug discovery with a knowledge-enhanced generative model. 2024. ArXiv. arXiv:2402.08790v2
17. Bastikar V, Bastikar A, Gupta P. Quantitative structure–activity relationship-based computational approaches. In: Computational approaches for novel therapeutic and diagnostic designing to mitigate SARS-CoV-2 infection. 2022;191–205. https://doi.org/10.1016/B978-0-323-91172-6.00001-7.
18. Verma J, Khedkar VM, Coutinho EC. 3D-QSAR in drug design—a review. Curr Top Med Chem. 2010;10:95–115. https://doi.org/10.2174/156802610790232260.
19. Ajmani S, Jadhav K, Kulkarni SA. Group-b sed QSAR (G-QSAR): mitigating interpretation challenges in QSAR. QSAR Comb Sci. 2009;28:36–51. https://doi.org/10.1002/qsar.200810063.
20. Kapetanovic IM. Computer-aided drug discovery and development (CADDD): in silico-chemico-biological approach. Chem Biol Interact. 2008;171:165–176. https://doi.org/10.1016/j.cbi.2006.12.006
21. Agustriawan D, Parikesit AA, Nurdiansyah R. Big data computation of drug design: from the natural products to the transcriptomic-based molecular development. In: Application of omics, AI and blockchain in bioinformatics research. World Scientific; 2019. p. 59–86. https://doi.org/10.1142/9789811203589_0005.
22. Roy K, Ka S, Das RN, Roy K, Kar S, Das RN. Background of QSAR and historical developments. In: Understanding the basics of QSAR for applications in pharmaceutical sciences and risk assessment. Elsevier; 2015. p. 1–46. https://doi.org/10.1016/B978-0-12-801505-6.00001-6.
23. Vamathevan D, Clark D, Czodrowski D, Dunham I, Ferran E, Lee G, Li B, Madabhushi A, Shah P, Spitzer M, Zhao S. Applications of machine learning in drug discovery and development. Nat Rev Drug Discov. 2019;18:463–77. https://doi.org/10.1038/s41573-019-0024-5.
24. Zanni R, Gálvez-Llompart M, Gálvez J, Garc a-Domenech R. QSAR multi-target in drug discovery: a review. Curr Comput Aided Drug Des. 2014;10:129–136. https://doi.org/10.2174/1573409910021407081051 2.
25. Taye MM. Understanding of machine learning with deep learning: architectures, workflow, applications and future directions. Computers. 2023;12:91. https://doi.org/10.3390/computers12050091.
26. Özçelik R, van Tilbog D, Jiménez-Luna J, Grisoni F. Structure-based drug discovery with deep learning. ChemBioChem. 2023;24:e202200776. https://doi.org/10.1002/cbic.202200776.
27. Gupta A, Müllr AT, Huisman BJH, Fuchs J, Schneider P, Schneider G. Generative recurrent networks for De Novo drug design. Molecul Inform. 2018;37:1700111. https://doi.org/10.1002/MINF.201700111.
28. Altalib MK, Salim N. Similarity-based virtual screen using enhanced siamese deep learning methods. ACS Omega. 2022;7:4769–86. https://doi.org/10.1021/acsomega.1c04587.
29. Ochiai T, Inuka T, Akiyama M, Furui K, Oue M, Matsumori N, Inuki S, Uesugi M, Sunazuka T, Kikuchi K, Kakeya H, Sakakibara Y. Variational autoencoder-based chemical latent space for large molecular structures with 3D complexity. Commun Chem. 2023;6:1–14. https://doi.org/10.1038/s42004-023-01054-6.
30. Hu C, Li S, Yag C, Chen J, Xiong Y, An G, Liu H, Hong L. ScaffoldGVAE: scaffold generation and hopping of drug molecules via a variational autoencoder based on multi-view graph neural networks. J Cheminform. 2023;15:91. https://doi.org/10.1186/s13321-023-00766-0.
31. Yu H, Li K, Shi J. DGANDDI: double generative adversarial networks for drug-drug interaction prediction. IEEE/ACM Trans Comput Biol Bioinform. 2023;20:1854–1863. https://doi.org/10.1109/TCBB.2022.319883.
32. Tripathi S, Augustin AI, Dunlop A, Sukuma R, Dheer S, Zavalny A, Haslam O, Austin T, Donchez J, Tripathi PK, Kim E. Recent advances and application of generative adversarial networks in drug discovery, development, and targeting. Artif Intell Life Sci. 2022;2:100045. https://doi.org/10.1016/j.ailsc.2022.100045.
33. Hussain S, Anees A, Das A, Nguyen BP, Marzuki M, Lin S, Wright G, Singhal A. High-content image generation for drug discovery using generative adversarial networks. Neural Netw. 2020;132:353–63. https://doi.org/10.1016/j.neunet.2020.09.007.

34. Vanhaelen Q, Lin Y-C, Zhavoronkov A. The advent of generative chemistry. ACS Med Chem Lett. 2020;11:1496–1505. https://doi.org/10.1021/acsmedchemlett.0c00088.
35. Bajorath J. Chemical language models for molecular design. Mol Inform. 2024;43:e202300288. https://doi.org/10.1002/minf.202300288.
36. Choi SR, Lee M. Transformer architecture and attention mechanisms in genome data analysis: a comprehensive review. Biology (Basel). 2023;12:1033. https://doi.org/10.3390/biology12071033.
37. Bajorath J. Chemical language models for molecular design. Mol Inf. 2024;43:e202300288. https://doi.org/10.1002/minf.202300288.
38. Maziaka Ł, Pocha A, Kacmarczyk J, Rataj K, Danel T, Warchoł M. Mol-CycleGAN: a generative model for molecular optimization. J Cheminform. 2020;12:2. https://doi.org/10.1186/s13321-019-0404-1.
39. Li C, Tang H, Zhu Y, Yamanishi Y. A reinforcement learning-driven transformer GAN for molecular generation. 2025. http://arxiv.org/abs/2503.12796. https://doi.org/10.48550/arXiv.2503.12796.
40. Buehler MJ. In situ graph reasoning and knowledge expansion using graph-PRefLexOR. Adv Intell Disc 202500006. https://doi.org/10.1002/aidi.202500006.
41. Deep Generative Models in De Novo Drug Molecule Generation | Request PDF. ResearchGate; 2025. https://doi.org/10.1021/acs.jcim.3c01496.
42. Dibha AF, Wahyuningsih S, Kharisma VD, Anori ANM, Widyananda MH, Parikesit AA, Rebezov M, Matrosova Y, Artyukhova S, Kenijz N, Kiseleva M, Jakhmola V, Zainul R. Biological activity of kencur (*Kaempferia galanga* L.) against SARS-CoV-2 main protease. Int J Health Sci. 2022;6:468–480. https://doi.org/10.53730/IJHS.V6NS.4779.
43. Wahyuni DK, Wacharasindhu S, Bankeeree W, Unnapayak H, Purnobasuki H, Junairiah J, Ansori AN, Kharisma VD, Parikesit AA, Suhargo L, Prasongsuk S. Molecular simulation of compounds from n-hexane fraction of *Sonchus arvensis* L. leaves as SARS-CoV-2 antiviral through inhibitor activity targeting strategic viral protein. J Pharm Pharmacogn Res. 2022;10:1126–1138. https://doi.org/10.56499/JPRES22.1489_10.6.1126.
44. Ansori ANM, Kharisma VD, Fadholly A, Tacharina MR, Antonius Y, Parikesit AA. Severe acute respiratory syndrome Coronavirus-2 emergence and its treatment with alternative medicines: a review. Res J Pharm Technol. 2021;14:5551–5557. https://doi.org/10.52711/097-360X.2021.00967.
45. Tambunan USF, Parikesit AA. Cracking the genetic code of human virus by using open source bioinformatics tools. Malays J Fundam Appl Sci. 2010;6. https://doi.org/10.11113/mjfa.v6n1.175.
46. Usman SFT, Parikesit AA. Molecular mechanisms of oseltamivir and other lead compounds towards influenza a virus based on proteomics and transcriptomics approaches. In: Parker R, editors. Proteomics: advances in research and applications. Hauppauge, New York: Nova Science Publishers; 2019. p. 136. https://novapublishers.com/shop/proteomics-advances-in-research-and-applications/
47. Parikesit AA. Introductory chapter: the emerging corner of the omics studies for rational drug design. In: Parikesit AA, editor. Drug design: novel advances in the omics field and applications. London: IntechOpen; 2021. p. 4. https://doi.org/10.5772/INTECHOPEN.95544.
48. Parikesit AA. Introductory chapter: the contribution of bioinformatics as blueprint lead for drug design. In: Glavic I, editor. Molecular insight of drug design. London: InTech; 2018. p. 7. https://doi.org/10.5772/intechopen.79664.
49. Valeska MD, Adisuja GP, Bernard S, Wijay R, Aldino M, Parikesit AA. The role of bioinformatics in personalized medicine: your future medical treatment. Cermin Dunia Kedokteran. 2019; 46:785–788. https://doi.org/10.5281/zenodo.4460835.
50. Temml V, Kutil Z. Structure-based molecular modeling in SAR analysis and lead optimization. Comput Struct Biotechnol J. 2021;19:1431–1444. https://doi.org/10.1016/j.csbj.2021.02.018.
51. Magoola M, Mariam Z, Magoola M. Optimizing lead compounds: the role of artificial intelligence in drug discovery. 2024. https://www.preprints.org/manuscript/202404.0055/v1. https://doi.org/10.20944/preprints202404.0055.v1.

52. Doron G, Genway S, Roberts M, Jasti S. Generative AI: driving productivity and scientific breakthroughs in pharmaceutical R&D. Drug Discov Today. 2025;30:104272. https://doi.org/10.1016/j.drudis.2024.104272.
53. Thomas M, Bender A, de Graaf C. Integrating structure-based approaches in generative molecular design. Curr Opin Struct Biol. 2023;79:102559. https://doi.org/10.1016/j.sbi.2023.10559.
54. Wang M, Wang Z, Sun H, Wang J, Shen C, Weng G, Chai X, Li H, Cao D, Hou T. Deep learning approaches for de novo drug design: an overview. Curr Opin Struct Biol. 2022;72:135–44. https://doi.org/10.1016/j.sbi.2021.1.001.
55. Liu Y, Xu C, Yang X, Zhang Y, Chen Y, Liu H. Application progress of deep generative models in de novo drug design. Mol Divers. 2024;28:2411–27. https://doi.org/10.1007/s11030-024-10942-5.
56. Malheiro V, Santos B, Figueiras A, Mascarenhas-Melo F. The potential of artificial intelligence in pharmaceutical innovation: from drug discovery to clinical trials. Pharmaceuticals. 2025;18:788. https://doi.org/10.3390/ph18060788.
57. Ferreira FJN, Carni AS. AI-driven drug discovery: a comprehensive review. ACS Omega. 2025. https://doi.org/10.1021/acsomega.5c00549.
58. Tang X, Dai H, Knight E, Wu F, Li Y, Li T, Gerstein M. A survey of generative AI for de novo drug design: new frontiers in molecule and protein generation. Brief Bioinform. 2024;25:bbae338. https://doi.org/10.1093/bib/bbae338.
59. Li Y, Pei J, Lai L. Structure-based de novo drug design using 3D deep generative models. Chem Sci. 12:13664–13675. https://doi.org/10.1039/d1sc04444c.
60. Hakaman H, Surya F, Geradine GE, Honggo H, Paikesit A. In-Silico molecular docking analysis of monoclonal antibodies, approved inhibitors, and plant-based inhibitors targeting extracellular and intracellular HER2 receptor. Berkala Penelitian Hayati. 2025;31:33–48. https://doi.org/10.23869/bphjbr.31.1.20256
61. Lorell J, Gautama NRP, Tanzil TS, Subagja M, Parikesit AA: In silico study of RNA polymerase inhibitor drugs for Rift Valley fever virus using RdRp protein as the target. J Pharm Pharmacogn Res. 2025;13:1–15. https://doi.org/10.56499/jppres24.1967_13.1.1.
62. Wan M, Li S, Wang J, Zhang O, Du H, Jiang D, Wu Z, Deng Y, Kang Y, Pan P, Li D, Wang X, Yao X, Hou T, Hsieh C-Y. ClickGen: directed exploration of synthesizable chemical space via modular reactions and reinforcement learning. Nat Commun. 2024;15:10127. https://doi.org/10.1038/s41467-024-54456-y.
63. Bilodeau C, Jin W, Jaakkola T, Barzilay R, Jensen KF. Generative models for molecular discovery: recent advances and challenges. WIREs Comput Molec Sci. 2022;12:e1608. https://doi.org/10.1002/wcms.1608.
64. Tabana Y, Babu D, Fahlman R, Sirki AG, Barakat K. Target identification of small molecules: an overview of the current applications in drug discovery. BMC Biotechnol. 2023;23:44. https://doi.org/10.1186/s12896-023-00815-4.
65. Zhou G, Rusnac D-V, Park H, Canzani D, Nguyen HM, Stewart L, Bush MF, Nguyen PT, Wulff H, Yarov-Yarovoy V, Zheng N, DiMaio F. An artificial intelligence accelerated virtual screening platform for drug discovery. Nat Commun. 2024;15:7761. https://doi.org/10.1038/s41467-024-52061-7.
66. Rehman AU, Li, Wu B, Ali Y, Rasheed S, Shaheen S, Liu X, Luo R, Zhang J. Role of artificial intelligence in revolutionizing drug discovery. Fundam Res. 2025;5:1273–1287. https://doi.org/10.1016/j.fmre.2024.04.021.
67. Heveer KE, Pesavento R, Ren J, Lee H, Ratia K, Johnson ME. Chapter twelve: Hit-to-lead: hit validation and assessment. In: Lesburg CA, editor. Methods in enzymology. Academic Press; 2018. p. 265–309. https://doi.org/10.1016/bs.mie.2018.09.022.
68. Schapin N, Majewski M, Varela-Rial A, Arranz C, Fabritiis GD. Machine learning small molecule properties in drug discovery. Artif Intell Chem. 2023;1:100020. https://doi.org/10.1016/j.aichem.2023.10000.
69. Singh S, Gupta H, Sharma P, Sahi S. A vances in Artificial Intelligence (AI)-assisted approaches in drug screening. Artif Intell Chem. 2024;2:100039. https://doi.org/10.1016/j.aichem.2023.10009.

70. Ivanenkov Y, Zagribelnyy B, Malyshev A, Evteev S, Terentiev V, Kamya P, Bezrukov D, Aliper A, Ren F, Zhavoronkov A. The Hitchhiker's guide to deep learning driven generative chemistry. ACS Med Chem Lett. 2023;14:901–15. https://doi.org/10.1021/acsmedchemlett.3c00041.
71. Fromer JC, Coley CW. Computer-aided multi-objective optimization in small molecule discovery. Patterns. 2023;4:100678. https://doi.org/10.1016/j.patter.2023.100678.
72. Bleicher LS, Daelen T, Honeycutt J, Hassan M, Chandrasekhar J, Shirley W, Tsui V, Schmitz U. Enhanced utility of AI/ML methods during lead optimization by inclusion of 3D ligand information. Front Drug Discov. 2022;2. https://doi.org/10.3389/fddsv.2022.1074797.
73. Chakraborty C, Bhattacharya M, Pal S, Islam MA. Generative AI in drug discovery and development: the next revolution of drug discovery and development would be directed by generative AI. Ann Med Surg (Lond). 2024;86:6340–6343. https://doi.org/10.1097/MS9.0000000000002438.
74. Li Q, Ahadpour M, Rocca M, Huang S-M. Clinical pharmacology regulatory sciences in drug development and precision medicine: current status and emerging trends. AAPS J. 2021;23:54. https://doi.org/10.1208/s12248-021-00563-3.
75. Hardman TC, Aitchison R, Scaife R, Edwards J, Slater G. The future of clinical trials and drug development: 2050. Drugs Context. 2023;12:2023-2-2. https://doi.org/10.7573/dic.2023-2-2.
76. Alowais SA, Alghamdi SS, Alsuhebany N, Alqahtani T, Alshaya AI, Almohareb SN, Aldairem A, Alrashed M, Bin Saleh K, Badreldin HA, Al Yami MS, Al Harbi S, Albekairy AM. Revolutionizing healthcare: the role of artificial intelligence in clinical practice. BMC Med Educ. 2023;23:689. https://doi.org/10.1186/s12909-023-04698-z.
77. Zhang K, Yang X, Wang Y, Yu Y, Huang N, Li G, Li X, Wu JC, Yang S. Artificial intelligence in drug development. Nat Med. 2025;31:45–59. https://doi.org/10.1038/s41591-024-03434-4.
78. Wang J, Liu Y, Li P, Lin Z, Sindakis S, Aggarwal S. Overview of data quality: examining the dimensions, antecedents, and impacts of data quality. J Knowl Econ. 2023;1–20. https://doi.org/10.1007/s13132-022-01096-6.
79. Herráiz-Gil S, Nygren-Jiménez E, Acosta-Alonso DN, León C, Guerrero-Aspizua S. Artificial intelligence-based methods for drug repurposing and development in cancer. Appl Sci. 2025;15:2798. https://doi.org/10.3390/app15052798.
80. Nguyen-Vo T-H, Teesdale-Spittle P, Harvey JE, Guyen BP. Molecular representations in bio-cheminformatics. Memetic Comp. 2024;16, 519–536. https://doi.org/10.1007/s12293-024-00414-6.
81. Mswahili M, Jeong Y-S. Transformer-based models for chemical SMILES representation: a comprehensive literature review. Heliyon. 2024;10:e39038. https://doi.org/10.1016/j.heliyon.2024.e39038.
82. Hassija V, Chaola V, Mahapatra A, Sinhal A, Goel D, Huang K, Scardapane S, Spinelli I, Mahmud M, Hussain A. Interpreting black-box models: a review on explainable artificial intelligence. Cogn Comput. 2024;16:45–74. https://doi.org/10.1007/s12559-023-10179-8.
83. Vamathevan J, Clark D, Czodrowski P, Dunham I, Ferran E, Lee G, Li B, Madabhushi A, Shah P, Spitzer M, Zhao S. Applications of machine learning in drug discovery and development. Nat Rev Drug Discov. 2019;18:463–477. https://doi.org/10.1038/s41573-019-0024-5.
84. Flores-Ernandez H, Martinez-Ledesma E. A systematic review of deep learning chemical language models in recent era. J Cheminform. 2024;16:129. https://doi.org/10.1186/s13321-024-00916-y.
85. Yang J, Cai Y, Zhao K, Xie H, Chen X. Concepts and applications of chemical fingerprint for hit and lead screening. Drug Discovery Today. 2022;27:103356. https://doi.org/10.1016/j.drudis.2022.103356.
86. Gao W, Coley CW. The synthesizability of molecules proposed by generative models. J Chem Inf Model. 2020;60:5714–23. https://doi.org/10.1021/acs.jcim.0c00174.
87. Li B, Wang Z, Liu Z, Tao Y, Sha C, He, Li X. DrugMetric: quantitative drug-likeness scoring based on chemical space distance. Brief Bioinform. 2024;25:bbae321. https://doi.org/10.1093/bib/bbae321.

88. Adelusi TI, Oyedele A-QK, Boyenle ID, Ogunlana AT, Adeyemi RO, Ukachi CD, Idris MO, Olaoba OT, Adedotun IO, Kolawole OE, Xiaoxing Y, Abdul-Hammed M. Molecular modeling in drug discovery. Inform Med Unlocked. 2022;29:100880. https://doi.org/10.1016/j.imu.2022.100880.
89. Ko YS, Parkinson J, Liu C, Wang W. TUnA, an uncertainty-aware transformer model for sequence-based protein–protein interaction prediction. Brief Bioinform. 2024;25:bbae359. https://doi.org/10.1093/bib/bbae359.
90. Son A, Park J, Kim W, Lee W, Yoon Y, Ji J, Kim H. Integrating computational design and experimental approaches for next-generation biologics. Biomolecules. 2024;14:1073. https://doi.org/10.3390/biom14091073.
91. Kaswan KS, Dhatterwal JS, Malik K, Baliyan A. Generative AI: a review on models and applications. In: 2023 international conference on communication, security and artificial intelligence (ICCSA). 2023. p. 699–704. https://doi.org/10.1109/ICCSAI59793.2023.10421601.
92. Bandi A, Adapa PSR, Kuchi YEV: The power of generative AI: a review of requirements, models, input–output formats, evaluation metrics, and challenges. Fut Internet. 2023;15:260. https://doi.org/10.3390/fi15080260.
93. Kolici V, Xhafa F, Barolli L, Lala A. Scalability, memory issues and challenges in mining large data sets. In: 2014 international conference on intelligent networking and collaborative systems. 2014. p. 268–73. https://doi.org/10.1109/INCoS.2014.50.
94. Sanka AI, Cheung RCC. A s systematic review of lockchain scalability: issues, solutions, analysis and future research. J Netw Comput Appl. 2021;195:103232. https://doi.org/10.1016/j.jnc.2021.103232.
95. Tynchenko VV, Malashin I, Kurashkin SO, Tynchenko V, Gantimurov A, Nelyub V, Borodulin A. Multi-criteria genetic algorithm for optimizing distributed computing systems in neural network synthesis. Fut Internet. 2025;17:215. https://doi.org/10.3390/fi17050215.
96. Rane NL, Rane J, Alick SK, Kaya Ö. Artificial intelligence, machine learning, and deep learning in cloud, edge, and quantum computing: a review of trends, challenges, and future directions. Deep Sci Publ. 2024. https://doi.org/10.70593/978-81-981271-0-5.
97. Tang B, Ewalt J, Ng H-L. Generative AI models for drug discovery. In: Saxena AK, editor. Biophysical and computational tools in drug discovery. Cham: Springer International Publishing; 2021. p. 221–243. https://doi.org/10.1007/7355_2021_24.
98. Robinson S, Brooks R. Assumptions and simplifications in discrete-event simulation modelling. J Simulat. 1–18. https://doi.org/10.1080/17477778.2024.2407369.
99. Robinson S. Exploring the relationship between simulation model accuracy and complexity. J Oper Res Soc. 2023;74:1992–2011. https://doi.org/10.1080/0160682.2022.2122740.
100. Kim E, Isozaki I, Sirkin N, Robson M. Generative artificial intelligence reproducibility and consensus. 2024. http://arxiv.org/abs/2307.01898. https://doi.org/10.48550/arXiv.2307.01898.
101. Aal E Ali, R.S., Meng, J., Khan, M.E.I., Jiang, X.: Machine learning advancements in organic synthesis: A focused exploration of artificial intelligence applications in chemistry. Artificial Intelligence Chemistry. 2, 100049 (2024). https://doi.org/10.1016/j.aichem.2024.1000 9.
102. Nada H, Meanwell NA, Gabr MT. Virtual screening: hope, hype, and the fine line in between. Exp Opin Drug Disc. 2025;20:145–162. https://doi.org/10.1080/17460441.2025.245666.
103. Reddy S. Generative AI in healthcare: an implementation science-informed translational path on application, integration, and governance. Implement Sci. 2024;19:27. https://doi.org/10.1186/s13012-024-01357-9.
104. Appel G, Neelbauer J, Schweidel DA. Generative AI has an intellectual property problem. https://hbr.org/2023/04/generative-ai-has-an-intellectual-property-problem.
105. Bonomi L, Huang Y, Ohno-Machado L. Privacy challenges and research opportunities for genomic data sharing. Nat Genet. 2020;52:646–54. https://doi.org/10.1038/s41588-020-0651-0.
106. Hartung T, Whean M, Tong W, Califf RM. Is regulatory science ready for artificial intelligence? npj Digit Med. 2025;8:1–5. https://doi.org/10.1038/s41746-025-01596-0.

107. Mariam Z, Nia SK, Magoola M. Unlocking the future of drug development: generative AI, digital twins, and beyond. BioMedInformatics. 2024;4:1441–1456. https://doi.org/10.3390/biomedinformatics4020079.
108. Fu C, Chen Q. The future of pharmaceuticals: artificial intelligence in drug discovery and development. J Pharm Anal. 2025;101248. https://doi.org/10.1016/j.jpha.2025.101248.
109. Mumtaz H, Riaz MH, Wajid H, Saqib M, Eeshan MH, Khan SE, Chauhan YR, Sohail H, Vohra LI. Current challenges and potential solutions to the use of digital health technologies in evidence generation: a narrative review. Front Digit Health. 2023;5:1203945. https://doi.org/10.3389/fdgth.2023.1203945.
110. Niazi K, Mariam Z. Artificial intelligence in drug development: reshaping the therapeutic landscape. Therap Adv Drug Saf. 2025;16:20420986251321704. https://doi.org/10.1177/20420986251321704.
111. Balasubramaniam S, Chirchi V, Kadry S, Agoramoorthy M, Gururama SP, Satheesh KK, Sivakumar TA. The road ahead: emerging trends, unresolved issues, and concluding remarks in generative AI—a comprehensive review. Int J Intell Syst. 2024;2024:4013195. https://doi.org/10.1155/202/4013195.
112. Lopez V, Hoang L, Martinez-Galindo M, Fernánde-Díaz R, Sbodio ML, Ordonez-Hurtado R, Zayats M, Mulligan N, Bettencourt-Silva J. Enhancing foundation models for scientific discovery via multimodal knowledge graph representations. J Web Semant. 2025;84:100845. https://doi.org/10.1016/j.websem.2024.100845.
113. Piazza L, Srinivasan S, Tuccinardi T, Ajorath J. Transforming molecular cores, substituents, and combinations into structurally diverse compounds using chemical language models. Eur J Med Chem. 2025;291:117615. https://doi.org/10.1016/j.ejmech.225.117615.
114. Suri S. Defining our future with generative AI. Nat Comput Sci. 2024;4:641–3. https://doi.org/10.1038/s43588-024-00694-5.
115. Romanelli C, Cerchia C, Lavecchia A. Unlocking the potential of generative artificial intelligence in drug discovery. In: Lyu Z, editor. Applications of generative AI. Cham: Springer International Publishing; 2024. p. 37–63. https://doi.org/10.1007/978-3-031-4638-2_3.
116. .Ismi DP, Pulungan R. Deep learning for protein secondary structure prediction: pre and post-AlphaFold. Comput Struct Biotechnol J. 2022;20:6271–6286. https://doi.org/10.1016/j.csbj.2022.11.012.
117. Alquraishi M. AlphaFold at CASP13. Bioinformatics. 2019;35:4862–5. https://doi.org/10.1093/bioinformatics/btz422.
118. Bordukova M, Makarov N, Rodriguez-Esteban R, Schmich F, Menden MP. Generative artificial intelligence empowers digital twins in drug discovery and clinical trials. Exp Opin Drug Discov. 2024;19:33–42. https://doi.org/10.1080/17460441.2023.2273839.
119. Lu J, Choi K, Eremeev M, Gobburu J, Goswami S, Liu Q, Mo G, Musante CJ, Shahin MH. Large language models and their applications in drug discovery and development: a primer. Clin Trans Sci. 2025;18:e70205. https://doi.org/10.1111/cts.70205.
120. Chakraborty C, Bhattacharya M, Pal S, Chatterjee S, Das A, Lee S-S. AI-enabled language models (LMs) to large language models (LLMs) and multimodal large language models (MLLMs) in drug discovery and development. J Adv Res. 2025. https://doi.org/10.1016/j.jare.2025.02.011.
121. Blanco-González A, Cabezón A, Seco Gonzále A, Conde-Torres D, Antelo-Riveiro P, Piñeiro Á, Garcia-Fandino R. The role of AI in drug discovery: challenges, opportunities, and strategies. Pharmaceuticals (Basel). 2023;16:891. https://doi.org/10.3390/ph16060891.
122. Vogt M. Exploring chemical space—generative models and their evaluation. Artif Intell Life Sci. 2023;3:100064. https://doi.org/10.1016/j.ailsc.2023.100064.
123. Thomas M, O'Boyle NM, Bender A, De Graaf C. MolScore: a scoring, evaluation and benchmarking framework for generative models in de novo drug design. J Cheminform. 2024;16:64. https://doi.org/10.1186/s13321-024-00861-w.
124. Abeer ANMN, Urban NM, Weil MR, Alexander FJ, Yoon B-J. Multi-objective latent space optimization of generative molecular design models. Patterns. 2024;5:101042. https://doi.org/10.1016/j.patter.2024.101042.

125. Askr H, Elgeldawi E, Aboul Ella H, Elshaier YAMM, Gomaa MM, Hassanien AE. Deep learning in drug discovery: an integrative review and future challenges. Artif Intell Rev. 2023;56:5975–6037. https://doi.org/10.1007/s10462-022-10306-1.
126. Krakowski S. Human-AI agency in the age of generative AI. Inf Organ. 2025;35:100560. https://doi.org/10.1016/j.infoandorg.2025.10050.
127. Chen W, Song C, Leng L, Zhang S, Chen S. The application of artificial intelligence accelerates G protein-coupled receptor ligand discovery. Engineering. 2024;32:18–28. https://doi.org/10.1016/j.eng.2023.09.011.

Chapter 2
Computational Design of mRNA-Based Cancer Immunotherapies

Dženan Kovačić, Selma Ibrahimpašić Huseinbegović, Azra Karić Muratović, and Melissa Jahibašić

Abstract Messenger RNA (mRNA) has rapidly advanced from a platform for prophylactic vaccination to a credible modality for cancer immunotherapy. Yet its successful translation to oncology is constrained by tumor heterogeneity, immune suppression, and the pharmacology of mRNA-lipid nanoparticle (LNP) systems, as well as modest in vitro-to-in vivo translatability during preclinical development. This chapter synthesizes a computationally driven framework for end-to-end design of mRNA cancer immunotherapies that couples multi-omics with modern machine learning and mechanistic modeling. We first outline the biological determinants of antigen processing and presentation across MHC classes, emphasizing how tumor mutational burden, clonal architecture, and microenvironmental cues shape neoepitope visibility. We then survey integrated pipelines for neoantigen discovery that combine immunopeptidomics-trained predictors with RNA/DNA evidence and HLA context, using toolchains such as pVACtools as exemplars while noting their current limits. Next, we detail sequence-level engineering of therapeutic mRNAs, codon and UTR optimization, linker/junction control, trafficking motifs, and structure, expression trade-offs, highlighting emerging AI models that jointly optimize translation efficiency and structural stability. On the delivery side, we review data-driven LNP design, including graph- and transformer-based models for lipid/formulation search and multiscale molecular dynamics to interrogate self-assembly, internal architecture, and endosomal escape. Finally, we describe systems biology methods, gene-regulatory network modeling, agent-based simulations, and quantitative systems pharmacology (QSP), to connect molecular design choices to predicted cellular and tissue-level responses, optimizing dosing, scheduling, and combination strategies. Across these layers, we propose practical integration points that reduce candidate attrition, shorten iteration cycles, and improve physiological relevance.

D. Kovačić (✉) · M. Jahibašić
Department of Genetics and Bioengineering, International Burch University, Sarajevo, Bosnia and Herzegovina
e-mail: dzenan.kovacic@stu.ibu.edu.ba

S. Ibrahimpašić Huseinbegović · A. Karić Muratović
BiOptimizers, Reno, USA

A. A. Parikesit and A. N. M. Ansori (eds.), *Recent Computational Techniques in De Novo Drug Design*, Bio-IT and AI, https://doi.org/10.1007/978-3-032-20428-8_2

Keywords mRNA · mRNA vaccines · Cancer immunotherapy · mRNA immunotherapy

2.1 Introduction

The past decade has seen an extraordinary evolution in RNA-based vaccines and therapeutics. Though initially validated as a potent platform technology for prophylactic vaccines, messenger (m)RNA has emerged as a promising new approach to cancer immunotherapy with encouraging clinical results. Such a surge of interest is not surprising, as the mRNA platform offers tunable immunogenicity, scalability and personalization; all of which are crucial criteria to meet for successful deployment in clinical oncology. However, designing an effective mRNA immunotherapy remains a formidable challenge, stemming from cancer biology, immunobiology and the pharmacological aspects of mRNA-based vaccines themselves. Furthermore, the most utilized mRNA delivery vectors—lipid nanoparticles (LNPs)—though tremendously promising and versatile, are hallmarked with non-trivial delivery, specificity and toxicity challenges. To further complexify the issue, there is a modest in vitro-to-in vivo translatability during the pre-clinical phases of development, necessitating the usage of costly animal studies or sophisticated and complex tissue culture models for design validation or proof-of-concept. With the central challenge in cancer immunotherapy being precision, designing an mRNA vaccine construct capable of eliciting a durable, targeted and effective immune response necessitates a technically challenging and integrative approach, informed heavily by high-quality cancer-centric multi-omics data.

Cancer is not a single disease; rather a landscape of heterogenic cellular ecosystems, each with its own set of mutational, transcriptional and regulatory landscapes, which hurdles efforts at designing stable and immunologically potent mRNA constructs. High throughput sequencing methods have given us unprecedented insight into the molecular dynamism of cancer, resulting in multi-dimensional data at a massive scale that is challenging to curate, integrate and analyze, particularly for the purpose of cancer therapy development.

Computational methods, from established pipelines to novel artificial intelligence (AI)-centric architectures, are becoming indispensable for turning these data into therapeutically actionable insights. Implementation of computational tools for mRNA cancer immunotherapy can certainly be a force multiplier if implemented properly, as it allows for the determination of patient and population-specific cancer neoantigens capable of stimulating an immune response, thus streamlining neoepitope search and selection. Furthermore, not all neoepitopes are created equal. Thorough understanding of the translational, mechanistic and regulatory context of the corresponding source protein(s) harboring a particular neoantigen is crucial, as not every neoantigen will necessarily be available to the immune system, either due to conditional expression, mutation profile or simply its localization. To this end, multi-omics computational analyses of high-resolution data may be efficiently integrated with sophisticated disease modeling platforms, whereby resolved expression networks and their response to simulated perturbations are used to interrogate the complex behavior of the cellular ecosystem of cancer; either to study basic cancer biology or simulate treatment response. Though such sophisticated data integrations are by no means technically simple to implement, they offer a substantial time and cost-saving advantage, particularly for high-throughput immunotherapy development. Simply stated, implementation of computational methods in mRNA immunotherapy development can help reduce the chance of candidate failure or underperformance across multiple meaningful metrics during development.

This chapter begins with an overview of the biological principles behind mRNA immunotherapy and its key components. We then explore how computational pipelines and 'AI-based' models may be utilized to overcome numerous design challenges, emphasizing data integration, construct optimization and systems-level computational simulations. We also discuss current limitations and future opportunities for the field, with emphasis on realistic implementation of AI and systems biology within today's experimental and clinical frameworks.

2.2 The Biological Basis of Antibody-Based and mRNA-Based Immunotherapies

The fundamental premise of mRNA immunotherapy is deceptively simple, involving systemic delivery of an mRNA construct whose immunogenic protein product stimulate an immune response to a particular neoantigen or a set of neoantigens, using either conventional or self-amplifying constructs (Fig. 2.1). Such an approach has certainly enjoyed noteworthy success in the domain of infectious disease prophylaxis in recent years. However, translating this success to cancer, while implementable in principle, hinges on numerous non-trivial factors rooted in cancer biology. Characterization of infectious agents and the corresponding host immunodynamics is comparatively straightforward by today's standards for many infections, which is rarely the case for cancer.

Depending on DNA repair mechanism efficiency, tumors can either have a high or low tumor mutational burden (TMB), resulting in phenotypes that are broadly referred to as immunologically hot or cold respectively (Fig. 2.2) [1–4]. Immunologically hot tumors are generally more responsive to immunotherapy checkpoint inhibitors, while cold tumors are less responsive to such regiments, at least without additional therapeutic modalities [5]. This is not at all surprising, as high TMB usually results in a broader pool of available immunogenic neoantigens, thereby leading to dense immune cell infiltration. Additionally, cold tumors can become hot if the corresponding DNA repair mechanisms fail or due to therapy-induced cellular damage [5, 6]. While these categories are certainly an oversimplification given cancer's heterogenic nature, they are nevertheless crucial when designing immunotherapies. For

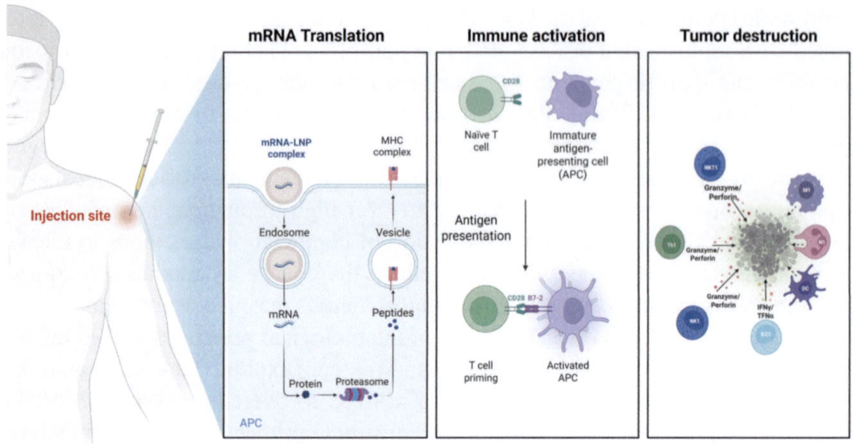

Fig. 2.1 Illustrative depiction of the underlying principles behind mRNA immunotherapies. Once an mRNA is internalized by an antigen presenting cell (APC), ideally a dendritic cell (DC), the APC migrates to lymphoid organs in order to initiate immune priming and cloncal expansion of the adaptive immune system. After this, professional immune cells migrate to the site of the tumor and initiate tumor killing. Figure was generated using BioRender

instance, roughly 80% of colorectal cancers are microsatellite stable (MSS), meaning that their functional mismatch repair (MMR) proteins are functional, resulting in low TMB [7]. Such tumors generate fewer frameshift mutations, thereby limiting the number of immunogenic neoantigens and diminishing the clinical utility of immune checkpoint inhibitors outside of hypermutated subgroups with DNA polymerase epsilon (POLE) mutations.

Considering this, a fundamental question arises: why are antibody-based immunotherapies effective against immunologically hot tumors and what are their mechanisms of action? While, indeed, immunologically hot tumors are replete with immune cell infiltrates, these immune cells are often hallmarked with functional exhaustion due to chronic antigen exposure by the tumor. This state is characterized by diminished cytotoxic capacity of effector T-cells, reduced cytokine production and chronic, low-grade inflammation driven by the exhausted infiltrates. This persistent inflammation paradoxically favors recruitment of suppressive immune cell phenotypes, including regulatory T cells (Tregs), myeloid derived suppressive cells (MDSCs), M2-like tumor associated macrophages (TAMs) and fibroblasts, which reinforce and perpetuate the exhausted cell states [4–6, 8–10]. Antibody-based immunotherapies can not only salvage exhausted infiltrates, but also initiate new T-cell responses, bridge innate-adaptive toxicity, expand cell repertoire diversity and reprogram immunosuppressive niches [11, 12]. For instance, immune checkpoint inhibitors targeting programmed cell death protein 1 (PD-1), PD-ligand 1 (PD-L1), cytotoxic lymphocyte associated protein 4 (CTLA-4) and lymphocyte activation gene-3 (LAG-3) reverse exhaustion while enabling new priming and

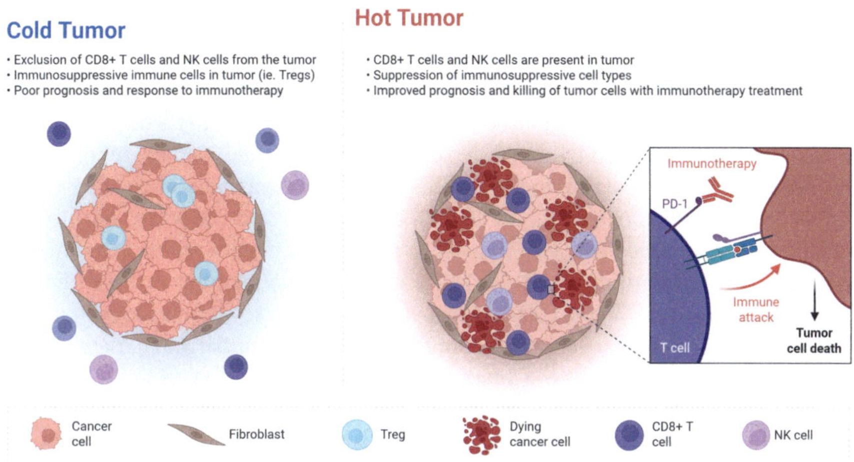

Fig. 2.2 Illustrative depiction of immunologically hot and immunologically cold tumors, highlighting the immunological besis behind immune checkpoint inhibitor effectiveness against hot tumors. Figure was adopted from BioRender

repertoire expansion, while tumor-directed antibodies (anti-human epidermal growth factor receptor 2 (HER2), anti-epidermal growth factor receptor (EGFR), anti-CD38, anti-CD-20, anti-B-cell maturation antigen (BCMA)) mediate antibody-dependent cellular toxicity (ADCC) and complement-dependent cytotoxicity (CDC) via antigen binding. Tumor specific antibodies enable dendritic cells (DCs) to cross-present tumor antigens on major histocompatibility complex class I (MHC-I), thereby effectively broadening the immune response as to recruit and prime new cytotoxic T-cell clones.

While antibody-based immunotherapies are certainly effective against immunologically hot tumors, this is significantly less pronounced in cold tumors, as they provide little-to-no immunobiological substrate to support immune infiltration. Simply stated, there is nothing for antibody-based immunotherapies to rescue or stimulate in this case, thereby necessitating strategies for initial immune priming prior to immunotherapy inclusion. Indeed, current therapeutic modalities such as oncolytic viruses, radiation and chemotherapy have their place in supporting immune priming by directly damaging or stressing tumors, however these can reflect rather poorly on overall quality of life due to off-target effects or toxicity profiles. This posits mRNA technology as an incredibly attractive platform for immunotherapy development, as it allows not only for priming of de novo, neoantigen-specific immune responses, but also the delivery of neoantigen-encoding mRNA directly to the tumor to facilitate a distinct immunological signature. In this way, mRNA vaccines complement antibody-based immunotherapies: while antibodies salvage and expand existing infiltrates, mRNA vaccines can potentially convert immunologically cold tumors into inflamed, immunotherapy sensitive states. This notion is not without merit, as was demonstrated by phase 1 clinical trial results of the pancreatic ductal adenocarcinoma (PDAC) personalized mRNA-lipoplex vaccine, autogene cevumeran (BNT122, RO7198457), published by BioNTech and Genentech, with encouraging data on multi-year de novo $CD8^+$ T-cell longevity [13]. The phase 1 trial entailed surgery, followed by one dose of anti-PD-L1 immunotherapy, upon which mRNA-induced immune priming was initiated through 8 intravenous doses. Patients further received chemotherapy and an additional single vaccine boost, resulting in much longer recurrence-free survival (RFS) in responders without reaching median RFS and with a hazard ration of ~ 0.14. After priming alone, the newly generated T-cell clones had a median estimated lifespan of ~ 1.1 years, wherein the administration of the booster dose rose this median by ~ 7.7 years. Furthermore, many clones persisted at significant levels ~ 3 years post-vaccination.

The results of this trial are of tremendous importance to the mRNA immunotherapy field, particularly since PDAC is a notoriously lethal and immunologically cold type of adenocarcinoma. By no means is this an isolated occurrence, as this vaccination platform was also tested in animal models of esophageal squamous cell carcinoma with encouraging results in terms of T-cell responses and response duration [14]. This work focused on utilizing a multi-epitope mRNA vaccine design, whereas the Genentech/BioNTech mRNA vaccine was a personalized single epitope construct. Although much work lies ahead in further pioneering these therapies, the crucial proof-of-concept step has resulted in encouraging in vitro and in vivo data.

2.3 Foundations of mRNA Immunotherapy Design: Bridging Concept with Computation

2.3.1 Neoantigens and Their Computational Prediction

The therapeutic success of mRNA vaccines—whether single epitope or multi epitope—critically depends on intracellular antigen processing efficiency of the corresponding protein and downstream antigen presentation [15–17]. When antigen-presenting cells (APCs)—primarily DCs—are transfected by the mRNA vaccine, the encoded antigen is synthesized endogenously and processed via the proteasome-transporter associated with antigen processing (TAP) pathway for MHC-I loading and subsequent presentation to $CD8^+$ T-cells. Additional lysosomal targeting motifs may be added to the mRNA sequence to bias the antigen towards MHC-II compartments, which enables $CD4^+$ T-cell priming [15, 16, 18, 19]. Furthermore, cross-presentation enables exogenous mRNA-delivered antigens to be shuttled into MHC-I processing pathways upon APC internalization, adding yet another avenue of immunogenicity should there be off-target mRNA delivery. Though conceptually simple, ensuring efficient antigen processing of mRNA-encoded immunogenic proteins is by no means a trivial task. Firstly, antigen processing is an intrinsically inefficient process, and is constrained by the multiplicity of steps required, the limitations of the immune system's repertoire to recognize foreign/neoantigenic peptides along with the complex regulatory landscape characterizing competent adaptive immune response biology [20]. This is sensical, after all, as such inefficiencies are necessary for the immune system to competently distinguish self from non-self-peptide motifs when priming a potent, yet tightly controlled, immune response. Further conditioning adaptive immune response competence is the repertoire of MHC alleles (where the term human leukocyte antigen (HLA) refers to human MHC proteins) carried by an individual. The structure of MHC proteins dictates which peptide fragments are presented to T cells, thus shaping the landscape of immune recognition. Polymorphisms in MHC class I (HLA-A, -B, -C) and class II (HLA-DR, -DP, -DQ) loci alter the structure of the so-called peptide binding groove, thus impacting peptide binding affinity and specificity (Fig. 2.3). Thus, allelic differences in these protein structures dictate whether critical viral, bacterial or tumor-derived epitopes can be presented to naïve adaptive immune cells at all for priming. While some HLA variants result in broad antigen-binding capacity, enabling the presentation of diverse epitopes, others exhibit narrow specificity or epitope preferences that leave gaps in immune surveillance [21]. HLA variants, therefore, contribute to individual immune response variations to immunization, infection, cancer and immunotherapy. Thus, selection of epitopes encoded by immunotherapeutic mRNA should be informed by their corresponding affinity for HLA variants, the likeliness of them being processed by the proper antigen-processing machinery and their actual presentation in a biologically plausible context.

Tumor antigens originate from diverse sources, such as non-synonymous single nucleotide polymorphisms (SNVs)/indels found in non-canonical ORFs (including frameshifts), gene fusions, aberrant splicing events (exon skipping, intron retention),

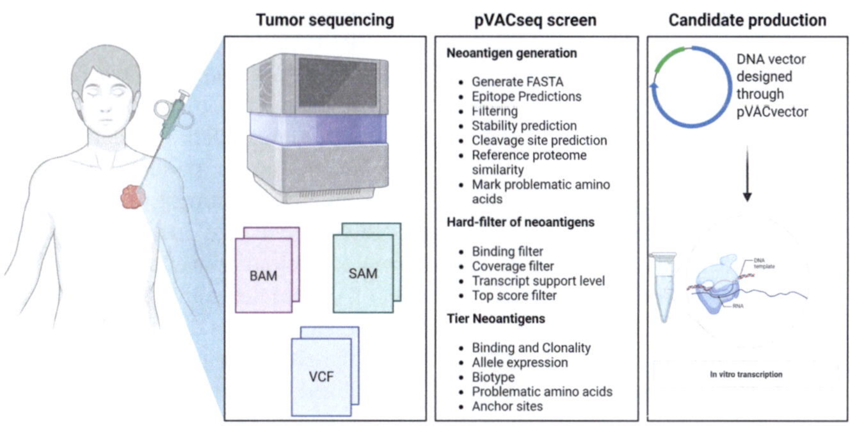

Fig. 2.3 Illustrative depiction of how patient tumor biopsies may be used in conjunction with pVACseq to develop personalized multi-epitope mRNA immunotherapies. Figure was generated using BioRender

non-canonical translations (upstream ORFs, cryptic/antisense ORFs, endogenous retroelements) and post-translationally modified peptides [3, 22]. Antigen supply available to the immune system is governed by protein abundance and turnover, where nonsense-mediated decay escapees and ubiquitin–proteasome targeting by tumor-altered E3 ligases all modulate the peptide precursor pool, thereby affecting tumor immunogenicity [23–25]. Furthermore, copy-number changes and aneuploidy can amplify or dilute antigen sources, with transcriptional program landscapes further contributing to proteome reshaping that influences the epitope landscape. For MHC I presentation, precursor proteins are degraded by the 20S/26S proteasome. In tumors exposed to interferon gamma (IFN-γ), for example, immunoproteasome subunits PSMB8/LMP7, PSMB9/LMP2, PSMB10/MECL-1) replace constitutive catalytic sites, which biases cleavage towards C-terminal residues that favor MHC-I anchoring [26, 27]. Proteasomes may also generate spliced peptides (cis/trans proteasome splicing), adding further diversity but complexifying epitope prediction. Cytosolic peptidases further shape intermediates before transport via TAP1/2 proteins into the endoplasmic reticulum (ER), where ER aminopeptidases ERAP1/ERAP2 trim N-termini to optimal lengths, meaning that polymorphisms of ERAP can overly trim and delete otherwise presentable epitopes [28–30]. When shuttled to the ER, the peptide-loading complex (PLC) (calnexin, calreticulin, Erp57, tapasin) edits and optimizes binding onto MHC-I heavy chain/β2-microglobulin (B2M) heterodimers. Dependence on tapasin varies by allele, thereby influencing peptide repertoires and surface stability during presentation [31–33]. After this, loaded pHLA-I traffics to the surface, where the complex half-life determines steady-state abundance and immune visibility, explaining why stability can outperform raw affinity as a correlate of presentation. This axis is commonly compromised in cancer, where B2M loss of function, HLA-I heavy chain mutations, HLA loss or heterozygosity, along with aberrant epigenetic changes can reduce surface pHLA-I [23]. Disruptions in IFN-γ—Janus kinase (JAK)—signal transducer and activator of transcription 1 (STAT1)—interferon regulatory factor 1 (IRF1) signaling axis blunt induction of antigen processing machinery (APM) genes, collapsing antigen flow [23, 26]. Oncogenic pathways, such as β-catening/Wnt signaling can block the recruitment of conventional dendritic cells (cDCs) and cross priming, which indirectly selects against antigenic tumors [34, 35]. Additional factors that can perturb proteostasis and ER function can originate from the tumor microenvironment (TME), where hypoxia, acidosis, reactive oxygen species (ROS), lipid peroxidation can rewire peptide processing and glycosylation. Furthermore, myeloid-derived suppressor cells (MDSCs) and tumor-associated macrophages (TAMs), release interleukin (IL)-10/transforming growth factor β (TGFβ) and arginase/indoleamine dioxygenase, dampening IFN signaling and T-cell effector function, thereby reducing selective pressure to maintain antigen presentation [36–38]. Specialized DCs capture tumor debris or vesicles, route antigens through cytosolic or vacuolar pathways and present them via their own HLA-I molecules to naïve T-cells—a step particularly sensitive to TME cues. For MHC II processing, professional APCs internalize tumor antigens via endocytosis or autophagy. In APC endo-lysosomes, cathepsin proteases generate peptides that load onto HLA-II after invariant chain removal and HLA-DM-mediated editing

[39]. Tumor cells themselves seldom express robust HLA-II without potent cytokine induction, but when this occurs, $CD4^+$ T-cells assist in the enhancement of $CD8^+$ responses, thereby shaping epitope hierarchies [40]. Autophagy flux, lysosomal pH and cathepsin balance influence the class-II peptidome and aberrations here can both create potent tumor-specific epitopes and erase them [41, 42].

For both MHC classes, immunoediting sculpts the epitope landscape. Tumor cell clones bearing highly visible neoepitopes are counter-selected, while antigen-poor or preservation-defective clones expand [43, 44]. This creates clonality gradients: clonal, truncal neoepitopes are more promising immunotherapy targets than subclonal passengers [45, 46]. Herein, the importance of abundance is twofold: at the mRNA/protein level (antigen source supply), and the pHLA level (surface presentation density and complex stability). These and other biological realities both offer clues regarding prediction hurdles and limit current computational pipelines for tumor antigen identification, making design of effective epitope-based mRNA immunotherapies rather difficult. Computational tools for antigen prediction strongly depend on high-resolution experimental data, and failure to recapitulate the subtleties and complexities lead to algorithms with coarse predictive power.

Nevertheless, a variety of in vitro methods exist to determine antigen binding affinity and processing, all of which have substantially informed the development of in silico tools that are now indispensable in cancer epitope discovery, a bulk of which are based on artificial neural network (ANN) architectures. Early work with enzyme-linked immunosorbent-based stabilization assays, fluorescence polarization and radiolabeled competition assays provided large-scale, systematically curated datasets of peptide-HLA binding affinities, often summarized and ranked by IC50 values. This resulted in the emergence of popular databases like the Immune Epitope Database (IEDB), the growth of which enabled the development of machine learning algorithms such as NetMHC and NetMHCpan, which leverage thousands of affinity measurements for binding strength predictions across numerous HLA alleles—including rare ones [47, 48]. IEDB itself makes available numerous other tools for the prediction of MHCI/II epitopes, including linear and discontinuous B cell epitopes. It further hosts convenient tools for analyzing population coverage of HLA alleles upon epitope binding prediction screens. However, binding affinity alone is an insufficient metric to accurately predict candidate epitopes, as many peptides with substantiated in vitro data are not presented in vivo [49, 50]. Considering that legacy epitope prediction algorithms are based on such data, this necessitated the development of more functional assays of processing and stability, investigating the half-life of peptide-MHC complexes, proteasome digestion and TAP transport. The resulting data from these efforts gave rise to processing-aware prediction algorithms, such as NetChop for proteasomal cleavage, TAPPred for transporter activity, and NetMHC-stab/NetMHCstabPan for peptide-HLA complex stability. These tools have also been made available on IEDB [49, 51, 52].

For computational immuno-oncology in particular, such processing improvements helped filter out strong binders with little-to-no in vivo translation, thereby increasing the chances of successful epitope identifications [53, 54]. Development and incorporation of mass spectrometry-based immunopeptidomics further refined immunoinformatics, generating datasets revealing anchor residue patterns, allele-specific presentation events and epitope length preferences; metrics unavailable through binding assays alone. Consequently, computational models such as MixMHCpred [55], HLAthena [56], DeepLigand [57] and MHCquant [58] were developed, offering prediction models of higher physiological relevance for use in cancer research that significantly outperform affinity-only approaches. Lastly, functional immunology assays such as tetramer staining, enzyme-linked ImmunoSpot (ELISPOT) and cytotoxicity studies contributed data pertaining to epitope immunogenicity, adding a critical layer of empirical evidence for the development of more sophisticated immunoinformatic workflows [59]. For instance, immunogenicity predictors such as DeepImmuno, DeepHLApan and iNeoEpp [60] capture probabilities of T-cell recognition upon antigen processing and presentation. However, it should be noted that the predictive power of these algorithms remains limited due to the complex immunodynamics underlying immune recognition, particularly in the case of cancer. While these tools are of modest utility when used in isolation, their integration into complex, multilayered epitope screening workflows has cascaded into a myriad of computational tools that attempt to reflect each stage of antigen presentation as to inform immunotherapy design. We summarize popular epitope prediction tools in Table 2.1.

To string this together with an example from computational immuno-oncology, we shall discuss a powerful and integrative open-source computational toolkit for personalized neoantigen prediction—pVACtools—with a particular emphasis on the pVACseq (personalized Variant Antigens by Cancer Sequencing) immunotherapy pipeline [69]. Namely, pVACtools is a Python-based command-line utility developed by Parker Institute/Van Allen Lab, designed specifically for neoantigen prediction using cancer sequencing data. Notably, it integrates binary alignment map (BAM)-derived phased and unphased variant call format (VCF) files with RNA expression data (RNA sequencing BAM files) offering analysis modules such as pVACseq (identifies neoantigens from tumor variants), pVACbind (runs peptide-HLA binding predictions), pVACfuse (handles fusion-derived epitopes), pVACvector (designs DNA-based constructs) and pVACview (web graphical user interface (GUI) for simple visualization). The pipeline successfully integrates affinity predictors (NetMHC, NetMHCpan, MHCflurry, PickPocket), stability predictors (NetMHCstab, NetMHCstabPan), processing predictors (NetChop, NetCTLpan) and ligandome-trained predictors (MixMHCpred, MHCflurry, HLAthena) to reflect the layered nature of antigen processing and presentation and increase the likelihood of biologically plausible antigen prediction (Table 2.2). Specifically, the way in which epitopes are filtered via pVACtools is by integrating the outputs of these prediction algorithms, accompanied by variant phasing (when available), allele-level epitope specificity and corresponding protein expression levels.

The pVACvector module further functionalizes the pipeline, wherein a ranked list of epitope candidates (usually from pVACseq or pVACbind) is transformed into an optimized linear sequence of epitopes arranged into DNA vectors. Importantly, the algorithm minimizes junctional epitopes (which may create novel, unintended

Table 2.1 Popular epitope prediction algorithms and their functions

Algorithm/s	Class	Function/s	Limitations
NetMHC/ NetMHCpan (4.x)	I	ANN models predicting peptide–MHC I binding; trained on large binding affinity (IC50) sets plus some MS ligands (pan-allelic coverage)	Strong for common alleles but biased by training data; less reliable for rare alleles; binding ≠ immunogenicity
MHCflurry [61]	I	ML/NN ensemble predicting binding and presentation; incorporates ligandome MS data; GPU-optimized	Still an indirect proxy for true presentation; does not model T-cell recognition
HLAthena	I	Trained on > 185k MS-identified ligands; outputs calibrated presentation likelihood	Limited immunogenicity info; weaker for alleles underrepresented in MS datasets
ImmuneApp [62]	I	Interpretable DL trained on extensive MS datasets; improves neoepitope prioritization	Newer tool; ecosystem less mature; still presentation focused
EpiScan Predictor [63]	I	Uses targeted immunopeptidomics; useful for antibody-specific epitopes	Scope tied to targeted protocols; limited allele generalizability
NetCTL/NetCTLpan [64, 65]	I	Combines binding (NetMHCpan), proteasomal cleavage (NetChop), and TAP transport	Processing models simplified; gains modest compared to ligandome-trained predictors
NetChop [66]	I	Proteasome cleavage predictor (processing step)	Limited data; immunoproteasome and context effects not fully captured
IEDB MHC-I (meta server)	I	Wraps multiple MHC I engines (NetMHC, NetMHCpan, MHCflurry, NetCTL) with unified outputs	Output only as good as underlying models; no immunogenicity
NetMHCII/ NetMHCIIpan [67]	II	Neural network predictors for MHC II binding; MA training adds MS ligandome data	Peptide register ambiguity; weaker for DP/D Q alleles; presentation ≠ immunogenicity
MARIA [68]	II	Deep learning model for MHC II presentation; integrates ligand MS, expression, and cleavage signals	Coverage uneven; performance tied to input expression data; no T-cell recognition model

(continued)

Table 2.1 (continued)

Algorithm/s	Class	Function/s	Limitations
MixMHC2pred	II	Trained on MS-eluted ligands; deconvolves cores for MHC II presentation	Core assignment can be noisy; DP/DQ coverage less complete
IEDB MHC-II (meta server)	II	Wraps MHC II binding predictors (NetMHCIIpan, NN-align, SMM-align)	Limited by underlying models; weaker accuracy for rare alleles and register-shifting peptides

Abbreviations ANN artificial neural network, *ML* machine learning, *MS* mass spectrometry, *NN* neural network

Table 2.2 Overview of different modules available from pVACtools and their corresponding epitope prediction algorithms

Module	MHC I Algorithms	MHC II Algorithms
pVACseq	NetMHC, NetMHCpan, PickPocket, MHCflurry, MHCnuggets-I, NetCTL, NetCTLpan	NetMHCII, NetMHCIIpan, SMM-align, NN-align, MHCnuggets-II
pVACbind	NetMHC, NetMHCpan, PickPocket, MHCflurry, MHCnuggets-I, NetCTL, NetCTLpan	NetMHCII, NetMHCIIpan, SMM-align, NN-align, MHCnuggets-II
pVACfuse	NetMHC, NetMHCpan, PickPocket, MHCflurry, MHCnuggets-I, NetCTLpan	NetMHCII, NetMHCIIpan, SMM-align, NN-align, MHCnuggets-II
pVACvector	Any epitope list generated by upstream modules	
pVACviz	Visualization only	

peptide epitopes) by efficient peptide ordering, where all possible junctional peptides are modelled, followed by prediction of which ones might be immunologically problematic. The sequences are further optimized by addition of spacer sequences (also known as linkers), aimed at minimizing the likelihood of junctional peptides occurring. For mRNA immunotherapies, this is an important functionality that streamlines DNA vector designs for in vitro transcription (IVT) of candidate constructs (Fig. 2.3).

Despite the comprehensiveness of pVACtools as an impressive open-source computational immuno-oncology toolkit, its limitations must be addressed. Namely, pVACtools sits at the crossroads between older immunoinformatics models and the modern demands of mRNA cancer vaccine design, resulting in limitations that stem from biological scope and software architecture. Biologically, pVACtools still treats binding affinity as a surrogate for immunogenicity, where NetMHC, NetMHCpan, MHCflurry and NetMHCIIpan focus on IC50 values or ligandome likelihood. Furthermore, processing remains a weak point, despite the incorporation of NetChop and NetCTLpan. These algorithms still offer coarse approximations of proteasome cleavage and TAP transport, with the suite itself being naïve to post-translational modifications (PTMs), alternative splicing or cryptic open reading frames (ORFs) contributing to tumor immunopeptidomes. Bias towards common alleles is yet another issue, where predictions on rare allelic variants remain unreliable despite pan-allelic prediction models. Tumor biology itself is also oversimplified, as only linear epitopes from coding variants are considered, which omits HLA heterozygosity or loss, tumor microenvironment (TME) suppressiveness and immunodominance hierarchies that determine epitope relevance.

In terms of software architecture, pVACtools inherits issues stemming from its integrated engines, as much of the core development took place between 2016 and 2019. Though the suite undergoes updates to remain compatible with newer Python releases, it is still primarily a wrapper whose successful implementation depends on external binaries which are not containerized by default, nor maintained in concert with pVACtools versions. As this can lead to installation issues and version conflicts, it makes the pipeline somewhat fragile when leveraging all available predictors within different prediction modules. Considering this pipeline was made with single patient sample analysis in mind, it can be computationally intensive for batch prediction strategies and overall lacking in direct GPU acceleration options. The NN-based modules (NetMHC, NetMHCpan, NetMHCII and NetMHCIIpan, NetChop) represent the most modern elements of pVACtools, though they are constrained by the composition of their training datasets. By contrast, methods such as PickPocket (structure-based), SMM-align (matrix method) are legacy algorithms with narrower predictive power. Limitations aside, pVACtools unifies powerful prediction algorithms into a single pipeline, leaving few options of its caliber for streamlined open-source immune-informatics. Ultimately, the utility of this wrapper will depend on the maintenance and optimization of the core prediction engines it unifies, which currently holds a bright future given the eruption and democratization of high-throughput experimental methods for cancer epitope discovery.

We shall now discuss the usage of the pVACseq module for personalized antigen discovery, with a particular focus on input preparation. Ideally, upon whole-exome sequencing of a patient's tumor and matched normal tissues, accompanied by whole-transcriptome RNA sequencing to provide expression-level evidence and allele-specific transport support, somatic VCFs are generated. VCF files are then annotated via the Ensembl Variant Effector Predictor (VEP), using the full suite of plugins that are required by the pVACseq module (consult tool documentation). This step ensures precise mapping of nucleotide changes to altered peptide sequences. Using bam-readcount in conjunction with vcf-readcount-annotator available from VAtools, sample DNA and RNA read support metrics (eg. AD, DP, AF for DNA and RAD, RDP, RAF for RNA) are appended, enabling pVACseq to filter variants by allelic depth, fraction, and RNA mutant read support. Further enrichment of the VCF file with transcript per million (TPM) or fragments per kilobase of transcript per million (FPKM) can be achieved using the vcf-expression-annotator, which populates the GX and TX fields corresponding to gene and transcript abundance, respectively. At runtime, the user may specify filtering thresholds for coverage cutoffs (DNA and RNA variant allele frequencies), including tumor purity values (for variant clonality calculations based on variant allele frequencies), along with maximum transcript support levels. When available, a separate phased VCF may be incorporated to correctly model compound variants. These annotated data layers enable pVACseq to extend beyond mere theoretical antigenic peptide generation, retaining only epitopes with genomic validity, phasing support and transcriptional evidence. Once these filters are applied, the generated antigenic peptides are scored across user-defined HLA alleles using a myriad of available predictors for binding (NetMHCpan, MHCflurry, PickPocket), stability (NetMHCstab) and processing (NetChop, NetCTLpan). The ranked list of epitopes can then be passed on to the pVACvector module for DNA vector design.

Another powerful resource for neoepitope prediction is the Caner Epitope Database and Analysis Resource (CEDAR) which, similarly to IEDB, integrates and democratizes powerful prediction engines accessible through the web [70]. While not overly dissimilar from IEDB in terms of MHC-I/II prediction algorithms, the analysis resource contains other tools, such as the Mutated Peptide Generator, Axel-F [71], pepX [72] and TCRmatch [73]. Axel-F, for instance, considers mRNA expression levels of the source proteins, where neoantigenic peptides, their predicted HLA binders and corresponding source protein TPM values may be added. PepX, on the other hand, takes a peptide as input and identifies from which proteins the specific peptide can be derived, returning estimated expression level of the source protein based on publicly available databases. Lastly, the Mutated Peptide Generator consider SNPEff-anotated VCFs and generates predictions for neoantigenic peptides alongside their corresponding wildtype (WT) peptides. Each of these tools can be reasonably integrated into a pVACtools-centric vaccine design pipeline for further filtering and evaluation of predicted candidates, although in a post hoc fashion. Given the open-source nature of many of these tools—pVACtools included—this enables pipeline overhauls beyond what they currently offer. While such undertakings will certainly be technically challenging, the future of neoantigen prediction tools in cancer immunotherapy design is quite promising.

2.3.2 *Computational Design of Immunotherapeutic mRNA Constructs: Sequence Engineering*

Regardless of the current benefits and pitfalls of computational neoantigen prediction, once a list of biologically plausible peptides has been generated, they must be incorporated into a synthetic mRNA construct that permits efficient translation while minimizing unintended immunogenicity. Beyond challenges of delivery and molecular stability, a central obstacle lies in the intrinsic innate immunostimulatory properties of mRNA itself, which can trigger pattern-recognition receptors such as toll-like receptor 3 (TLR3), TLR7/8, retinoic acid-inducible gene 1 RIG-1 and melanoma differentiation-associated protein 5 (MDA5) [74, 75]. Although this response can be acceptable, or even beneficial, for prophylactic vaccines administered to immunocompetent individuals, it may compromise the efficiency of therapeutic immunotherapies by promoting premature transcript degradation or systemic inflammation. Consequently, rational mRNA construct design is by no means a trivial engineering task, regardless of neoantigen quality. At this stage, emphasis shifts to computational sequence design and optimization, ensuring that the encoded polyepitope is correctly processed and robustly translated in APCs, particularly DCs. The following section outlines theoretical and practical considerations in mRNA sequence engineering, including codon optimization, secondary-structure minimization, untranslated-region (UTR) design and chemical modification strategies. Furthermore, we discuss newly emerging AI-driven platforms that have shown promise in mRNA construct design and optimization.

In practice, a bulk of mRNA sequence design efforts are centered around the ORF that yields reliable expression, routes the antigen/s to the proper processing pathway, avoids unintended junctional epitopes and manufacturing issues, resulting in a construct that doesn't overstimulate, yet still stably translates (Fig. 2.4) [15, 17, 19, 74, 75]. Chemically, a junctional epitope is a novel peptide that spans the C-terminus of one epitope and the N-terminus of the next, which may cause adaptive immunological consequences, including tolerance. To this end, a typical mRNA construct starts with a strong and recognizable Kozak context (gccRccaugG) immediately upstream of the START codon, followed by a secretion or routing sequence motif (e.g., signaling peptide that facilitates ER entry), along with the epitope cassette arranged to minimize immunodominant junctions via linker inclusion (AAY linkers bias towards MHC-I proteasomal cleavage, GPGPG favors MHC-II pathways), and optional signal trafficking motifs for further MHC biasing [15–17, 75]. Spatially, these linkers separate out individual epitopes, thereby improving antigen processing. Trafficking motifs include, but are not limited to, tissue plasminogen activator (tPA), C-terminal MHC-I targeting domain (MITD), ubiquitin tags (for MHC-I presentation biasing), or a LAMP1/invariant-chain endosomal targeting motif to bias MHC-II [14–16, 76, 77]. Notably, epitope ordering can also be optimized to further minimize junctional immunogenicity, an optimization approach where computational sequence optimization tools offer tremendous benefit. Tools such as pVACvector, become quite valuable here, as it allows for epitope order permutations and linker insertions only where needed, explicitly scoring all junctional k-mers against the corresponding HLA alleles as to avoid the creation of super-binders at junctional sites. Completion of ORF cassette design is typically followed by inclusion of a STOP codon. For multi-ORF strategies, however, 2A peptides may be theoretically included to facilitate ribosomal skipping, though such design philosophies currently remain underexplored [78]. Engineering single-ORF construct is preferable given the status of the field, as it helps keep translation efficient and predictable.

Incorporation of UTRs flanking the coding sequence are critical for mRNA translation and stability, wherein the 5' UTR is typically designed to mediate ribosomal recruitment and scanning. An optimal 5' UTR is often taken from highly expressed human genes or engineered to avoid stable secondary structure formation. Similarly, the 3' UTR is chosen to prolong mRNA half-life and further support translation. Common 3' UTR strategies include incorporation of elements from human β-globin or albumin 3' UTRs to stabilize the transcript. Proper UTR optimization, however, is best achieved in concert with codon optimization, as such sequence elements are

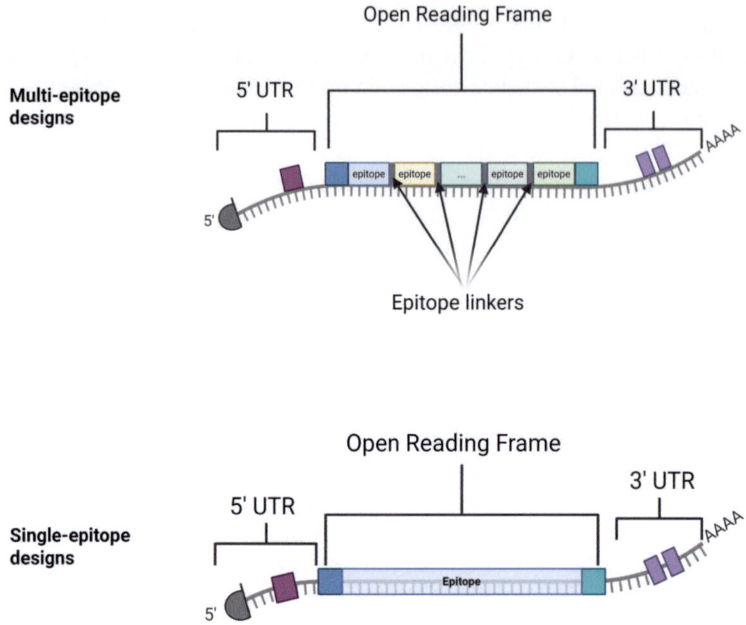

Fig. 2.4 Illustrative depiction of multi-epitope and single-epitope mRNA vaccine designs. Figure was generated using BioRender

interdependent in facilitating interactions between eukaryotic translation machinery and mRNA constructs. Lastly, the mRNA construct is given a 5' cap, a modification achieved during IVT, and a poly(A) tail (~100–120 nucleotides), to mimic full processed eukaryotic mRNA. Addition of poly(A) tails may be done directly during IVT, or enzymatically achieved as a post hoc production step, provide that polyadenylation signals are added directly to the construct itself during design of the final DNA template.

2.3.2.1 Advancements in AI-Driven Tools for mRNA Construct Design and Optimization

Recent advances have produced several noteworthy AI-based platforms for optimizing mRNA vaccine sequences, offering unprecedented speed in design and reducing the likelihood of therapeutic failure. Transformer and language-model architectures have been applied to codon optimization and UTR design, while generative models enable de novo mRNA sequence generation. While such tools vary in input requirements and specific targets, they do share the common goal of tuning sequence features via machine learning. Remarkably, the design outputs of some of these tools have been validated either in vivo or in vitro, encouraging their further development and ultimate extension towards comprehensive, multi-layered pipelines for mRNA vaccine design. At the present, their implementation in mRNA construct design particularly shines in the context of sequence optimization, and we shall discuss relevant novel sequence engineering tools, along with their architectures and limitations.

Transformer-Driven Codon Optimization

In brief, codon optimization represents a technique in mRNA sequence engineering tasked with replacing rare codon regions with synonymous, canonically abundant codons as to increase translation efficiency without modifying a protein's amino acid sequence. This is a crucial step, after all, as ensuring efficient translation without sacrificing crucial immunostimulatory amino acid motifs contributes to therapeutic efficiency. Conceptually this is simple enough to achieve, as the technique relies on identifying an organism's preferred codon usage patterns to make mRNA a favorable substrate for proteins orchestrating translation. Computationally, on the other hand, creation of codon-optimizing neural networks entails training on vast quantities of data pertaining to mRNA translation efficiency, along with corresponding codon contexts, all derived from high-quality wet lab data. It should be noted that organismal favoring of specific codons is cell, tissue and context dependent, influenced by aspects such as cell function, differentiation states, transport RNA (tRNA) abundance and overall cellular energetics [79–82]. Given the compromised health status of oncological patients, particularly those undergoing multi-modal therapeutic regiments, such factors are not trivial [83, 84]. Furthermore, and as discussed previously, mRNA delivery should optimally be APC specific, necessitating careful codon optimization that favors, for instance, efficient DC translation. Such optimization efforts may, in principle, be leveraged further to ensure poor mRNA translation in non-target tissue, while facilitating substantial APC translation [76, 85, 86]. Numerous AI-driven approaches have recently emerged for codon and UTR optimization, some of which have demonstrated promising in vivo translatability.

CodonTransformer, for instance, is a bidirectional encoder representation for transformers (BERT)-like encoder model trained on ~ 1 million coding-sqeuence/protein pairs from 164 species [87]. Its input combines tokens from organism ID, amino acid identity, and masked codons in a custom scheme ("STREAM"), allowing the model to predict host-specific DNA codons given an amino acid sequence. Practically, CodonTransformer-generated sequences have "natural-like codon distribution profiles" and very few deleterious motifs. During the fine-tuning process, the model was trained on the top tenth percentile of genes with highest codon similarity in each organism, which yielded optimized sequences with native-like GC content and minimal negative cis-regulatory elements.

Similarly, CodonBERT, a 12-layer BERT large transformer pretrained on > 10 million full-length mRNA sequences from diverse organisms, learns biologically meaningful codon embeddings and can be fine-tuned to predict mRNA properties such as expression, stability and immunogenicity. Uniquely, it tokenizes at the codon level, meaning that each input token is a codon rather than a single nucleotide.

In published tests, CodonBERT outperformed previous mRNA models on various prediction tasks, positioning it as a useful tool for evaluating candidate sequences in its learned latent space. It should be noted, however, that CodonBERT does not generate actual codons. Another published (preprint) BERT-based architecture, High-Codon, takes raw codon sequences and predicts optimal codon substitutions at each position [88]. It incorporates an expression-weighted loss so that highly expressed reference genes influence learning. This architecture was benchmarked on 100 test proteins, leading to outperformance of standard frequency-based optimizers in predicted expression.

RNN and LSTM Models for Codon Optimization

Beyond transformers, recurrent neural networks (RNNs) have also been leveraged for codon optimization. For example, ICOR, a codon optimizer using a bidirectional LSTM, was trained on ~ 7000 highly expressed *E. coli* genes to learn sequential codon biases [89]. Given a novel protein sequence, ICOR can predict the optimal codon sequence, one amino acid at a time. Computational metrics demonstrated that ICOR-optimized sequences achieve a higher codon adaptation index (CAI), balanced GC content, and fewer negative motifs than naïve frequency-based designs. While, indeed, this architecture was leveraged for bacterial codons, the work by Jain and colleagues showcased how LSTMs may be leveraged for capturing local and global context in codon choice.

For human vaccines in particular, Ravi and colleagues trained BiLSTM models on human expression data to derive tissue and protein-specific codon rules [90]. The group collected the top-expressed genes from brain, liver and muscle tissues and trained separate RNNs using BiLSTM layers to predict codon biases. Luciferase reporter assays re-encoded by these models produced much higher protein levels in vitro than the native or generically optimized sequences. Notably, the model trained on liver-expressed genes resulted in the largest improvement in expression across all tested cell types, suggesting that DL can learn human codon usage nuances that general algorithms tend to miss. Summarily, RNN/LSTM-based tools hold tremendous potential for learning sequential codon dependencies from real-world transcriptome data, yielding optimized mRNAs that experimentally translate more protein.

Generative and Multi-objective Optimization of mRNA Sequences

Some deep-learning tools treat codon design as a generative or multi-objective optimization task. For instance, RiboDecode is a deep generative framework that directly learns from ribosome profiling data to generate optimized mRNA sequences [91]. As input, it takes an amino acid sequence and systematically searches the codon space using a learned model that predicts translation efficiency in different cellular contexts. In silico, RiboDecode demonstrated robust prediction accuracy on novel genes and environments, whilst in vitro and in vivo experiments validated the generated mRNA sequences, reporting dramatically more protein production than from conventional sequences. For instance, an influenza HA mRNA sequence optimized by RiboDecode elicited ~ tenfold higher neutralizing antibody titers in mice than the unoptimized sequence versions. These results clearly demonstrate that data-driven generative approaches can far surpass mere rule-based optimization strategies for encoding vaccine antigens.

RNop is another multi-objective design framework, trained on > 3 million coding sequences and uses specialized loss functions to enforce multiple criteria, including sequence fidelity, high CAI, high tRNA adaptation (tAI) and favorable mRNA structure (MFE), simultaneously [92]. RNop achieved high throughput (tens of optimized sequences per second) and is validated computationally and experimentally. The authors reported that RNop-designed mRNAs significantly increased protein output of several test genes compared to controls or standard optimizers. This approach demonstrates how deep learning can directly balance translation efficiency against potential drawbacks, such as rare codons or strong structure, via tailored loss terms.

Implications for Cancer Vaccine Polyepitope Design

Regardless of whether multi-epitope or single-epitope mRNA designs are being considered, the role of efficient codon optimization remains paramount, as it significantly influences mRNA translation efficiency and overall stability. While, indeed, legacy codon optimization tools have been useful at achieving this, the implementation of transformers, reinforcement learning and deep learning (among other methods) offers tremendous promise in streamlining and accurizing mRNA sequence optimization efforts. This is particularly important for multi-epitope mRNA cancer vaccine designs, where balanced expression of every antigen is a key therapeutic metric. In clinical reality, mRNA translation efficiency and stability relate to dosing requirements, along with immunostimulatory potential. This should not be overlooked, particularly when factoring in issues with mRNA delivery mechanisms (LNPs), whose structural constituents (eg. polyethylene glycol (PEG)) may elicit antagonistic immune responses towards the delivery vector itself, thereby potentially reducing the efficiency of mRNA-LNP complexes with each subsequent dose [93–96]. This issue, however, is under debate, and it is not yet clear what sort of consequences this would have in the context of mRNA immunotherapy outside of facilitating faster LNP clearance. While not a codon optimization issue, LNPs exhibit significant organ tropism. Thus, when an immunotherapeutic mRNA does reach its intended target—a DC—its translation should be as efficient possible to minimize the need for overly repetitive dosing regiments. Proper sequence optimization leads to stronger overall immune activation without losing epitopes to poor expression. Deep codon optimization tools, with their numerous architectures, provide a novel computational arsenal for designing mRNA cancer vaccines by maximizing protein yield whilst minimizing off-target innate immune liabilities. We succinctly summarize some useful mRNA sequence optimization tools in Table 2.3.

Table 2.3 A list of useful mRNA sequence optimization/design tools with encouraging benchmarking data

Tool	Focus	Validation
LinearDesign [97]	Joint optimization of RNA secondary structure (low MFE) and codon optimality (high CAI); balances transcript stability with translation efficiency	Validated in vitro and in vivo; improved antigen expression and up to 128 × higher antibody titers
mRNAid [98]	Multi-parameter design: codon usage, GC content, mRNA stability, immunogenic motif removal	Validated in vitro; comparable or superior translation and stability vs. commercial designs
CodonTransformer [87]	Host-aware codon design; learns native-like codon distributions and avoids deleterious cis-elements	In silico only; generates context-aware codon sequences with natural-like bias and low motif burden
CodonBERT	Learns codon–context relationships from transcriptomic data; predicts optimal synonymous codons for high expression	In silico only; outperforms prior models in predictive benchmarks
ICOR	ML-driven codon substitution; emulates host codon usage, optimizes CAI and GC balance	In silico validation; ~ 42% CAI improvement and reduced rare codon frequency
VaxPress [99]	End-to-end mRNA vaccine design; integrates codon and mRNA stability optimization (via iCodon model)	Computational validation; mRNA vaccine–specific modular framework
DERNA [100]	Generates Pareto-optimal coding sequences maximizing both stability (MFE) and translation efficiency (CAI)	In silico validation; provides trade-off landscape for codon–structure design
mRNArchitect [101]	Integrates codon, GC, and structure optimization with immunogenicity constraints	No experimental validation; in silico demonstration
Legacy Tools (OPTIMIZER [102], JCat [103], ATGme [104], CDSfold [105])	Frequency-based codon replacement or structural folding optimization	No experimental validation; foundational rule-based codon optimization used prior to DL-based approaches

2.3.2.2 Predicting mRNA Structure and Translation Efficiency

The structure of mRNA is considered at two levels: the secondary (2D) structure, defined by intramolecular base-pairing (stems, loops, etc.), and the tertiary structure (3D), which is the full 3D folding of the RNA chain. For mRNA vaccines, understanding the formation of these structures is critical, as these structural features can impact mRNA stability, immunogenicity and translation efficiency. A variety of open-source tools exist to predict 2D/3D structures and to estimate (and optimize) translation efficiency. These computational approaches can be categorized as thermodynamics-based algorithms (physics-based) and machine learning algorithms. Below, we shall review these tools, their underlying methods, scope and limitations, alongside their relevance to mRNA immunotherapy design.

RNA Secondary Structure Prediction Tools

Classic RNA secondary structure prediction tools leverage dynamic programming with experimentally derived nearest-neighbor (NN) free energy parameters to estimate the minimum free energy fold, with pioneering examples including Mfold and the Vienna RNAfold program [106–109]. These tools assign a thermodynamic stability score to candidate structures and identify the lowest free energy configuration in the form of a base-pairing pattern, primarily in dot-bracket notation in terms of output format. Similarly, the RNAstructure software uses free-energy minimization with updated NN parameters [110]. Such physics-based approaches effectively predict simple nested base-pair structures and possess the capabilities to compute ensemble properties, leveraging partition functions to recover base-pairing probabilities and alternative folds. Pseudoknot structures, however, are typically neglected by these approaches unless special heuristics are included due to the non-nested pair complexities underlying pseudoknot formation [109]. At a higher compute cost, tools like HotKnots, IPknot and KnotFold introduce heuristic or integer programming strategies to factor in pseudoknots [111–114]. Furthermore, classical algorithms often struggle to account for non-canonical base-pairs outside of typical G-U wobbles, often assuming a single structure despite RNA being capable of folding into an ensemble of structures within its conformational space [115]. Furthermore, newer algorithms address the issue of scalability by improving computational speed for long sequences, which is particularly important for therapeutic mRNAs that may extend up to several kilobases. For instance, LinearFold leverages a linear-time approximate parsing algorithm to predict secondary structure for very long RNAs within seconds, though it sacrifices some accuracy to achieve these short timescales [116]. Another tool, LinearPartition, computes base-pairing probabilities in linear time and aids in the identification of well-defined versus ambiguous regions in the structure [117].

Recently, data-driven methods have significantly improved the accuracy of mRNA 2D structure predictions by learning from resolved RNA structures. Numerous approaches use deep learning neural networks to predict base-pair probabilities or structure directly from sequence information alone. Noteworthy examples include

SPOT-RNA, which uses an ensemble of deep networks for transfer learning and UFold, which leverages a U-net CNN that treats the base-pair matrix as an image to predict RNA pairing [118, 119]. Another novel tool, MXfold2, integrates deep learning with thermodynamic information, training a neural network to output 'pseudo-free energies' for base-pairing, effectively learning folding parameters while penalizing deviations from known thermodynamic rules [120]. Overall, the implementation of transformers in RNA 2D structure predictions has led to marked improvements in predictions, including improved sensitivity to non-canonical motifs and effective handling of pseudoknot structures. It should be noted, however, that generalizability remains a challenge: many models will perform rather well on RNAs similar to those in their training set but may falter or novel families [121]. Efforts to mitigate this include incorporation of supplementary information; for example, some models may integrate evolutionary covariation data, akin to those in protein folding, or enforce physics-based constraints. UFold can, for example, factor in constraints from a coarse structural model, like CDPfold [122], to inform its predictions, while MXfold2 adds a thermodynamic regularization term. One particularly sophisticated tool, BPfold, creates a comprehensive library of base-pair motifs in the form of short helix neighbors with precomputed energies via 3D modeling (based on BRIQ [123]), using a specialized neural network to incorporate these motif energies in 2D predictions. BPfold achieves high accuracy and, importantly, better generalization to unseen RNA families than purely data-driven models.

2D structure prediction algorithms are highly applicable in the design process of mRNA therapeutics. mRNA stability correlates with its structure, where highly structured mRNAs can resist nuclease degradation, thereby having a longer half-life. This principle was confirmed in the study that showcased the LinearDesign algorithm, where the SARS-CoV-2 mRNA vaccine sequence was re-engineered to achieve greater structural stability. The lattice parsing approach thermodynamic algorithm of LinearDesign optimized the secondary structure of the ORF while respecting codon usage, with the resulting optimized mRNA sequence achieving up to 128-fold antibody titer in mice compared to conventionally codon-optimized mRNA [97]. Such experimental validation substantiated the notion that more structured mRNA persists longer, thereby yielding more antigen. On the other hand, the presence of too many secondary structural motifs risk impeding translation, especially if stable hairpins form near the 5' UTR or the START codon. Thus, 2D structure prediction algorithms can be advantageously leveraged to avoid deleterious constraints.

RNA Tertiary Structure Prediction Tools

The role of 3D modeling in mRNA therapeutic design is currently complementary and exploratory, with the primary focus being given to 2D predictions. Unlike protein structures, whose folding predictions hinge on the biological reality that proteins have an incredibly limited conformational space that confers proper function, the sole purpose of mRNA is to translate. This means that, endogenously, there is a certain degree of tolerance for structural changes that mRNA can undergo when inside the cytosol without sacrificing translation potential. This is not to say, however, that 3D predictions do not have their own value for mRNA design. For instance, certain mRNAs may be engineered to include riboswitches or aptamers that adopt specific 3D folds to regulate translation in response to a ligand, an approach that has garnered much attention in recent years. Similarly, understanding the formation of pseudoknot structures, which may lead to ribosomal frameshifting, contain specific 3D junction geometries. Furthermore, prediction of 3D structure can inform on the presence of immunogenic structures detectable by cytosolic RNA sensors. While 2D prediction algorithms are useful in flagging long helices, 3D modeling may also reveal if two distant regions have potential to interact together to create pseudo double-stranded structure or a G-quadruplex. It should be noted, however, that direct experimental validation of entire 3D mRNA structures is quite rare due to their size and stability, while interrogation of local domains within mRNA structures have been increasingly validated by cryo-EM or X-ray analysis. Actual prediction of the full 3D conformation of RNA, especially mRNAs, is a long-standing challenge in structural bioinformatics. This is unsurprising considering that RNA 3D structure depends on the interactions of secondary structure elements via tertiary contacts, such as coaxial stacking of helices, loop-loop interactions, pseudoknots and complex long-range interactions. Nonetheless, several computational tools do exist for predicting the 3D structure of RNA, implementing either template-based methods de novo strategies or, more recently—learning algorithms.

Template-based modeling methods effectively use resolved RNA structures as templates to model new RNAs which share similarities in terms of sequence or secondary structure. An example of this is ModeRNA, which can take an input sequence with a proposed secondary structure and map it onto fragments of known RAN structures [124]. If a homologous RNA or common motif does exist and is available to these algorithms, they can leverage template-based modeling to produce reasonably accurate structures rather quickly. In actual practice, however, template-based methods are quite limited given the modest size and diversity of the Protein Data Bank (PDB) RNA library. Therapeutic mRNA candidates sometimes contain structural motifs like internal ribosomal entry sites (IRES) or pseudoknots with no known templates, where tools such as the VfoldLA pipeline can be particularly useful as it implements such motif templates in 3D predictions [125].

In terms of de novo prediction, such methods build the RNA 3D structure from scratch, which typically involves initial secondary structure prediction (or using a provided one), followed by exploration of 3D conformational space via conformational sampling and energy minimization [126]. For brevity, we summarize some powerful de novo prediction tools in Table 2.4. Collectively, de novo methods often predict local 3D motifs correctly, yet for large RNAs like full length mRNA sequences, these methods struggle significantly, owed in part due to computational complexity, and due to the fact that a single mRNA may not have one fixed 3D conformation. Simply stated, it is quite difficult to generate accurate 3D structures for mRNA due to its complex topology. In practice, these modeling approaches can be

Table 2.4 A summary of popular de novo 3D structure prediction tools for RNA

Tool name	Prediction method	Accuracy
FARFAR2 [127]	Fragment assembly (Monte Carlo) refined by high-resolution Rosetta potential	Often near-native for short RNAs (< 20 nt; < 2 Å RMSD in ~ 44% of tests); ~ 6.3 Å average RMSD on blind RNA-Puzzles targets
SimRNA [128]	Coarse-grained replica-exchange Monte Carlo with a knowledge-based statistical potential	~ 3.9 Å mean RMSD on small RNAs (Das&Baker set); ~ 17.1 Å on ~ 50–100-nt RNAs (average of first models)
RNAComposer [129]	Knowledge-based 3D assembly guided by secondary structure templates	Requires secondary structure input; ~ 17.4 Å RMSD on ~ 50–100-nt RNAs (no models < 4 Å in one benchmark)
3dRNA [130]	Two-step assembly of secondary structure elements (hairpins, junctions, etc.) into full 3D models	~ 3.7 Å mean heavy-atom RMSD on 300 RNAs (up to 100 nt); e.g. 5.9 Å for a 101-nt ribosomal RNA domain
Vfold	Hierarchical folding pipeline: 2D structure prediction (thermodynamic/DCA, with optional SHAPE data) followed by motif-based 3D assembly	Fast but moderate accuracy—e.g. ~ 8–9 Å RMSD in automated tests (Vfold3D/LA pipeline)
RhoFold +	Transformer-based RNA language model (RNA-FM) with MSA features; end-to-end neural network predicts 3D coordinates via invariant point attention	State-of-the-art accuracy: ~ 4.0 Å average RMSD on RNA-Puzzles targets (outperforms next-best by ~ 2.3 Å)
trRosettaRNA [126]	Hybrid deep learning + physics: transformer predicts base–base distances and orientations from MSA & predicted secondary structure, followed by restrained energy minimization	~ 8.5 Å average RMSD on independent RNAs (~50–150 nt); ~ 20% of models < 4 Å RMSD (none < 4 Å for SimRNA/RNAComposer)
DeepFoldRNA [131]	Attention-based deep neural network (uses MSA and secondary structure to predict distance/orientation constraints) integrated with an all-atom refinement	Consistently high global accuracy—achieved lowest median RMSD among methods in one broad benchmark (outperforms classical sampling tools in RMSD and TM-score)
iFoldRNA [132]	Discrete molecular dynamics simulation (coarse-grained) with optional secondary-structure restraints	Lower accuracy on larger RNAs without restraints (e.g. ~ 21 Å for tRNA with sequence-only input), improving to ~ 11 Å when given correct secondary structure

leveraged to predict stable subdomains of an mRNA molecule and model those individually, rather than modeling entire kilobase-length predictions in single prediction runs.

Inspired by the remarkable success of AlphaFold2 for protein structure prediction, efforts have been made to apply deep learning towards RNA 3D structure prediction. Typically, these models predict inter-nucleotide distances or contact probabilities, followed by coordinate reconstruction, which is analogous to protein contact-map predictions. For instance, trRosettaRNA implements a transformer network to predict 1D/2D geometric constraints and then folds the RNA with an energy minimization procedure [126]. In the CASP15 and RNA-Puzzles blind tests, trRosettaRNA was competitive with human expert models for natural RNAs [126]. Another modeling approach, RhoFold+, leverages an LLM pre-trained on millions of RNA sequences and generates structural embeddings that, in spite of limited data, aid in predicting 3D coordinates with high accuracy [133]. RhoFold+ remarkably outperformed previous methods, including purely computational and expert human groups, on several blind RNA structure targets, achieving the best accuracy on RNA-Puzzle challenges in its category [133]. Indeed, such performance indicates that learning algorithms can capture subtle folding rules through learning from sequence patterns even in the absence of abundant RNA structures for supervised learning. Impressiveness aside, the prediction accuracy of these models tends to be higher for smaller RNAs and helical regions, while resolving complex long-range tertiary contacts in large RNAs remains difficult.

Predicting mRNA Translation Efficiency

Influences on mRNA translation efficiency are multiple, ranging from sequence codon usage bias, 2D structural motifs, the presence of upstream ORFs or internal Shine-Delgarno-like sequences, nucleotide composition, to list a few. Historically, codon optimization has been the primary strategy to improve protein expression from mRNA, at little-to-no improvement on overall mRNA structure and stability [97]. Optimally, a balance between codon usage and structural stability should be achieved during sequence engineering. Purely codon-optimized sequences may translate fast initially, but ultimately suffer from a shorter half-life, though simple metrics such as CAI or tAI are still widely used as rough predictors of expression in spite of omitting secondary structure impacts. In mRNA vaccine development, translation prediction tools are used in concert with structure predictions. For example, a researcher might generate a large number of synonymous versions of an mRNA and use RNAfold to assess the presence of problematic 2D structures, followed by the usage of deep-learning tools to predict the protein yield of each variant.

With the emergence of high-throughput datasets such as those of ribosome profiling or massively parallel reporter assays with varying UTR sequences, data-driven models have become powerful in predicting translation efficiency. For example, a recent breakthrough, RiboNN, which is a deep learning multitask convolutional neural network, covers 140+ human/mouse cell types based on training using > 3800 ribosome profiling experiments [134]. It takes an mRNA sequence as input and learns features across the 5' UTR, ORF, and 3' UTR to quantitatively predict translation efficiency for different cell types. Furthermore, the tool integrates sequence features from across the mRNA, capturing effects like codon-dependent ribosomal processivity and the influence of tRNA abundance on elongation. Essentially, this approach moves beyond treating translation efficiency as an initiation-only problem, accounting for how codon choices throughout the ORF affect protein expression. Impressively, RibboNN's predictions matched experimentally measured translation efficiency across many cell types about twice as accurately as prior models. This is a major breakthrough, as it was also recently discovered (by the same group) that mRNA translation shares so called 'translation efficiency covariation' (TEC), meaning that mRNAs encoding proteins with related biological functions are translated at similar levels across different cell types, according to cell line data [135]. RiboNN can further generalize to predict effects of modified nucleotides, such as pseudouridine on translation. This is of immense importance, as most mRNA therapeutic candidates use modified nucleotides to reduce immunogenicity.

Another approach reported by the Seeling lab, Optimus 5' Prime, is a convolutional neural network trained on parallel reporter data to predict mean ribosome load (MRL) on a given 5' UTR [136]. In essence, it outputs how strongly a particular UTR will drive translation, explaining a large fraction of the variance in translation rates due to UTR sequence in held-out test data. Such an approach is powerful, as it allows researchers to screen or design novel 5' UTRs in sillico to facilitate greater ribosomal MRL. For example, Castillo-Hair and colleagues reported using a deep learning model to design novel 5' UTRs for an mRNA encoding a proprietary gene-editing enzyme (megaTAL™), achieving high protein expression and gene editing activity in human PBMCs [137]. The authors performed polysome profiling on thousands of random UTR variants in order to train the model to predict translation from UTR sequence, followed by the usage of a generative design algorithm (gradient descent on the model and a generative network) to generate optimized UTRs. The best of the designed UTRs outperformed the canonically used alpha-globin UTR in driving translation.

Predicting translation efficiency is a complex task and, while the discussed approaches are definitely reshaping the domain of mRNA therapeutic design, their inherent limitations must be addressed. Namely, current models largely focus on intrinsic sequence determinants of translation, failing to capture/predict mRNA sequestering into stress granules or effects of RNA interference (RNAi) unless such effects correlate strongly with sequence features in training data. Another limitation stems from training libraries, potentially impeding generalization to different genes or conditions without retraining. Additionally, certain models omit explicit modeling of secondary structure dynamics during translation. Considering that an mRNA structure may be partially unfolded by helicases during translation, resulting in a very stable hairpin that can delay the ribosome, thereby leaving the specific region unpaired for some time, such details may be important to capture [138]. Limitations aside, however, translation efficiency predictors, particularly those leveraging deep-learning methods, have become invaluable for mRNA therapeutic design.

2.3.3 Computational Design of Lipid Nanoparticle Systems for mRNA Delivery

LNPs currently stand at the forefront of nucleic acid delivery, despite the multitude of explored nanocarrier alternatives, as they are biocompatible, relatively simple to synthesize, and can be functionalized in numerous ways. While plagued with their own challenges, ranging from organ tropism, toxicity, innate and adaptive immunological reactions and scalable manufacturing, they nevertheless offer a reasonable balance between real-world industry utility and cost-effective translatable benchtop development. LNPs themselves are composed of structural lipids and helper lipids, with the former offering structural integrity while the latter ensures mRNA internalization along with contributing to structural integrity, biodistribution and cellular delivery (Fig. 2.5) [139].

Regardless of the challenges when it comes to LNPs, their remarkable utility as mRNA delivery vectors was particularly validated during the SARS-CoV-2 pandemic, and they continue to be engineered and explored as the gold standard for mRNA delivery, persisting in newly emerging clinical mRNA immunization candidates. Canonically, LNP engineering entailed laborious and time-consuming experimental efforts, particularly concerning optimal candidate lipid screenings during formulation, and the necessitation of in vivo validation in animal models for any meaningful translation metrics [95]. Considering the options for effective in vitro screenings are indeed limited for LNPs in terms of administration in vivo, there exists an urgent need to bridge this gap [140]. While emerging technologies such as microfluidics-based tissue culture offer a promising answer to this, much work lies

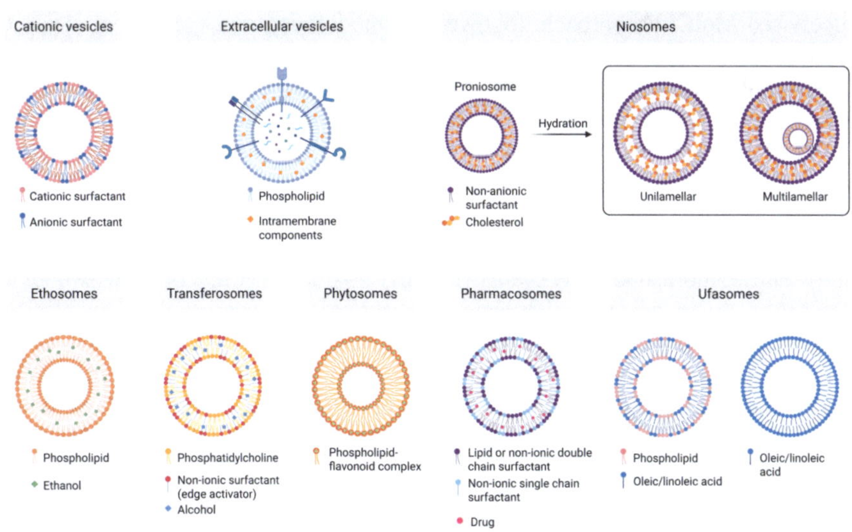

Fig. 2.5 Illustrative depiction of different varieties of lipid nanoparticles. Figure was adopted from BioRender

ahead in functionalizing such platforms towards efficient LNP screenings. Simply stated, it is quite difficult to predict the fate of an LNP vector in a living biological system based on currently available laboratory methods. In fact, while 95% of LNPs are readily internalized by host cells, ~ 5% or less actually manage to deliver translatable mRNA into the cytoplasm. Thus, endosomal entrapment upon internalization persists as a non-trivial issue that awaits resolution. Strategies for optimizing LNP designs are primarily centered around lipid chemistries, with additional functionalization molecular motifs recently emerging as promising enhancers, such as peptides and antibodies [139]. Nevertheless, research efforts have generated a swath of in vitro and in vivo data that may be leveraged for the development of learning algorithms tasked with optimizing LNP designs towards higher efficiency, reduced organ tropism and perhaps even enhanced endosomal escape. We shall briefly discuss some newly emerging and quite impressive computational models for LNP design and candidate lipid screening.

2.3.3.1 Machine Learning for Lipid Candidate Screening

Recent 'AI-driven' platforms are expediting the discovery of ionizable lipids, the key component of mRNA-LNP formulations, by virtually screening vast chemical libraries. Rather than empirically testing thousands of lipid structures, deep learning models can predict the mRNA delivery performance of a lipid *in sillico*, thus drastically narrowing the ionizable lipid search space. One noteworthy example is the AI-guided Ionizable Lipid Engineering (AGILE) platform leverages graph neural networks (GNNs) to evaluate ionizable lipid candidates from combinatorial chemistry libraries [141]. This model was pre-trained on millions of molecules via a self-supervised MolCLR approach, followed by fine-tunning on experimental datasets (derived from HeLa and RAW 264.7 cell lines), allowing it to learn nuanced structure–activity relationships for mRNA transfection on in vitro. Essentially, lipids were encoded as molecular graphs, where atoms were represented as network nodes while bonds were represented as network edges, allowing the GNN to recapitulate features such as head-group ionizability, tail length and steric bulk. Notably, this approach allowed AGILE to uncover cell-type-specific lipid preferences, underscoring the notion that different lipid chemistries may be tailored towards enhanced cell specificity. Furthermore, one study evaluated nearly 20 million virtual lipids in two iterations of AI-driven design, which yielded novel lipids that outperformed the clinical gold-standard cationic lipid DLin-MC3-DMA (MC3) for some generated designs, in vivo [142]. The underlying code used to generate the results in the aforementioned study, however, is not readily available; an unfortunate fact, given that in vivo translation is the single most important challenge plaguing benchtop LNP development. Nevertheless, the encouraging in vivo results are certainly a triumph for such computational approaches. Typically, such models predict properties like mRNA delivery efficiency through learning on experimental data proxies such as reporter protein luminescence, along with key physicochemical parameters such as pKa. Regardless of their current limitations, which primarily stem from the quality of experimental data used to train these models, such approaches promise to significantly streamline the generally time-consuming process of identifying optimal chemical permutations for rational LNP design. Whil AGILE, for instance, is quite a powerful example of leveraging learning methods for this purpose, it currently best serves as a tool to shorten in vitro screenings, given the lack of in vivo training data.

2.3.3.2 Generative Models for Ionizable Lipid Discovery

Beyond predictions on resolved lipid structures, generative models can be leveraged to generate entirely new structures with mRNA optimization delivery at their fulcrum. Deep generative models, including autoencoders, GANs and diffusion models, can learn the distributions of effective lipids and sample novel candidates. For instance, GAN-based approaches have been reported to produce structurally novel ionizable lipids that support high mRNA encapsulation efficiency. These models often operate on molecular graphs or SMILES strings, specifically tailored to lipid chemistry. A recent preprint reported that, by encoding lipid synthesis pathways as directed acyclic graphs (DAGs), a generator can be trained to output synthetic routes alongside structures [143]. Such an approach is of particular interest in novel lipid chemistries, as it holds potential to streamline synthesis by estimating synthetic feasibility and optimal synthesis routes, similar to how generative methods are used in small molecule synthesis route prediction. Optimizing such generative models is of particular importance, as they promise to significantly streamline the search for more rational and efficient lipid designs.

2.3.3.3 Multi-component Formulation Optimization

Considering that the efficiency of an LNP design does not depend on individual lipids alone, but rather their meticulous combination in precise ratios, algorithms designed for LNP formulation should, therefore, be holistic. A prime example of this is the Composite Material Transformer (COMET) model, which leverages transformer neural networks to encode entire LNP formulations and predict their performance [144]. With this transformer architecture, the molecular structure of each component is first converted into a learned embedding, and the ingredient list of the formulation (along with its molar ratios) is fed into a transformer as a token sequence. Inspired by large language model (LLM) architectures, it allows for flexible input of varying compositions and ingredients, including unconventional formulations with multiple ionizable lipids and polymers. The model itself was trained on LANCE, which is a dataset of over 3000 LNP formulations where each entry had corresponding mRNA delivery efficiency, quantified by luciferase expression. Learning from such multifactorial data allowed the transformer to capture subtle interactions between lipid components. For example, COMET was used in a virtual screen of ~ 50 million candidate LNP formulations, with varying components and lipid ratios, and successfully identified top-ranked LNPs that yielded high protein expression in vitro and in vivo during empirical validation. Impressively, COMET demonstrated the ability to predict LNP performance on cell lines not seen in training, along with predicting stability after lyophilization through the implementation of fine-tunning datasets modest in size. For certain in silico hits that underwent in vivo tests, COMET-optimized LNPs achieved > 40-fold and > fivefold bioluminescence compared to the MC3 and SM-102 clinical benchmarks, respectively. Such an advance underscores the relevance and capability of these models to effectively tackle complex lipid combinatorics and formulation design in optimizing maximal delivery efficiency.

2.3.3.4 Integration of Molecular Dynamics Simulations

Molecular dynamics (MD) simulations are becoming increasingly attractive as beneficial computational approaches for LNP design. When appropriately implemented, MD can provide useful insights into nanoparticle behavior that is quite challenging to interrogate experimentally, including lipid self-assembly with mRNA, the internal structure of LNP cores, particle stability and dynamic properties such as lipid exchange and deformity. Broadly, the MD family of physics-based modeling methods can be divided into atomistic and coarse-grained (CG), where the former represent each atom in the system and the latter uses a simpler schema for representing groups of atoms or molecules. Although sacrificing finer atomistic detail, CG simulations offer the advantage of longer simulation runtimes (and scales) at lower compute cost. Nevertheless, both MD approaches have been applied towards understanding LNP structure, formulation and behavior at the molecular level. Through leveraging forcefield families such as CHARMM and AMBER, atomistic MD has been used to capture specific interactions, such as hydrogen bonding between lipids and RNA, while CG simulations—often using the Martini forcefield—were used to simulate entire LNP structures. For instance, a recent study performed large-scale Martini CG simulations incorporating all LNP constituents, including ionizable lipids, distearoylphosphatidylcholine (DSPC), cholesterol, PEG and mRNA in aqueous environments, highlighting how CG enables full-size LNP modeling with millions of particles [145]. Parallel to this, atomistic models have been developed for novel lipids, such as a neutral DLin-MC3-DMA model to study its bilayer behavior, enabling insight into subtle details such as protonation effects [146]. However, accurately representing pH-dependent protonation of ionizable lipids in simulations remains quite challenging, as their pKa (~ 6—6.5) means that their charge state can shift between the neutral environment in the bloodstream and the acidic environment found in endosomes. Through a combination of such multiscale modeling approaches, mechanistic insights into LNP assembly, mRNA encapsulation and LNP behavior in biological environments, are becoming increasingly available with appreciable detail.

LNP Self-assembly, Encapsulation Dynamics and Stability Captured by MD

MD simulations can be leveraged to interrogate self-assembly mechanics of individual LNP components under various conditions, including those canonically used in laboratory synthesis. For instance, Zhang and colleagues investigated the ethanol dilution method used commonly in microfluidic-based LNP synthesis, via atomistic MD simulations, modeling mixtures of the ionizable lipid SM-102, DSPC, cholesterol and PEG, with and without mRNA [147]. Considering that the majority simulation works on LNPs use pure water in their simulation environments, this work showcased how MD can be leveraged to represent and study crucial, real-world LNP synthesis approaches. Their results demonstrated that under neutral pH, simulating initial lipid mixing in ethanol–water, lipids spontaneously self-assemble into a nanoparticle, concentrating ionizable lipids within the interior core. Once the environment was acidified to mimic low-pH buffer used for mRNA loading, the LNPs disassembled and rearranged as protonated cationic lipids electrostatically bound to mRNA. Specifically, protonated SM-102 wrapped around the mRNA, forming stable lipid-mRNA hydrogen bonds and effectively re-assembling the LNP with the mRNA inside. It was reported that the dynamic encapsulation process depended on the N/P (lipid amines to RNA phosphates) ratio and buffer conditions, where low N/P or stronger acidic environments led to large but more uniformly mixed LNPs. Directly mirroring experimental formulation knowledge helped illuminate key encapsulation mechanisms, indicating that, at low pH, ionizable lipids become cationic and drive mRNA packaging via electrostatic attraction, wherein subsequent neutralization to $pH = 7$ locks the mRNA into a stable LNP.

CG MD studies have also been leveraged to interrogate LNP assembly and structure under various solvent conditions. Along these lines, Guruge and colleagues combined Martini 2 CG simulations with X-ray scattering to examine mRNA-LNP formulations at acidic and physiological pH, capturing internal molecular structuring of LNPs, including spatial distribution of RNA and water in environments of $pH = 4$ and $pH = 7.4$ [148]. Initial conditions for most of the simulation runs entailed varying water content and PEG arrangements to determine how such factors influence final LNP morphology. The authors found that acidic conditions result in different internal LNP organization in contrast to physiological pH conditions, likely involving disordered, water-replete interiors that gradually re-arrange in concert with pH shifts, aligning elegantly with experimental small-angle X-ray scattering (SAXS) data. These data suggest that LNPs may undergo a sort of structural maturation when transferred from preparation buffer to bloodstream-like conditions. LNP self-assembly MD simulations underscore that pH-triggered LNP phase transitions are essential in LNP assembly. In an atomistic MD study reported in a preprint by Trollmann and Böckmann, it was demonstrated that a model LNP (Comirnaty formulation) without mRNA possesses and 'oil-like' fluid core at physiological pH and undergoes significant structural transition in response to ionizable lipid protonation [149]. Given these findings, this pH-mediated LNP phase transition is likely critical for endosomal escape, rendering MD as a powerful computational approach towards

LNP optimization, resulting from detailed understanding of fine molecular details underpinning LNP behavior under various therapeutically important conditions.

MD simulations have also provided detailed views of how mRNA interacts with LNP lipids during encapsulation. Generally, ionizable cationic lipids are the primary drivers of RNA binding. When protonated, they form strong electrostatic complexes with the negatively charged phosphate backbone of RNA, mediated by specific H-bonds and coordination with phosphate or ribose oxygens. In the simulated encapsulation of mRNA by SM-102, for instance, each strand was reported as encapsulated by multiple protonated SM-102 molecules through electrostatic interactions, forming stable H-bonds to the mRNA phosphate groups, effectively bridging adjacent RNA segments and neutralizing charge. DSPC, given its zwitterionic nature, played more of a structural role and displayed fewer direct contacts with RNA in the study. Cholesterol, on the other hand, given its hydrophobicity, mostly aided in encapsulation by condensing the lipid-RNA complexes by physically filling the space around them. The results of the study effectively confirmed that ionizable lipids cluster around RNA, whilst cholesterol primarily occupies nearby hydrophobic regions, with PEG-lipids having minimal direct contact with nucleic acids during encapsulation. Alongside such fine molecular details, the strength of lipid-RNA interactions is of paramount importance, as it can understandably either serve to effectively protect the RNA cargo from degradation, or impede its release into the cytosol if too stable [150, 151]. One instructive atomistic study contrasted 1,2-dioleoyl-sn-glycero-3-phosphoethanolamine (DOPE) with 1,2-dioleoyl-sn-glycero-3-phosphocholine (DOPC) in ionizable lipid bilayers, reporting that DOPE formed much stronger H-bond interactions with MC3's amine and carbonyl oxygens [146]. This tight association kept MC3 lipids at the bilayer interface and significantly reduced water penetration into the membrane. Such findings may be used to infer that LNPs formulated with DOPE have a more dehydrated, rigid core that strongly entraps RNA, thereby contributing to poor release and transfection, which is consistent with experimental observations of MC3-DOPE LNP formulations [152]. In contrast, using DOPC or distearoylphosphatidylcholine (DSPC) results in slightly looser packing and an overall more hydrated core, thereby facilitating protonation-driven structural changes and cargo release. These studies elegantly demonstrate the potential of MD as a tool for substantiating various formulation choices during LNP design, and overall aiding in the selection of lipid candidates.

MD as a Tool for Studying LNP Internal Structure and Lipid Organization

Arrangement of lipid components during LNP assembly remains a fundamental question in LNP research, one that is quite difficult to answer experimentally in a time-conscious and cost-effective way. Recent MD simulations have converged on a general model of LNPs, where these nanoparticles exhibit a core–shell architecture with distinct regions enriched for different lipids. Previously discussed atomistic MD simulations by Zhang and colleagues revealed a clear radial stratification, where PEGylated lipids (DMG-PEG2000) segregate to the outer surface, DSPC forms the next inward layer, cholesterol populates an intermediate zone, and SM-102 (protonated when complexed with mRNA) occupies the innermost core—an ordering that is further supported by CG simulations. For instance, Pastre and colleagues constructed LNP models at a lipid ratio that mimics Moderna's mRNA vaccine for SARS-CoV-2 (50% MC3 ionizable lipid, 38.5% cholesterol, 10% DSPC, 1.5% PEG) and reported spontaneous formation of a cholesterol/ionizable lipid-rich core surrounded by a coating of DSPC and PEG upon equilibration [145]. Based on their findings, it appears that cholesterol intermixes with the ionizable lipid within the core, creating a rather fluid hydrophobic domain, as also reported in Trollman's and Böckmann's CG MD work. Their model demonstrated that DSPC molecules orient with their hydrophilic headgroups towards the aqueous exterior, forming a half-bilayer shell surrounding the lipid core. This reported monolayer shell is consistent with cryo-EM and SAXS interpretations of LNPs as single-leaflet vesicle-like structures rather than solid lipid spheres.

Unsurprisingly, MD simulations have also shed some light on the physical state of lipids within LNPs under various conditions. For instance, despite DSPC generally being a phospholipid that normally forms a gel phase at room temperature, nuclear magnetic resonance (NMR)-integrated MD studies indicate that DSPC remains remarkably fluid within the LNP environment [153]. Schroder and colleagues combined solution and solid-state NMR with MD simulations for short interfering (si)RNAs-LNP complexes, observing that the DSPC in the LNP adopts a liquid crystalline-like disordered state even when at 25 °C [153]. It is likely that the presence of cholesterol and the ionizable lipid, with its polyunsaturated tails, disrupts the packing of DSPC and depresses its effective melting temperature, thereby preventing LNP core crystallization. Such core fluidity is essential for proper LNP function, as it permits structural rearrangement during endosomal fusion and may prevent premature solidification during storage. Furthermore, CG models implementing high hydration conditions sometimes show small water packets or 'blebs' that penetrate the LNP core, separated by lipid-rich cores, in agreement with recent SAXS-derived models [154]. While atomistic confirmation and study of these substructures is indeed challenging, it has become increasingly accepted that LNPs are heterogenous structures with regions differing in lipid composition and possibly the presence of small water-filled defects. This is not at all trivial to explore and understand, as such structural details may influence RNA distribution within the LNP, its accessibility and overall LNP stability, both in vivo and during synthesis/production.

Understanding LNP Endosomal Fusion and Intracellular Fate Using MD

The fine molecular details underpinning LNP internalization, endosome fusion and endosomal escape are challenging to study experimentally, though it is generally thought that protonation of the ionizable lipids in the acidifying endosome triggers LNP-membrane fusion or disruption, eventuating in mRNA release. Recent MD simulations have provided novel insights into this process. For instance, one study explicitly demonstrated that the internal nanostructure of LNPs correlates with fusion efficiency, as LNP-RNA organization within the nanoparticle can modulate the energetic cost of creating a fusion pore [155]. LNPs that were structured in a way to readily form a contiguous connection via internal lipid arrangements that facilitate the creation of a bilayer curvature or partial water channel, showed significantly easier fusion. This implies that, past the details regarding LNP composition, morphological LNP characteristics also influence endosomal escape in a non-trivial manner. For instance, an LNP with a non-lamellar core may fuse more readily than one with a rigid, well-ordered core. This was indeed reported by Garaizar and colleagues, where the authors were able to reduce the energy required for endosomal fusion by modifying specific internal topologies in CG MD simulations [156]. Their work demonstrated that intentionally introducing structural heterogeneities, and possibly fusogenic lipids that contribute to core destabilization, could potentially render an LNP more prone to merge with the endosomal membrane. Such findings are rather encouraging, as they indicate that beyond selection of a potent ionizable lipid, other architectural manipulations may be of benefit towards achieving higher endosomal escape.

2.4 The Role of Systems Biology and Omics in Computational Design of mRNA Immunotherapies

Systems biology and multi-omics approaches hold promise not only as tools for studying cancer biology (and immunobiology) in exhaustive detail but also in helping predict the outcomes of therapeutic regiments in patient computational clinical trial models. Indeed, high-throughput sequencing technology has enabled the discipline of bioinformatics to converge on sophisticated algorithms and methods for neoepitope identification, druggable pathway identification and overall cancer immunobiology. Furthermore, their integration into systems biology approaches holds tremendous, albeit untapped, potential. Bridging previously discussed methods on epitope screening, sequence design and optimization, along with LNP design, systems biology and its vast toolkit of methods offer tremendous promise in functionalizing computational data into predictive workflows for therapeutic development. Even at the mere genomic level, outside of digital representations of cells, bioinformatics alone has proven to be quite useful in uncovering druggable targets with encouraging success.

This was particularly exemplified in a recent study in which publicly available single-cell RNA (scRNA) sequencing data were used to identify master cell state regulators whose manipulation led to converting colorectal cancer cells into normal-like enterocytes, with encouraging in vitro and in vivo validation [157]. The underlying mathematical concepts in this study were certainly not new, but their integration led to the creation of the so-called BENEIN framework, which essentially reconstructs Boolean gene regulatory networks (GRNs), based on quantification of pre-mature and mature mRNA reads to separate the transcriptional status of different cell types. It achieves this by identifying switching points on a phase plot of intron and exon reads for each gene, effectively binarizing intron and exon gene expression in accordance with those switching points. The function behind Boolean network construction is based on applying the Quine-McCluskey algorithm for all the genes in a network designed, a priori, by clustering results from conditional mutual information (CMI) calculations performed on the scRNA sequencing data, where relationships between transcription factors (TFs) and their targets are factored. This is indeed an oversimplification of the framework but nevertheless serves the purpose of underscoring how existing datasets can be leveraged for new discoveries by creatively bridging different computational methods.

Multi-omics data originating from cancer patients, particularly those capturing the transcriptome through scRNA or bulk RNA sequencing, can be successfully leveraged to infer gene expression and co-expression networks, particularly when information about TFs, enhancers, genomic data, miRNA expression and protein expression data is available. While technically challenging to combine for the purpose of GRN construction, even interrogating transcriptomes alone can offer biologically useful insight when implemented properly, offering hypothesis generation that may inform relatively straightforward in vitro or in vivo validation. While every mathematical model is definitionally wrong, they can nevertheless be tremendously useful when constructed and applied properly and, most importantly, biologically translatable, as demonstrated in the case BENEIN and GRN reconstruction works. The field of bioinformatics is certainly replete with analytical tools, from R-based packages to Python-based workflows, each possessing benefits and drawbacks related to computational efficiency, capability and ease of use. With the popularization and hasty development of numerous biology-centric deep-learning methods, the existing massive quantities of multi-omics data will likely yield novel insights, ultimately paving the way for new discoveries that may shape the therapeutic philosophy behind mRNA immunotherapies and their clinical implementation.

2.4.1 Quantitative Systems Pharmacology Modeling

Among the most critical questions that arise during mRNA therapy development, particularly when an associated LNP is engineered for its delivery, is biodistribution. Physiologically based pharmacokinetic (PBPK) models and associated quantitative systems pharmacology (QSP) approaches have emerged as powerful tools to simulate the systemic delivery of drugs, including mRNA-LNP complexes, albeit more recently. Fundamentally, these models compartmentalize human systemic physiology into separate organs and tissue, such as blood, liver, spleen, tumor, etc., and encode mechanistic processes that a particular drug undergoes, derived commonly from the mathematical apparatus of pharmacokinetics and pharmacodynamics, solving ordinary differential equations (ODEs) [158–160]. These mechanistic properties include transport via circulation, diffusion into tissues, cellular uptake and, in the case of mRNA-LNPs, even endosomal escape, along with translation and subsequent protein degradation/elimination [161]. By incorporating such processes, PBK/QSP models can be used to predict critical outcomes such as biodistribution of the nanoparticle, the concentration–time profile of the mRNA (or protein) in plasma and tissues along with estimates on biological impact in the context of the immune system or tumor, provided that the PBK/QSP model is coupled with a disease model. The recent work by Miyazawa and colleague provides a very concrete, albeit simplistic, example of a minimal PBPK-QSP model for studying mRNA-LNP delivery, calibrated to preclinical data for a liver-targeted mRNA. The group's in silico sensitivity analysis illuminated factors that strongly influenced therapeutic protein levels, where the intrinsic stability of the mRNA, translation rate of the protein and cellular uptake of LNPs were identified as the most sensitive determinants of protein exposure. By contrast, delivery factors such as the bulk influx of LNPs into the target tissue had comparatively lower impact on end-point protein levels, suggesting that mRNA biochemical properties such reduced sensitivity to nucleases may yield greater efficiency than mere dosage increase to the target site. Furthermore, the authors reported that, at a given LNP degradation half-life, enhancing endosomal escape led to near-linearly higher protein production, emphasizing the importance of engineering LNPs for enhanced endosomal escape.

Beyond explaining experimental observations, dosing strategies may be optimized via systemic mRNA-LNP models. For instance, QSP models can be run on virtual cohorts of patients or animal models through model parameter sampling, to simulate trials of different dosing regiments. In the aforementioned study, virtual animal populations were used to interrogate dosing frequency and dose levels required to maintain efficacy, which revealed dosing schedules that could reduce injection frequency while maintaining therapeutically meaningful protein levels. This approach elegantly demonstrates how PBPK/QSP simulations may function as translational platforms, assisting in the prediction of mRNA immunotherapy performance. Given the in vitro-to-in vivo translation gap issue currently impeding the field of mRNA therapeutics and LNPs, sophisticated development of such tools offers a reasonable promise to bridge this gap.

In real-world industry applications, the pharmaceutical industry has been increasingly applying QSP to guide drug development trajectories, reflected in the increased presence of QSP models in applications to the Food and Drug Administration (FDA), predominantly for phase 2 trial dose selection, though this has likely increased since 2021 [162]. This not only reflects the increased recognition of such modelling approaches in the broader scientific community, but the enormous potential that such methodologies hold for complex biologics like mRNA vaccines. However, it should be noted that standardization of PBPK/QSP methods is a crucial step towards their broader adoption, and that there currently exist few QSP options for mRNA-LNPs. This is unsurprising, as both fields are merely juvenile in comparison to well-stablished legacy methodologies rooted in in vitro and in vivo work. The current gap in standardization of these computational methods, particularly the absence of benchmarked and truly capable models for mRNA therapies, represents an exciting opportunity for the field.

2.4.2 Systemic Immune Response Simulations of mRNA Immunotherapies

Apart from implementing computational approaches to inform mRNA sequence design and LNP engineering, it is important to understand (and predict) the complex immune response that the resulting protein will elicit upon translation. Given the complexity of the immune response, and the non-trivial immunobiological nuances that must be factored in for cancer immunotherapies, systems immunology models have emerged as useful in simulating the immune response to immunotherapies in a quantitative manner [158, 163]. A common approach to achieving this is by using ODE-based QSP models to represent immune cell population dynamics and cytokine production over time, factoring in systemic circulation and different physiological compartments. These models are typically encoded with known biology of cancer immunobiology, including tumor antigen release, DC activation, T-cell priming in lymph nodes, immune cell infiltration of tumours and cytotoxic effects; aspects which are directly influenced by immunotherapeutic agents. Differing from empirical models, such methods allow testing of how different relevant parameters, such as vaccine antigenicity, dosing regiments or TME dynamics, influence overall therapy response. Critically, such simulation approaches can operate at the whole-system level, mirroring the complexity of tumor biology and the immune response to it. An instance where this is useful to model, for example, is in intramuscular injections of an mRNA vaccine, where local APCs would primarily uptake the mRNA-LNP complex, followed by migration to lymphoid organs to activate T cells. QSP models are well suited to capture this coordinative behaviour, representing lymph-node priming and systemic circulation to the tumor site, and faithfully recapitulate local tumor-immune interactions, including TME dynamics. There are numerous examples of this for a range of cancers, such as models of melanoma, lung and breast cancer, tailored towards immune checkpoint inhibitor therapies.

Similarly to PBPK/QSP models, a key challenge and focus for immune response simulations is validation against experimental benchmarks. The past decade has certainly seen a rise in studies rigorously comparing in silico immune simulations with in vivo and in vitro immune data to assess their predictive potential. An illustrative example of this comes from a recent mechanistic modeling study of mRNA cancer vaccine efficacy by Voutouri and colleagues, where the group developed a detailed computational model for tumor-immune interactions under mRNA vaccine treatment, tested on 1600 simulated patients [164]. The model's predictions strongly agreed with real-world clinical data where, for instance, the model correctly anticipated trends observed in mouse melanoma and breast cancer studies such as the degree of tumor shrinkage and T cell infiltration achievable through vaccination. Furthermore, the authors reported a 45% reduction in tumor size and significant increase in $CD8^+$ T-cell infiltration in responding tumors, aligning with experimental results. Strikingly, the simulation identified certain biomarkers of response with subsequent empirical verification. For instance, a higher density of APCs in lymph nodes (above ~ 500 cells/mm^3) was predicted to boost vaccine response rates by

> 50%, with a > 20% T-cell fraction in tumors being associated with a 60% reduction in tumor volume. This work exemplifies how properly calibrated systemic immune response simulations may serve as valuable methods for testing immunotherapy strategies. They synthesize disparate experimental data into a coherent framework that allows researchers to probe therapeutic aspects that would otherwise be too costly and time-consuming to explore empirically.

Of note is that immune simulations range in their scale and simulation approach. While some researchers employ agent-based models (ABMs) or stochastic simulations to capture high-resolution spatial data and cell–cell interactions, others use deterministic ODE models for the purpose of traceability and ease of standardization. Additionally, specialized platforms for immune response simulations exist, such as the C-ImmSim algorithm, which can model the immune response to a peptide-based antigen by stimulating B-cell and T-cell clonal expansion, antibody production and cytokine release over multiple hypothetical exposures [165, 166]. Such tools have been extensively used in computational papers dealing with vaccine design, including mRNA vaccines [14, 77, 167, 168]. While such tools heavily rely on theoretical rules and require meticulous validation, they contribute to a growing and important ecosystem of immune response simulations. As experimental benchmarks improve, iterative refinement of these models will become more achievable, positioning them as indispensable tools in the arsenal of mRNA vaccine design.

2.4.3 Multimodal Agent-Based Modelling Approaches

While whole-body and immune system models capture systemic dynamics of immunotherapy, an equally relevant scale is that of the tumor and its environment in a three-dimensional fashion. Solid tumors in particular are quite heterogenous, structured tissues where tumor cells, stromal cells, immune cells and the extracellular matrix dynamically interact with biologically relevant consequences. Such spatial complexity can strongly influence how well an immunotherapy works. For instance, even when an abundant number of T-cells are in circulation, they may fail to perform their effector functions on the tumor if they cannot penetrate the dense tumor tissue or if locally present immunosuppressive cells inhibit them. Such spatial and cellular details may be addressed with the implementation of ABMs, where cells (and even molecules) are represented as individual agents, with rules governing their behavior and mutual interactions. This approach enables the study of emergent biological phenomena in tumors, with examples including formation of an immune-infiltrated vs. immune-excluded tumor zones, the evolution of therapy-resistant subclones or the dynamics of cytokine gradients that can attract, repel or functionally exhaust immune cells. Over the last decade, ABMs have been developed and applied to cancer immunology questions, including how tumors respond to checkpoint inhibitors, adoptive T-cell transfers, oncolytic viruses and other immunotherapies, providing valuable insights that complement continuum models such as QSP [158, 169–172].

Conveniently, ABMs may be coupled with QSPs in order to bring a higher degree of biological realism towards therapeutic research.

A distinctive advantage of ABMs is their ability to incorporate tumor cell heterogeneity explicitly, along with recapitulating the finer biological consequence of it, including modeling regions of low and high antigen expression, variations in nutrient and oxygen levels along with immune cell infiltrates. While ODE-based models typically average over such spatial variations, ABMs can capture, for instance, distinct pockets of tumor cells that fail to present a particular antigen, or the presence of necrotic foci that are difficult for T-cells to infiltrate. Studies have indeed reported that spatial heterogeneity can qualitatively alter therapy outcomes. For example, spatial simulations have demonstrated how gradients of tumor antigens or immunosuppressive factors may "trap" immune cells in a particular tumor region, thereby allowing non-infiltrated tumor regions to evade immune recognition or killing, by using the PhysiCell open-source physics-based ABM platform [173–175]. What the authors essentially reported was that in tumors with a highly immunogenic core but poorly immunogenic rim, T cell infiltrates may sequester to the core while missing the cancer cells present at the periphery. Such emergent behaviors matching experimental observations of immune-excluded phenotypes only become apparent once spatial structure is modeled.

Integrating system-wide QSP models with the spatial resolution of ABMs is another interesting simulation avenue with tremendous potential that has successfully been implemented for cancer in recent works. A notable example is the spQSP-IO platform introduced by Gong and colleagues, coupling a whole-patient QSP ODE model of immunotherapy with an embedded 3D ABM representing tumor tissue [163]. Within their simulation framework, the QSP component was designed to track global dynamics such as pharmacokinetics and circulating immune cells, while the AB component specifically simulates the tumor, capturing cancer-immune cell interactions in a virtual lattice. This model was applied to non-small cell lung cancer (NSCLC) treated with anti-PD-1 checkpoint inhibitor, demonstrating the ABMs capability to capture the spatial distribution of cells in the tumor during treatment, such as region-specific T-cell infiltrates. The QSP module itself was adapted from a prior NSCLC immunotherapy model, thus ensuring baseline tumor and immune dynamics were grounded in clinical data [176]. Only a fraction of the tumor is specifically represented in the 3D space, as it is not feasible to simulate the entirety of the tumor cell-by-cell, thus resulting in a simulation representation of a so-called Region of Interest (ROI) within the tumor instead. In essence, at the ROI, it runs a high-granularity ABM simulation, whereas the whole-tumor simulation is course-grained. Through parameter sensitivity and virtual biomarker analyses, the model was further functionalized to identify mechanistic factors and potential pretreatment response indicators. Specifically, by systematically varying simulation parameters, the authors were able to identify which aspects, such as baseline effector/regulatory T cell ratio in the tumor, or the expression level of PD-L1 on the surface of cancer cells, predict patient response to checkpoint inhibitors. This model was extended by Ruiz-Martinez and colleagues to triple-negative breast cancer (TNBC), coupling a QSP ODE model with a spatial ABM, also encompassing compartments for central blood, peripheral tissues, the primary tumor and the tumor-draining lymph node analogous to the NSCLC model [158]. The authors leveraged digital pathology data to focus on the tumor's invasive front (IF), informing the spatial arrangement of cells at the tumor margin, a clinically relevant feature of TNBC. Apart from tumor and immune cells, both ABMs incorporate cytokine signaling, MDSC dynamics and mechanistic immune checkpoint biology. While tailored towards antibody-based immune checkpoint inhibitors, designing hybrid ABMs that factor in mRNA-LNP QSP parameters is evidently becoming an increasingly feasible approach, and one that will certainly have a complex yet likely fruitful future.

Given the rather complex nature of building and implementing ABMs, considering that the majority of the legacy codebase for such simulation engines was written in compiled languages such as C and C++, there is generally an entry-barrier to novice biologists with minimal-to-no coding experience. Numerous platforms have been developed for more streamlined ABM development, each with their own degree of user-friendliness. Popular examples include Repast Simphony, CompuCell3D, BioDynaMo and PhysiCell. PhysiCell in particular is interesting because of its user interface versatility and availability of plugins, manageable either through code or through the PhysiCell Studio, which places a relatively intuitive graphical user interface (GUI) for constructing ABMs. Essentially, all of the functionalities of PhysiCell are made available through PhysiCell studio, from defining agents, their parameters, behaviors, interactive rules, and even associated plugins. One particularly attractive feature of PhysiCell is its PhysiBoss plugin, which offers a tremendous benefit in bridging omics-derived Boolean gene regulatory networks (GRNs) derived for transcriptome data, with stochastic agent-based simulation. GRNs essentially model gene and protein networks as binary nodes ($0 = \text{OFF}$, $1 = \text{ON}$) with logical rules governing their interactions. Traditional Boolean models often use synchronous or asynchronous updating in discrete timesteps, but MaBoSS (Markovian Boolean Stochastic Simulator), extends this by introducing stochastic, continuous-time dynamics and is at the core of the PhysiBoss plugin. MaBoSS treats state changes in a network as random events in continuous time, where each node's switching is governed by a probability rate, rendering the model a continuous-Markov process. Effectively, this allows simulation of time-evolution of state probabilities and trajectories, reflecting heterogenous cellular responses over real time. Practically, transition rates are associated with each variable, thus setting the characteristic time scale for a particular node's activation or deactivation. Remarkably, initial states can also be probabilistic, such that one can specify the probability of each note starting at either 0 or 1. This framework provides a convenient way to simulate GRNs under biological variability and uncertainty, capturing a cell's regulatory stochasticity during transitioning between states over time, rather than simulating single-path trajectories. PhysiBoss embeds MaBoSS within agent-based modeling of cell populations using the underlying PhysiCell engine, where each individual agent is embedded with its own instance of the Boolean network, effectively adding a "digital" twin of the cell regulatory/signaling machinery to the agent. In terms of implementation for ABM, this integration enables two-way communication between the cell's internal GRN state and its external environment and behavior. PhysiBoss enables feedback loops between intracellular signaling and multicellular dynamics, enabling the simulated cells to sense their environment and neighbor signals, internally processing them via the Boolean network (with stochastic outcomes), followed by phenotype alterations or behavior via output node effects. It should be noted that other platforms implement similar approaches for incorporating Boolean or ODE networks, and that PhysiCell natively lacks the ability to represent different physiological compartments without significant code overhauls. Nevertheless, these platforms are quite promising for enabling high-resolution computational assessment of mRNA immunotherapy response with reasonable biological translatability.

While powerful, ABMs face significant validation hurdles. Direct comparison with experimental endpoints is a non-trivial task given the high dimensionality of the outputs of complex and truly meaningful ABMs. To date, validation has focused more on qualitative patterns, such as immune cell infiltration/accumulation at different tumor regions, or through proxy metrics such as tumor volume over time, often contrasted with animal data. Nonetheless, ABMs constructed based on substantiated biology can faithfully recapitulate biological phenomena seen with in vivo experiments alongside offering predictions for new ones. Another noteworthy example of this is seen in the work of Surendran and colleagues, where the short-term response of glioblastoma to a combination of chemotherapy (temozolomide) and an immune checkpoint inhibitor, was faithfully recapitulated [169]. To conclude, ABM and hybrid multiscale simulations are promising computational avenues for studying tumor response to mRNA-LNP therapies, complementing systemic models and design efforts discussed earlier. They embody the "bottom-up" systems biology philosophy, where starting from first principles of cell behavior and known (or modeled) molecular interactions, followed by simulation to assess which outcome would result in a macro, high-dimensional scale as a consequence of property emergence. When combined with top-down data, such as omics and imaging for calibration, these models may become predictive "digital twins" of tumors, useful for testing of immunotherapy strategies prior to reaching the laboratory bench or costly in vivo studies. Fueled by advances in big data and computational power, the concept of digital medical twins (DT) has emerged as a novel and promising paradigm in precision medicine. The concept is simple enough, entailing a virtual replica of a particular patient via a computational model continuously synchronized with clinical and molecular patient data [177–180]. In essence, the goal of such systems is to mirror a patient's health status in real time. For oncological applications, this modelling approach represents the promise of simulating tumor evolution and treatment response for each patient, effectively enabling clinicians to perform a sort of therapeutic rehearsal before in vivo application. Such convergence of systems biology with digital twin technology is indeed a natural progression, where deep mechanistic insights and multi-scale models from systems biology form the foundation for building faithful and predictive digital cancer patient twins.

For therapeutic development, utilization of patient digital twins may hold significant ethical and safety hurdles. Simply stated, it is far simpler to build a 'good enough' general purpose simulation stack, rather than navigating data privacy regulatory landscapes pertaining to medical records to use patient-specific DTs at the development stage. Instead of being a patient avatar, a general-purpose DT entails a reusable, modular simulation stack implementable to new biology, modalities and assays. Such approaches couple broad, pre-trained foundation models (for sequences, structure, chemistry and images) with mechanistic simulators (physics-based/ABM, QSP, absorption, distribution, metabolism, excretion and toxicity (ADMET)) and closed-loop experimentations so that the model learns from data it helps generate. Development of mRNA immunotherapies is indeed a systems problem, where sequence and LNP engineering along with host immunity and cancer biology form an intertwined stack of computational layers that benefit from this DT philosophy across the development continuum. Current design pipelines largely embody parts of such a DT framework, and their integration promises to significantly improve current development efforts in the field. Each of the modelling and simulation approaches discussed in this work can, ideally, converge on this concept, offering researchers unprecedented ability to test therapeutic candidates with critical systemic variables in mind (Fig. 2.6). Technical hurdles aside, these are likely going to be crucial tools in drug discovery, provided that efforts towards standardization and generalization are made.

Limitations and open questions remain, such as the merging.

2.5 Conclusions and Future Perspectives

Cancer treatment, whether using conventional chemotherapies or novel biologics like mRNA, continues to be a tumultuous and ever-evolving field, whose pace is equally defined by actionable insights into cancer itself as it is with the field's ability to design effective and precise drugs. Considering the massive financial and time expenditures that have historically characterized drug development, a tremendous opportunity exists to merge computational methods, including AI-enabled ones, for the process of biologics development. As mRNA vaccines are, in essence, validated technologies with satisfactory safety profiles, it is not at all difficult to see how implementation of computational methods in their design may lead to faster turnaround time for their development, as it already has in a primarily academic environment. The same can be said for LNP design, where for the first time in decades, researchers are being equipped with powerful computational tools for lipid candidate screenings, encapsulation simulation, and overall in vivo efficiency prediction. Remarkably, some of the tools and algorithms discussed in our review have seen in vivo validation, a fact that promises triumph over the current issue plaguing the drug development world, which is relatively poor in vitro-to-in vivo translation. The complex and efficiency-defining interactions and environments that mRNA-LNPs undergo upon systemic administration are simply too difficult to interrogate at the bench, or even in vivo, and that is without factoring in actual tumor biology and a tumor's ability to adapt to therapeutic stressors. Indeed, in recent decades we've seen a surge in high-quality, multi-omics data, which are replete with information that awaits careful and meticulous analysis, promising the emergence of new insights into tumor biology and its

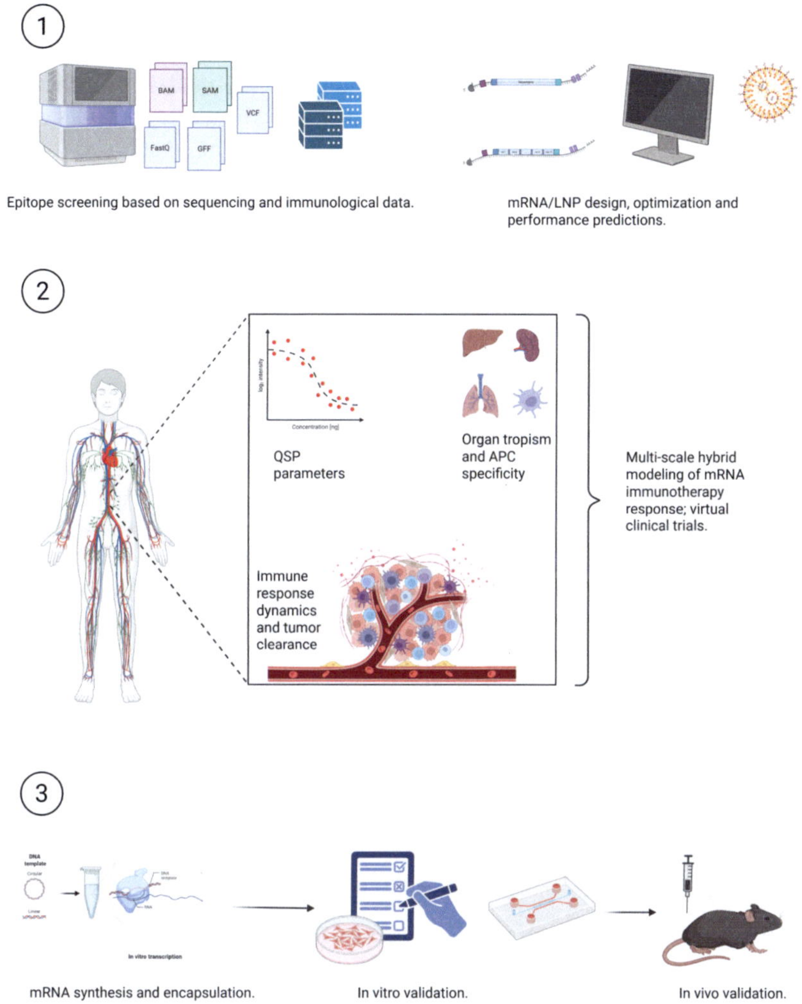

Fig. 2.6 Illustrative depiction of a computationally enabled mRNA immunotherapy development workflow/stack, representing a kind of general-purpose digital twin. (1) Epitope screening for mRNA construct development may be done via numerous tools, including pVACseq, followed by sequence building and its codon optimization, translation efficiency predictions, 2D/3D structure predictions and LNP encapsulation predictions via molecular dynamics (MD). Additionally, LNP design may be guided via tools for formulation optimization to increase transfection efficiency potential. (2) Once an mRNA-LNP achieved encouraging in silico results, key parameters from such simulations may be fed into a hybrid, multi-scale whole-patient tumor model, where each key compartment is faithfully recapitulated through a combination of quantitative systems pharmacology (QSP) and agent-based modeling (ABM). Such models may be functionalized by omics-derived gene regulatory networks (GRNs) to enable tracking of dynamic intracellular changes within the tumor and immune cell populations. In essence, these represent virtual clinical trials that inform choices for downstream candidate synthesis and laboratory validation. (3) Laboratory synthesis and validation of mRNA-LNP candidates using either static in vitro models or sophisticated microphysiological systems (MPS). Candidates with substantiated in silico and in vitro results pose a reduced risk of therapeutic failure for in vivo studies. Figure was generated using BioRender

relationship with its host. Remarkably, omics can not only be used to identify key pathways underlying tumor biology, whether it is regarding treatment, adaptation or therapy resistance, but also leveraged in virtual tumor models that effectively functionalize each simulated cell with its own metabolic/regulatory networks for the purpose of running high-dimensional virtual clinical trials. The combination of omics data and multi-scale ABM effectively brings us closer to a long hypothesized reality where biology can realistically be modelled in silico, thereby allowing researchers to reduce the chance of therapeutic candidate failure during development, and even gain insights into emergent properties that define therapy response, previously unavailable through classical methods alone, at least at a modest time and financial cost. Not only are we currently capable of identifying cancer-specific neoepitopes from multi-omics data and understanding their stochastic origins, but current computational methods enable streamlined design of mRNA constructs along with reasonably translatable predictions of key biochemical metrics that underpin their stability, translation and immunogenicity profiles. Combining the sequence and vector engineering tools discussed with omics-functionalized multi-scale disease models are not only within reach, but also necessary, to reduce development costs while also increasing development speed. Indeed, the current status of computational biology is both unstandardized and fragmented, and specialized multi-disciplinary teams are required to bring such pipelines to fruition, which is nevertheless within reach. Additionally, it is not difficult to see how insights from properly calibrated disease models can be used to train AI-enabled systems to rapidly predict mRNA vaccine efficiency, safety and even finer molecular details pertaining to its interaction with host physiology. Indeed, implementing any of these tools, including AI-enabled ones, hinges on high quality and high-resolution real-world data, which further must be properly implemented during model training, as without factoring in such biological finesse, we risk underutilizing these truly powerful technologies. Hurdles and challenges aside, these are indeed exciting times for computational biology in mRNA vaccine design.

Author Contributions The authors contributed equally to this work.

Conflict of Interest Statement The authors declare no conflicts of interest.

References

1. Harris RJ, et al. Tumor-infiltrating B lymphocyte profiling identifies IgG-biased, clonally expanded prognostic phenotypes in triple-negative breast cancer. Cancer Res. 2021;81(16):4290. https://doi.org/10.1158/0008-5472.CAN-20-3773.
2. Stockmann C, Schadendorf D, Klose R, Helfrich I. The impact of the immune system on tumor: angiogenesis and vascular remodelling. Front Oncol. 2014;4:69. https://doi.org/10.3389/FONC.2014.00069/BIBTEX
3. Apavaloaei A, Hardy MP, Thibault P, Perreault C. The origin and immune recognition of tumor-specific antigens. Cancers (Basel). 2020;12(9):2607. https://doi.org/10.3390/cancers12092607.

4. Chen DS, Mellman I. Elements of cancer immunity and the cancer-immune set point. Nature. 2017;541(7637):321–30. https://doi.org/10.1038/nature21349.
5. Sun S, et al. The role of neoantigens and tumor mutational burden in cancer immunotherapy: advances, mechanisms, and perspectives. J Hematol Oncol. 2025;18(1):1–40. https://doi.org/10.1186/S13045-025-01732-Z/FIGURES/3.
6. Wu B, Zhang B, Li B, Wu H, Jiang M. Cold and hot tumors: from molecular mechanisms to targeted therapy. Signal Transduct Target Ther. 2024;9(1):1–65. https://doi.org/10.1038/S41392-024-01979-X;SUBJMETA.
7. Boland CR, Goel A. Microsatellite Instability in Colorectal Cancer. Gastroenterology. 2010;138(6):2073. https://doi.org/10.1053/J.GASTRO.2009.12.064.
8. Greten FR, Grivennikov SI. Inflammation and cancer: triggers, mechanisms, and consequences. Immunity. 2019;51(1):27–41. https://doi.org/10.1016/J.IMMUNI.2019.06.025.
9. Uribe-Querol E, Rosales C. Neutrophils in cancer: two sides of the same coin. J Immunol Res. 2015;2015(1):983698. https://doi.org/10.1155/2015/983698.
10. Schnellhardt S, et al. Tumour-infiltrating inflammatory cells in early breast cancer: an underrated prognostic and predictive factor? Int J Mol Sci. 2020;21(21):1–17. https://doi.org/10.3390/IJMS21218238.
11. Mc Neil V, Lee SW. Advancing cancer treatment: a review of immune checkpoint inhibitors and combination strategies. Cancers. 2025;17(9):1408. https://doi.org/10.3390/CANCERS17091408.
12. Arafat Hossain M. A comprehensive review of immune checkpoint inhibitors for cancer treatment. Int Immunopharmacol. 2024;143(Pt 2). https://doi.org/10.1016/J.INTIMP.2024.113365.
13. Lopez J, et al. Autogene cevumeran with or without atezolizumab in advanced solid tumors: a phase 1 trial. Nat Med. 2025;31(1):152–64. https://doi.org/10.1038/S41591-024-03334-7;SUBJMETA.
14. Liang H, Qian L, Shen N, Zhao H, Lu Z, Zhang H. A multi-epitope mRNA vaccine for effective immunotherapy in esophageal squamous cell carcinoma. J Clin Oncol. 2025;43(16_suppl). https://doi.org/10.1200/JCO.2025.43.16_SUPPL.E16007.
15. Schlake T, Thess A, Fotin-Mleczek M, Kallen KJ. Developing mRNA-vaccine technologies. RNA Biol. 2012;9(11):1319. https://doi.org/10.4161/RNA.22269.
16. Pardi N, Hogan MJ, Porter FW, Weissman D. mRNA vaccines—a new era in vaccinology. Nat Rev Drug Discov. 2018;17(4):261–279. https://doi.org/10.1038/nrd.2017.243.
17. Sayour EJ, Boczkowski D, Mitchell DA, Nair SK. Cancer mRNA vaccines: clinical advances and future opportunities. Nat Rev Clin Oncol. 2024;21(7):489–500. https://doi.org/10.1038/s41571-024-00902-1.
18. Schoenmaker L, et al. MRNA-lipid nanoparticle COVID-19 vaccines: Structure and stability. Int J Pharm. 2021;601:120586. https://doi.org/10.1016/J.IJPHARM.2021.120586.
19. Linares-Fernández S, Lacroix C, Exposito JY, Verrier B. Tailoring mRNA vaccine to balance innate/adaptive immune response. Trends Mol Med. 2020;26(3):311–23. https://doi.org/10.1016/j.molmed.2019.10.002.
20. Pishesha N, Harmand TJ, Ploegh HL. A guide to antigen processing and presentation. Nat Rev Immunol. 2022;22(12):751–64. https://doi.org/10.1038/S41577-022-00707-2;SUBJMETA.
21. Crux NB, Elahi S. Human Leukocyte Antigen (HLA) and immune regulation: how do classical and non-classical HLA Alleles modulate immune response to human immunodeficiency virus and Hepatitis C virus infections? Front Immunol. 2017;8:832. https://doi.org/10.3389/FIMMU.2017.00832.
22. Minati R, Perreault C, Thibault P. A roadmap toward the definition of actionable tumor-specific antigens. Front Immunol. 2020;11:583287. https://doi.org/10.3389/fimmu.2020.583287.
23. Reeves E, James E. Antigen processing and immune regulation in the response to tumours. Immunology. 2016;150(1):16. https://doi.org/10.1111/IMM.12675.
24. Jian Y et al. Current advance of immune evasion mechanisms and emerging immunotherapies in renal cell carcinoma. Front Immunol. 2021;12. https://doi.org/10.3389/fimmu.2021.639636.

25. Chen B, Zhu H, Yang B, Cao J. The dichotomous role of immunoproteasome in cancer: friend or foe? Acta Pharm Sin B. 2023;13(5):1976–89. https://doi.org/10.1016/j.apsb.2022.11.005.
26. Jorgovanovic D, Song M, Wang L, Zhang Y. Roles of IFN-γin tumor progression and regression: a review. Biomark Res. 2020;8(1):1–16. https://doi.org/10.1186/S40364-020-00228-X/FIGURES/5.
27. Cerruti F, et al. Enhanced expression of interferon-γ-induced antigen-processing machinery components in a spontaneously occurring cancer. Neoplasia. 2007;9(11):960. https://doi.org/10.1593/NEO.07649.
28. Zhang S, et al. Polymorphisms in ERAP1 and ERAP2 genes are associated with tuberculosis in the Han Chinese. Front Genet. 2020;11:566190. https://doi.org/10.3389/FGENE.2020.566190/BIBTEX.
29. Yao Y, Liu N, Zhou Z, Shi L. Influence of ERAP1 and ERAP2 gene polymorphisms on disease susceptibility in different populations. Hum Immunol. 2019;80(5):325–34. https://doi.org/10.1016/j.humimm.2019.02.011.
30. Li C, et al. Polymorphisms in endoplasmic reticulum aminopeptidase genes are associated with cervical cancer risk in a Chinese Han population. BMC Cancer. 2020;20(1):341. https://doi.org/10.1186/S12885-020-06832-2.
31. Thuring C, et al. HLA class I is most tightly linked to levels of Tapasin compared with other antigen-processing proteins in glioblastoma. Br J Cancer. 2015;113(6):952–62. https://doi.org/10.1038/BJC.2015.297;SUBJMETA.
32. Belicha-Villanueva A, McEvoy S, Cycon K, Ferrone S, Gollnick SO, Bangia N. Differential contribution of TAP and tapasin to HLA class I antigen expression. Immunology. 2008;124(1):112. https://doi.org/10.1111/J.1365-2567.2007.02746.X.
33. Shionoya Y, et al. Loss of Tapasin in human lung and colon cancer cells and escape from tumor-associated antigen-specific CTL recognition. Oncoimmunology. 2017;6(2):e1274476. https://doi.org/10.1080/2162402X.2016.1274476.
34. Stefanski CD, Prosperi JR. Wnt-independent and Wnt-dependent effects of APC loss on the chemotherapeutic response. Int J Mol Sci. 2020;21(7844):2020. https://doi.org/10.3390/IJMS21217844.
35. Stamos JL, Weis WI. The β-catenin destruction complex. Cold Spring Harb Perspect Biol. 2013;5(1):a007898. https://doi.org/10.1101/CSHPERSPECT.A007898.
36. Yang C, et al. Arginine deprivation enriches lung cancer proteomes with cysteine by inducing arginine-to-cysteine substitutants. Mol Cell. 2024;84(10):1904-1916.e7. https://doi.org/10.1016/j.molcel.2024.04.012.
37. Albaugh VL, Pinzon-Guzman C, Barbul A. Arginine metabolism and cancer. J Surg Oncol. 2016;115(3):273. https://doi.org/10.1002/JSO.24490.
38. Lasser SA, Ozbay Kurt FG, Arkhypov I, Utikal J, Umansky V. Myeloid-derived suppressor cells in cancer and cancer therapy. Nat Rev Clin Oncol. 2024;21(2):147–164. https://doi.org/10.1038/S41571-023-00846-Y;SUBJMETA.
39. Roche PA, Furuta K. The ins and outs of MHC class II-mediated antigen processing and presentation. Nat Rev Immunol. 2015;15(4):203. https://doi.org/10.1038/NRI3818.
40. Johnson AM, et al. Cancer cell-intrinsic expression of MHC Class II regulates the immune microenvironment and response to anti–PD-1 therapy in lung adenocarcinoma. J Immunol. 2020;204(8):2295. https://doi.org/10.4049/JIMMUNOL.1900778.
41. Wang H, et al. Autophagy in tumor immune escape and immunotherapy. Mol Cancer. 2025;24(1):1–21. https://doi.org/10.1186/S12943-025-02277-Y/TABLES/3.
42. Jiang GM et al. The relationship between autophagy and the immune system and its applications for tumor immunotherapy. Mol Cancer 2019;18(1). https://doi.org/10.1186/S12943-019-0944-Z.
43. Martincorena I. Somatic mutation and clonal expansions in human tissues. Genome Med. 2019;11(1):1–3. https://doi.org/10.1186/S13073-019-0648-4/FIGURES/1.
44. Greaves M, Maley CC. Clonal evolution in cancer. Nature. 2012;481(7381):306–13. https://doi.org/10.1038/NATURE10762;SUBJMETA.

45. Lee KL, et al. Efficient tumor clearance and diversified immunity through neoepitope vaccines and combinatorial immunotherapy. Cancer Immunol Res. 2019;7(8):1359. https://doi.org/10.1158/2326-6066.CIR-18-0620.
46. Schlom J, et al. Hypothesis: the generation of T cells directed against neoepitopes employing immune-mediating agents other than neoepitope vaccines. J Immunother Cancer. 2024;12(7):e009595. https://doi.org/10.1136/JITC-2024-009595.
47. Vita R, et al. The immune epitope database (IEDB) 3.0. Nucleic Acids Res. 2015;43(D1):D405–12. https://doi.org/10.1093/NAR/GKU938.
48. Vita R, et al. The immune epitope database (IEDB): 2018 update. Nucleic Acids Res. 2019;47(D1):D339–43. https://doi.org/10.1093/nar/gky1006.
49. Lawrence PJ, Ning X. Improving MHC class I antigen-processing predictions using representation learning and cleavage site-specific kernels. Cell Reports Methods 2022;2(9). https://doi.org/10.1016/j.crmeth.2022.100293.
50. Chaves FA, Lee AH, Nayak JL, Richards KA, Sant AJ. The utility and limitations of current web-available algorithms to predict peptides recognized by CD4 T cells in response to pathogen infection. J Immunol. 2012;188(9):4235. https://doi.org/10.4049/JIMMUNOL.1103640.
51. Nielsen M, Andreatta M. NetMHCpan-3.0; improved prediction of binding to MHC class I molecules integrating information from multiple receptor and peptide length datasets. Genome Med. 2016;8(1). https://doi.org/10.1186/S13073-016-0288-X.
52. Jurtz V, Paul S, Andreatta M, Marcatili P, Peters B, Nielsen M. NetMHCpan 4.0: improved peptide-MHC class I interaction predictions integrating eluted ligand and peptide binding affinity data. J Immunol. 2017;199(9):3360. https://doi.org/10.4049/JIMMUNOL.1700893.
53. Mahafujul Alam SS, Mir SA, Samanta A, Nayak B, Ali S, Hoque M. Immunoinformatics based designing of a multi-epitope cancer vaccine targeting programmed cell death ligand 1. Sci Rep. 2025;15(1):1–26. https://doi.org/10.1038/S41598-025-87063-Y;SUBJMETA.
54. Hundal J, et al. PVAC-Seq: a genome-guided in silico approach to identifying tumor neoantigens. Genome Med. 2016;8(1):1–11. https://doi.org/10.1186/S13073-016-0264-5/FIGURES/3.
55. Tadros DM, Racle J, Gfeller D. Predicting MHC-I ligands across alleles and species: how far can we go? Genome Med. 2025;17(1):1–13. https://doi.org/10.1186/S13073-025-01450-8/FIGURES/4.
56. Sarkizova S, et al. A large peptidome dataset improves HLA class I epitope prediction across most of the human population. Nat Biotechnol. 2019;38(2):199. https://doi.org/10.1038/S41587-019-0322-9.
57. Zeng H, Gifford DK. DeepLigand: accurate prediction of MHC class I ligands using peptide embedding. Bioinformatics. 2019;35(14):i278–83. https://doi.org/10.1093/BIOINFORMATICS/BTZ330.
58. Bichmann L, et al. MHCquant: automated and reproducible data analysis for immunopeptidomics. J Proteome Res. 2019;18(11):3876–84. https://doi.org/10.1021/ACS.JPROTEOME.9B00313.
59. Slota M, Lim JB, Dang Y, Disis ML. ELISpot for measuring human immune responses to vaccines. Expert Rev Vaccines. 2011;10(3):299. https://doi.org/10.1586/ERV.10.169.
60. Wang G, et al. INeo-Epp: a novel T-cell HLA Class-I immunogenicity or neoantigenic epitope prediction method based on sequence-related amino acid features. Biomed Res Int. 2020;2020:5798356. https://doi.org/10.1155/2020/5798356.
61. O'Donnell TJ, Rubinsteyn A, Laserson U. MHCflurry 2.0: improved pan-allele prediction of MHC Class I-presented peptides by incorporating antigen processing. Cell Syst. 2020;11(1):42-48.e7. https://doi.org/10.1016/j.cels.2020.06.010.
62. Xu H, et al. ImmuneApp for HLA-I epitope prediction and immunopeptidome analysis. Nat Commun. 2024;15(1):1–15. https://doi.org/10.1038/S41467-024-53296-0;TECHMETA.
63. Wang C, Wang J, Song W, Luo G, Jiang T. EpiScan: accurate high-throughput mapping of antibody-specific epitopes using sequence information. NPJ Syst Biol Appl. 2024;10(1):1–16. https://doi.org/10.1038/S41540-024-00432-7;SUBJMETA.

64. Stranzl T, Larsen MV, Lundegaard C, Nielsen M. NetCTLpan: pan-specific MHC class I pathway epitope predictions. Immunogenetics. 2010;62(6):357–68. https://doi.org/10.1007/S00251-010-0441-4.
65. Larsen MV, Lundegaard C, Lamberth K, Buus S, Lund O, Nielsen M. Large-scale validation of methods for cytotoxic T-lymphocyte epitope prediction. BMC Bioinformatics. 2007;8(1):1–12. https://doi.org/10.1186/1471-2105-8-424/TABLES/3.
66. Nielsen M, Lundegaard C, Lund O, Keşmir C. The role of the proteasome in generating cytotoxic T-cell epitopes: insights obtained from improved predictions of proteasomal cleavage. Immunogenetics. 2005;57(1–2):33–41. https://doi.org/10.1007/S00251-005-0781-7.
67. Jensen KK, et al. Improved methods for predicting peptide binding affinity to MHC class II molecules. Immunology. 2018;154(3):394–406. https://doi.org/10.1111/IMM.12889.
68. Chen B, et al. Predicting HLA class II antigen presentation through integrated deep learning. Nat Biotechnol. 2019;37(11):1332–43. https://doi.org/10.1038/S41587-019-0280-2;SUBJMETA.
69. Hundal J, et al. PVACtools: a computational toolkit to identify and visualize cancer neoantigens: pVACtools: toolkit for tumor neoantigen characterization. Cancer Immunol Res. 2020;8(3):409. https://doi.org/10.1158/2326-6066.CIR-19-0401.
70. Koşaloğlu-Yalçın Z, et al. The Cancer Epitope database and analysis resource (CEDAR). Nucleic Acids Res. 2022;51(D1):D845. https://doi.org/10.1093/NAR/GKAC902.
71. Koşaloğlu-Yalçın Z et al. Combined assessment of MHC binding and antigen abundance improves T cell epitope predictions. iScience. 25(2):103850 (2022). https://doi.org/10.1016/J.ISCI.2022.103850.
72. Frentzen A, Greenbaum JA, Kim H, Peters B, Koşaloğlu-Yalçın Z. Estimating tissue-specific peptide abundance from public RNA-Seq data. Front Genet. 2023;14:1082168. https://doi.org/10.3389/FGENE.2023.1082168/FULL.
73. Chronister WD, et al. TCRMatch: predicting T-cell receptor specificity based on sequence similarity to previously characterized receptors. Front Immunol. 2021;12:640725. https://doi.org/10.3389/FIMMU.2021.640725/BIBTEX.
74. Verbeke R, Hogan MJ, Loré K, Pardi N. Innate immune mechanisms of mRNA vaccines. Immunity. 2022;55(11):1993. https://doi.org/10.1016/J.IMMUNI.2022.10.014.
75. Chaudhary N, Weissman D, Whitehead KA. MRNA vaccines for infectious diseases: principles, delivery and clinical translation. Nat Rev Drug Discov. 2021;20(11):817–38. https://doi.org/10.1038/S41573-021-00283-5;SUBJMETA.
76. Jin L, Zhou Y, Zhang S, Chen SJ. MRNA vaccine sequence and structure design and optimization: advances and challenges. J Biol Chem. 2025;301(1):108015. https://doi.org/10.1016/J.JBC.2024.108015.
77. Shi J, et al. In silico designed novel multi-epitope mRNA vaccines against Brucella by targeting extracellular protein BtuB and LptD. Sci Rep. 2024;14(1):1–17. https://doi.org/10.1038/s41598-024-57793-6.
78. Wang X, Marchisio MA. Synthetic polycistronic sequences in eukaryotes. Synth Syst Biotechnol. 2021;6(4):254. https://doi.org/10.1016/J.SYNBIO.2021.09.003.
79. Fumagalli SE, et al. Uncovering codon usage patterns during murine embryogenesis and tissue-specific developmental diseases. Front Genet. 2025;16:1554773. https://doi.org/10.3389/FGENE.2025.1554773/BIBTEX.
80. Liu Y, Yang Q, Zhao F. Synonymous but not silent: the codon usage code for gene expression and protein folding. Annu Rev Biochem. 2021;90:375. https://doi.org/10.1146/ANNUREV-BIOCHEM-071320-112701.
81. Ran X, Xiao J, Cheng F, Wang T, Teng H, Sun Z. Pan-cancer analyses of synonymous mutations based on tissue-specific codon optimality. Comput Struct Biotechnol J. 2022;20:3567–80. https://doi.org/10.1016/J.CSBJ.2022.07.005.
82. Tomaz da Silva P, et al. Cellular energy regulates mRNA degradation in a codon-specific manner. Mol Syst Biol. 2024;20(5):506–20. https://doi.org/10.1038/S44320-024-00026-9.
83. Jiang SH, et al. Systemic regulation of cancer development by neuro-endocrine-immune signaling network at multiple levels. Front Cell Dev Biol. 2020;8:586757. https://doi.org/10.3389/FCELL.2020.586757/BIBTEX.

84. Swanton C, et al. Embracing cancer complexity: hallmarks of systemic disease. Cell. 2024;187(7):1589–616. https://doi.org/10.1016/J.CELL.2024.02.009.
85. Paremskaia AI, et al. Codon-optimization in gene therapy: promises, prospects and challenges. Front Bioeng Biotechnol. 2024;12:1371596. https://doi.org/10.3389/FBIOE.2024.1371596/FULL.
86. Lai CJ, et al. Viral codon optimization on SARS-CoV-2 Spike boosts immunity in the development of COVID-19 mRNA vaccines. J Med Virol. 2023;95(10):e29183. https://doi.org/10.1002/JMV.29183.
87. Fallahpour A, Gureghian V, Filion GJ, Lindner AB, Pandi A. CodonTransformer: a multispecies codon optimizer using context-aware neural networks. Nat Commun. 2025;16(1):1–12. https://doi.org/10.1038/S41467-025-58588-7;TECHMETA.
88. Li J, Dong X, Liu J. High-codon: a deep learning-based codon optimization tool for enhanced heterologous protein expression in *Escherichia coli.* bioRxiv 2025;2025.05.30.656984. https://doi.org/10.1101/2025.05.30.656984.
89. Jain R, Jain A, Mauro E, LeShane K, Densmore D. ICOR: improving codon optimization with recurrent neural networks. BMC Bioinf. 2023;24(1):1–13. https://doi.org/10.1186/S12859-023-05246-8/TABLES/1.
90. Ravi S, et al. A deep learning model trained on expressed transcripts across different tissue types reveals cell-type codon-optimization preferences. Nucleic Acids Res. 2025;53(6):233. https://doi.org/10.1093/NAR/GKAF233.
91. Li Y et al. Deep generative optimization of mRNA codon sequences for enhanced mRNA translation and therapeutic efficacy. bioRxiv. 2025;2024.09.06.611590. https://doi.org/10.1101/2024.09.06.611590.
92. Gong Z, Jiang Z, Gao W, Zhuo D, Ma L. A new deep-learning-based approach for mRNA optimization: high fidelity, computation efficiency, and multiple optimization factors. 2025. Accessed: 23 Oct 2025 [Online]. Available: https://www.arxiv.org/pdf/2505.23862
93. Albertsen CH, Kulkarni JA, Witzigmann D, Lind M, Petersson K, Simonsen JB. The role of lipid components in lipid nanoparticles for vaccines and gene therapy. Elsevier B.V.; 2022. https://doi.org/10.1016/j.addr.2022.114416.
94. Hassett KJ, et al. Optimization of lipid nanoparticles for intramuscular administration of mRNA vaccines. Mol Ther Nucleic Acids. 2019;15:1–11. https://doi.org/10.1016/j.omtn.2019.01.013.
95. Hou X, Zaks T, Langer R, Dong Y. Lipid nanoparticles for mRNA delivery. Nat Rev Mater. 2021;6(12):1078–94. https://doi.org/10.1038/S41578-021-00358-0;SUBJMETA.
96. Omata D, et al. Effect of anti-PEG antibody on immune response of mRNA-loaded lipid nanoparticles. Mol Pharm. 2024;21(11):5672–80. https://doi.org/10.1021/ACS.MOLPHARMACEUT.4C00628/ASSET/IMAGES/LARGE/MP4C00628_0006.JPEG.
97. Zhang H, et al. Algorithm for optimized mRNA design improves stability and immunogenicity. Nature. 2023;621(7978):396–403. https://doi.org/10.1038/S41586-023-06127-Z;TECHMETA.
98. Vostrosablin N et al. mRNAid, an open-source platform for therapeutic mRNA design and optimization strategies. NAR Genom Bioinform. 2024;6(1):lqae028. https://doi.org/10.1093/NARGAB/LQAE028.
99. ChangLabSNU/VaxPress. Codon optimizer for mRNA vaccine design. Accessed: 23 Oct 2025 [Online]. Available: https://github.com/ChangLabSNU/VaxPress
100. Gu X, Qi Y, El-Kebir M. DERNA enables Pareto optimal RNA design. J Comput Biol. 2024;31(3):179–96. https://doi.org/10.1089/CMB.2023.0283.
101. Budzinska MA et al. mRNArchitect: optimized design of mRNA sequences. bioRxiv. 2024;2024.12.03.626696. https://doi.org/10.1101/2024.12.03.626696.
102. Puigbò P, Guzmán E, Romeu A, Garcia-Vallvé S. OPTIMIZER: a web server for optimizing the codon usage of DNA sequences. Nucleic Acids Res. 2007;35(suppl_2):W126–W131. https://doi.org/10.1093/NAR/GKM219.
103. Grote A et al. JCat: a novel tool to adapt codon usage of a target gene to its potential expression host. Nucleic Acids Res. 2005;33(Web Server issue):W526. https://doi.org/10.1093/NAR/GKI376.

104. Daniel E, Onwukwe GU, Wierenga RK, Quaggin SE, Vainio SJ, Krause M. ATGme: open-source web application for rare codon identification and custom DNA sequence optimization. BMC Bioinform. 2015;16(1):1–6. https://doi.org/10.1186/S12859-015-0743-5/FIGURES/2.
105. Terai G, Kamegai S, Asai K. CDSfold: an algorithm for designing a protein-coding sequence with the most stable secondary structure. Bioinformatics. 2016;32(6):828–34. https://doi.org/10.1093/BIOINFORMATICS/BTV678.
106. Zuker M. Mfold web server for nucleic acid folding and hybridization prediction. Nucleic Acids Res. 2003;31(13):3406. https://doi.org/10.1093/NAR/GKG595.
107. Lorenz R, et al. ViennaRNA package 2.0. Algorith Molecul Biol. 2011;6(1):1–14. https://doi.org/10.1186/1748-7188-6-26/TABLES/2.
108. Gruber AR, Lorenz R, Bernhart SH, Neuböck R, Hofacker IL. The vienna RNA websuite. Nucleic Acids Res. 2008;36(Web Server issue):W70. https://doi.org/10.1093/NAR/GKN188.
109. Zhu H, Tang F, Quan Q, Chen K, Xiong P, Zhou SK. Deep generalizable prediction of RNA secondary structure via base pair motif energy. Nat Commun. 2025;16(1):1–13. https://doi.org/10.1038/S41467-025-60048-1;SUBJMETA.
110. Mathews DH. RNA secondary structure analysis using RNAstructure. In: Baxevanis AD et al, editors. Current protocols in bioinformatics, vol. 12, no 1. 2006. p. Unit. https://doi.org/10.1002/0471250953.BI1206S13.
111. Gong T, Ju F, Bu D. Accurate prediction of RNA secondary structure including pseudoknots through solving minimum-cost flow with learned potentials. Commun Biol. 2024;7(1):1–13. https://doi.org/10.1038/S42003-024-05952-W;SUBJMETA.
112. Sato K, Kato Y, Hamada M, Akutsu T, Asai K. IPknot: fast and accurate prediction of RNA secondary structures with pseudoknots using integer programming. Bioinformatics. 2011;27(13):i85–93. https://doi.org/10.1093/BIOINFORMATICS/BTR215.
113. Andronescu MS, Pop C, Condon AE. Improved free energy parameters for RNA pseudoknotted secondary structure prediction. RNA. 2010;16(1):26. https://doi.org/10.1261/RNA.1689910.
114. Ren J, Rastegari B, Condon A, Hoos HH. HotKnots: Heuristic prediction of RNA secondary structures including pseudoknots. RNA. 2005;11(10):1494. https://doi.org/10.1261/RNA.7284905.
115. Lin J, Chen Y, Zhang Y, Lin H, Ouyang Z. Deciphering the role of RNA structure in translation efficiency. BMC Bioinform. 2022;23(3):1–15. https://doi.org/10.1186/S12859-022-05037-7/FIGURES/6.
116. Huang L, et al. LinearFold: linear-time approximate RNA folding by 5'-to-3' dynamic programming and beam search. Bioinformatics. 2019;35(14):i295–304. https://doi.org/10.1093/BIOINFORMATICS/BTZ375.
117. Zhang H, Zhang L, Mathews DH, Huang L. LinearPartition: linear-time approximation of RNA folding partition function and base-pairing probabilities. Bioinformatics. 2020;36(Supplement_1):i258–i267. https://doi.org/10.1093/BIOINFORMATICS/BTAA460.
118. Singh J, Hanson J, Paliwal K, Zhou Y. RNA secondary structure prediction using an ensemble of two-dimensional deep neural networks and transfer learning. Nat Commun. 2019;10(1):1–13. https://doi.org/10.1038/S41467-019-13395-9;SUBJMETA.
119. Fu L, Cao Y, Wu J, Peng Q, Nie Q, Xie X. UFold: fast and accurate RNA secondary structure prediction with deep learning. Nucleic Acids Res. 2022;50(3):e14–e14. https://doi.org/10.1093/NAR/GKAB1074.
120. Sato K, Akiyama M, Sakakibara Y. RNA secondary structure prediction using deep learning with thermodynamic integration. Nat Commun. 2021;12(1):1–9. https://doi.org/10.1038/S41467-021-21194-4;SUBJMETA.
121. Chaturvedi M, Rashid MA, Paliwal KK. Transformers in RNA structure prediction: a review. Comput Struct Biotechnol J. 2025;27:1187. https://doi.org/10.1016/J.CSBJ.2025.03.021.
122. Zhang H et al. A new method of RNA secondary structure prediction based on convolutional neural network and dynamic programming. Front Genet. 2019;10:467. https://doi.org/10.3389/FGENE.2019.00467.

123. Xiong P, Wu R, Zhan J, Zhou Y. Pairing a high-resolution statistical potential with a nucleobase-centric sampling algorithm for improving RNA model refinement. Nat Commun. 2021;12(1):2777. https://doi.org/10.1038/S41467-021-23100-4.
124. Rother M, Rother K, Puton T, Bujnicki JM. ModeRNA: a tool for comparative modeling of RNA 3D structure. Nucleic Acids Res. 2011;39(10):4007. https://doi.org/10.1093/NAR/GKQ1320.
125. Li J, Zhang S, Zhang D, Chen SJ. Vfold-Pipeline: a web server for RNA 3D structure prediction from sequences. Bioinformatics. 2022;38(16):4042–3. https://doi.org/10.1093/BIOINFORMATICS/BTAC426.
126. Wang W, et al. TrRosettaRNA: automated prediction of RNA 3D structure with transformer network. Nat Commun. 2023;14(1):1–13. https://doi.org/10.1038/S41467-023-42528-4;SUBJMETA.
127. Watkins AM, Rangan R, Das R. FARFAR2: improved De Novo Rosetta prediction of complex global RNA folds. Structure. 2020;28(8):963-976.e6. https://doi.org/10.1016/j.str.2020.05.011.
128. Boniecki MJ, et al. SimRNA: a coarse-grained method for RNA folding simulations and 3D structure prediction. Nucleic Acids Res. 2016;44(7):e63. https://doi.org/10.1093/NAR/GKV1479.
129. Biesiada M, Pachulska-Wieczorek K, Adamiak RW, Purzycka KJ. RNAComposer and RNA 3D structure prediction for nanotechnology. Methods. 2016;103:120–7. https://doi.org/10.1016/J.YMETH.2016.03.010.
130. Wang J, Wang J, Huang Y, Xiao Y. 3dRNA v2.0: an updated web server for RNA 3D structure prediction. Int J Mol Sci. 2019;20(17):4116. https://doi.org/10.3390/IJMS20174116.
131. Pearce R, Omenn GS, Zhang Y. De Novo RNA tertiary structure prediction at atomic resolution using geometric potentials from deep learning. bioRxiv 2022;2022.05.15.491755. https://doi.org/10.1101/2022.05.15.491755.
132. Krokhotin A, Houlihan K, Dokholyan NV. IFoldRNA v2: folding RNA with constraints. Bioinformatics. 2015;31(17):2891–3. https://doi.org/10.1093/BIOINFORMATICS/BTV221.
133. Shen T, et al. Accurate RNA 3D structure prediction using a language model-based deep learning approach. Nat Methods. 2024;21(12):2287–98. https://doi.org/10.1038/S41592-024-02487-0;SUBJMETA.
134. Zheng D et al. Predicting the translation efficiency of messenger RNA in mammalian cells. Nat Biotechnol. 2025;1–14. https://doi.org/10.1038/S41587-025-02712-X;SUBJMETA.
135. Liu Y et al. Translation efficiency covariation identifies conserved coordination patterns across cell types. Nat Biotechnol. 2025;1–13. https://doi.org/10.1038/S41587-025-02718-5;SUBJMETA.
136. Castillo-Hair SM, Seelig G. Machine learning for designing next-generation mRNA therapeutics. Acc Chem Res. 2021;55(1):24–34. https://doi.org/10.1021/ACS.ACCOUNTS.1C00621.
137. Castillo-Hair S, et al. Optimizing 5'UTRs for mRNA-delivered gene editing using deep learning. Nat Commun. 2024;15(1):1–15. https://doi.org/10.1038/S41467-024-49508-2;TECHMETA.
138. Marintchev A. Roles of helicases in translation initiation: a mechanistic view. Biochim Biophys Acta. 2013;1829(8):799. https://doi.org/10.1016/J.BBAGRM.2013.01.005.
139. Xu S, Hu Z, Song F, Xu Y, Han X. Lipid nanoparticles: Composition, formulation, and application. Mol Ther Methods Clin Dev. 2025;33(2):101463. https://doi.org/10.1016/J.OMTM.2025.101463.
140. Lindsay S, Hussain M, Binici B, Perrie Y. Exploring the challenges of lipid nanoparticle development: the in vitro–in vivo correlation gap. Vaccines (Basel). 2025;13(4):339. https://doi.org/10.3390/VACCINES13040339/S1.
141. Xu Y, et al. AGILE platform: a deep learning powered approach to accelerate LNP development for mRNA delivery. Nat Commun. 2024;15(1):1–16. https://doi.org/10.1038/S41467-024-50619-Z;TECHMETA.

142. Wang W, et al. Artificial intelligence-driven rational design of ionizable lipids for mRNA delivery. Nat Commun. 2024;15(1):1–17. https://doi.org/10.1038/S41467-024-55072-6;TECHMETA.
143. Ou Y, Zhao J, Tripp A, Rasoulianboroujeni M, Hernández-Lobato JM. A deep generative model for the design of synthesizable ionizable lipids. 2024; Accessed: 23 Oct 2025 [Online]. Available: https://arxiv.org/pdf/2412.00928v1
144. Chan A et al. Designing lipid nanoparticles using a transformer-based neural network. Nat Nanotechnol. 2025;1–11. https://doi.org/10.1038/S41565-025-01975-4;SUBJMETA.
145. Pastre A, Brass O, Terreux R, Megy S. Coarse-grained simulations of RNA-loaded lipid nanoparticles: implications on RNA stability and antigenic protein expression in mRNA vaccines. J Phys Chem B. 2025;129(38):9643–57. https://doi.org/10.1021/ACS.JPCB.5C04980.
146. Ermilova I, Swenson J. DOPC versus DOPE as a helper lipid for gene-therapies: molecular dynamics simulations with DLin-MC3-DMA. Phys Chem Chem Phys. 2020;22(48):28256–68. https://doi.org/10.1039/D0CP05111J.
147. Zhang Z, et al. Molecular dynamics simulation of lipid nanoparticles encapsulating mRNA. Molecules. 2024;29(18):4409. https://doi.org/10.3390/MOLECULES29184409/S1.
148. Guruge AG, He Y, Clulow AJ, Pouton CW, Chalmers DK. Structuring of lipid nanoparticle mRNA formulations at acidic and neutral pH: X-ray scattering and molecular dynamics studies. Mol Ther Methods Clin Dev. 2025;33(4):101596. https://doi.org/10.1016/j.omtm.2025.101596.
149. Trollmann MFW, Böckmann RA. Mechanistic insight into pH-driven phase transition of lipid nanoparticles. bioRxiv. 2024;2024.11.27.625717. https://doi.org/10.1101/2024.11.27.625717.
150. Chatterjee S, Kon E, Sharma P, Peer D. Endosomal escape: a bottleneck for LNP-mediated therapeutics. Proc Natl Acad Sci U S A. 2024;121(11):e2307800120. https://doi.org/10.1073/PNAS.2307800120/ASSET/753B2757-1D43-4A2A-AC8D-8BD68E7140F8/ASSETS/IMAGES/LARGE/PNAS.2307800120FIG03.JPG.
151. Varkouhi AK, Scholte M, Storm G, Haisma HJ. Endosomal escape pathways for delivery of biologicals. 2011. https://doi.org/10.1016/j.jconrel.2010.11.004.
152. Schober GB, Story S, Arya DP. A careful look at lipid nanoparticle characterization: analysis of benchmark formulations for encapsulation of RNA cargo size gradient. Sci Rep. 2024;14(1):1–10. https://doi.org/10.1038/S41598-024-52685-1;SUBJMETA.
153. Schroder R, et al. Molecular dynamics of lipid nanoparticles explored through solution and solid-state NMR spectroscopy and MD simulations. J Phys Chem B. 2025;129(25):6200–19. https://doi.org/10.1021/ACS.JPCB.5C01399.
154. Li H, et al. Mesoscopic structure of lipid nanoparticles studied by small-angle X-Ray scattering: a spherical core-triple shell model analysis. Membranes (Basel). 2025;15(5):153. https://doi.org/10.3390/MEMBRANES15050153/S1.
155. Zheng L, Bandara SR, Leal C. Lipid nanoparticle topology regulates endosomal escape and delivery of RNA to the cytoplasm. bioRxiv. 2022;2022.05.20.492895. https://doi.org/10.1101/2022.05.20.492895.
156. Garaizar A, et al. Toward understanding lipid reorganization in RNA lipid nanoparticles in acidic environments. Proc Natl Acad Sci. 2024;121(45):e2404555121. https://doi.org/10.1073/PNAS.2404555121.
157. Gong JR et al. Control of cellular differentiation trajectories for cancer reversion. Adv Sci. 2024;2402132. https://doi.org/10.1002/ADVS.202402132.
158. Ruiz-Martinez A et al. Simulations of tumor growth and response to immunotherapy by coupling a spatial agent-based model with a whole-patient quantitative systems pharmacology model. PLoS Comput Biol. 2022;18(7). https://doi.org/10.1371/JOURNAL.PCBI.1010254.
159. Azer K et al. History and future perspectives on the discipline of quantitative systems pharmacology modeling and its applications. Front Physiol. 2021;12. https://doi.org/10.3389/FPHYS.2021.637999.

160. Androulakis IP. Towards a comprehensive assessment of QSP models: what would it take? J Pharmacokinet Pharmacodyn. 2022;51(5):521. https://doi.org/10.1007/S10928-022-09820-0.
161. Miyazawa K, Liu Y, Bazzazi H. Development of a minimal PBPK-QSP modeling platform for LNP-mRNA based therapeutics to study tissue disposition and protein expression dynamics. Front Nanotechnol. 2024;6:1330406. https://doi.org/10.3389/FNANO.2024.1330406/BIBTEX.
162. Bai JPF et al. FDA-Industry scientific exchange on assessing quantitative systems pharmacology models in clinical drug development: a meeting report, summary of challenges/gaps, and future perspective. AAPS J 2021;23(3). https://doi.org/10.1208/S12248-021-00585-X.
163. Gong C, Ruiz-Martinez A, Kimko H, Popel AS. A spatial quantitative systems pharmacology platform spqsp-io for simulations of tumor—immune interactions and effects of checkpoint inhibitor immunotherapy. Cancers (Basel). 2021;13(15). https://doi.org/10.3390/CANCERS13153751/S1.
164. Voutouri C, Munn LL, Stylianopoulos T, Jain RK. Biomarkers of mRNA vaccine efficacy derived from mechanistic modeling of tumor-immune interactions. PLoS Comput Biol. 2025;21(6):e1013163. https://doi.org/10.1371/JOURNAL.PCBI.1013163.
165. Rapin N, Lund O, Castiglione F. Immune system simulation online. Bioinformatics. 2011;27(14):2013–4. https://doi.org/10.1093/BIOINFORMATICS/BTR335.
166. Toubanaki DK, Allemailem KS. In silico development of a chimeric multi-epitope vaccine targeting *Helcococcus kunzii*: coupling subtractive proteomics and reverse vaccinology for vaccine target discovery. Pharmaceuticals. 2025;18(9):1258. https://doi.org/10.3390/PH18091258.
167. An Y, et al. CP91110P: a computationally designed multi-epitope vaccine candidate for tuberculosis via TLR-2/4 synergistic immunomodulation. Biology. 2025;14(9):1196. https://doi.org/10.3390/BIOLOGY14091196.
168. Ahammad I, Lira SS. Designing a novel mRNA vaccine against SARS-CoV-2: an immunoinformatics approach. Int J Biol Macromol. 2020;162:820–37. https://doi.org/10.1016/j.ijbiomac.2020.06.213.
169. Surendran A, et al. Agent-based modelling reveals the role of the tumor microenvironment on the short-term success of combination temozolomide/immune checkpoint blockade to treat glioblastoma. J Pharmacol Exp Ther. 2023;387(1):66–77. https://doi.org/10.1124/jpet.122.001571.
170. Jenner AL et al. Agent-based computational modeling of glioblastoma predicts that stromal density is central to oncolytic virus efficacy. iScience. 2022;25(6):104395. https://doi.org/10.1016/J.ISCI.2022.104395.
171. Macal CM. Everything you need to know about agent-based modelling and simulation. J Simulat. 2016;10(2):144–56. https://doi.org/10.1057/JOS.2016.7.
172. Halling-Brown M et al. ImmunoGrid: towards agent-based simulations of the human immune system at a natural scale. R Soc of Lond Philos Trans Math Phys Eng Sci 2010;368(1920):2799–2815. https://doi.org/10.1098/RSTA.2010.0067.
173. Ozik J, et al. High-throughput cancer hypothesis testing with an integrated PhysiCell-EMEWS workflow. BMC Bioinformatics. 2018;19(Suppl 18):483. https://doi.org/10.1186/S12859-018-2510-X.
174. Ghaffarizadeh A, Heiland R, Friedman SH, Mumenthaler SM, Macklin P. PhysiCell: an open source physics-based cell simulator for 3-D multicellular systems. PLoS Comput Biol. 2018;14(2):e1005991. https://doi.org/10.1371/JOURNAL.PCBI.1005991.
175. Smeriglio R, Bardini R, Savino A, Di Carlo S. Start & stop: a PhysiCell and PhysiBoSS 2.0 add-on for interactive simulation control. BMC Bioinf. 2025;26(1):1–17. https://doi.org/10.1186/S12859-025-06144-X/TABLES/1.
176. Jafarnejad M, et al. A computational model of neoadjuvant PD-1 inhibition in non-small cell lung cancer. AAPS J. 2019;21(5):1–14. https://doi.org/10.1208/S12248-019-0350-X/FIGURES/7.

177. Masison J, et al. A modular computational framework for medical digital twins. Proc Natl Acad Sci. 2021;118(20):e2024287118. https://doi.org/10.1073/PNAS.2024287118.
178. Niarakis A, et al. Immune digital twins for complex human pathologies: applications, limitations, and challenges. NPJ Syst Biol Appl. 2024;10(1):1–14. https://doi.org/10.1038/S41540-024-00450-5;SUBJMETA.
179. Akbarialiabad H, et al. Enhancing randomized clinical trials with digital twins. NPJ Syst Biol Appl. 2025;11(1):110. https://doi.org/10.1038/S41540-025-00592-0;SUBJMETA.
180. Shen S, et al. From virtual to reality: innovative practices of digital twins in tumor therapy. J Transl Med. 2025;23(1):1–31. https://doi.org/10.1186/S12967-025-06371-Z/TABLES/8.

Chapter 3
Protein and Peptide Design with Generative AI

David Agustriawan

Abstract The rational design of proteins and peptides is a longstanding challenge in molecular biology, with broad implications for biotechnology, therapeutics, and synthetic biology. Traditional methods—such as directed evolution and structure-guided engineering—are limited in scope and throughput, often failing to efficiently explore the vast combinatorial space of amino acid sequences. This work aims to comprehensively review and analyze the application of generative artificial intelligence (AI) models for the de novo design of proteins and peptides, highlighting recent advancements, dataset resources, generative methods, and their real-world implications. We examine state-of-the-art generative architectures, including Generative Adversarial Networks (GANs), Transformer-based language models, diffusion models, and reinforcement learning strategies. These methods are assessed based on their ability to model protein sequence-function relationships, generate novel sequences, and optimize specific biochemical properties. We also discuss the role of curated datasets (e.g., UniProt, APD3, PDB). Generative AI models demonstrate substantial capabilities in producing novel, diverse, and functional proteins and peptides with application-specific properties. Several studies show AI-generated candidates outperforming traditionally designed ones in stability, activity, and novelty. Integration with downstream pipelines—such as structure prediction (e.g., AlphaFold2) and in silico docking—further enhances their utility. Generative AI offers a paradigm shift in protein and peptide design, enabling rapid and scalable exploration of biomolecular sequence space. Its applications span industrial enzyme optimization, therapeutic peptide development, synthetic circuit design, and personalized medicine. As experimental validation pipelines mature, these models are poised to become foundational tools in computational biology and bioengineering.

Keywords Protein design · Peptide generation · Generative AI · Transformer models · Synthetic biology · Drug discovery · Deep learning · GANs

Never underestimate your capability because when you have put all your efforts persistently you will know nothing impossible in this world.

D. Agustriawan (✉)
Department of Informatics, Universitas Multimedia Nusantara, Tangerang, Banten, Indonesia
e-mail: david.agustriawan@umn.ac.id

A. A. Parikesit and A. N. M. Ansori (eds.), *Recent Computational Techniques in De Novo Drug Design*, Bio-IT and AI, https://doi.org/10.1007/978-3-032-20428-8_3

3.1 Introduction

Proteins and peptides are the cornerstone of biological systems. Proteins are macromolecules characterized by sequences of amino acids (i.e., > 50 amino acids) function as enzymes, structural frameworks, signal transducers, and immune effectors [1]. Proteins consist of one or more chains of amino acids and various secondary structure feature (e.g., alpha helices and beta sheets). In other hand, Peptides as short chains of amino acids (i.e., < 50 amino acids) are also performing diverse roles, such as antimicrobial defense, hormone signaling, and molecular targeting in therapy [2]. Designing novel protein or peptide sequences that can carry out specific tasks is a long-standing goal in the fields like biotechnology, drug discovery, and synthetic biology. Historically, De novo peptide and protein design has been started since 1980s with an aim to create novel peptide and protein components in the drug screening and discovery stage of the drug development pipeline [3]. It has been suggested that peptide-based/protein-based drugs are considered as target-specific therapeutics with fewer side effects and thereby are more suitable for personalized medicine than the traditional "small molecule" drugs [4, 5]. It has also been shown that artificial intelligence (AI) methods such as machine learning and deep learning algorithms are more efficient than other exiting approaches in the process of extracting high-level features from protein sequences [6]. Since the rise of AI technologies, researchers have continued making significant contributions in the interdisciplinary fields of machine learning, deep learning, generative chemistry, and computer-aided drug design and discovery [7–12]. In the emerging field of generative chemistry, the goal of computer computer aided AI methods such as machine learning and deep learning technologies is to accomplish in silico design of novel molecular chemical structures with desired biological and pharmacological properties [12, 13]. Similarly, in the process of drug design and discovery, the objective of computer-aided AI methods such as machine learning and deep learning technologies is to facilitate data-driven algorithms that are mainly capable of helping streamline the multiple stages of the drug development pipeline to reduce costs, including, but not limited to, drug target prediction, drug screening and discovery, preclinical trials, and clinical trials [8, 14, 15]. Up-to-date improvements in machine learning and deep learning technologies, especially deep learning techniques have displayed their favorable impacts in connection with generative chemistry and computer-aided drug design and discovery [16–20]. For example, deep learning techniques such as generative adversarial network (GAN) algorithms have been utilized to produce novel peptide and protein structures in generative chemistry for drug targets during the drug screening and discovery stage in the drug development pipeline [11]. Therefore, it has been indicated that deep learning techniques such as GAN algorithms play an essential role in the future of generative chemistry and computer-aided drug design and discovery as these advantageous approaches can be applied to numerous features in the arena of generative chemistry and computer-aided drug design and discovery [9, 21]. With recent advances in deep learning techniques, GAN algorithms have a great potential in de novo peptide and protein design [22–25]. Deep learning approaches utilize deep artificial neural networks with multiple layers to build up the hierarchical portrayals, instead of artificial neural networks with only single layer [26, 27]. In other words, deep learning techniques such as GAN algorithms commonly employ multiple layers of abstraction to establish hierarchical representations for the data [26, 28, 29]. For the purpose of handling the challenging difficulties we face nowadays in the fields of generative chemistry and computer-aided drug design and discovery, there is a tremendous need of utilizing deep learning techniques such as GAN algorithms for various drug development projects such as de novo peptide and protein design [8]. GAN algorithms are based on a generative method which trains deep artificial neural networks to model the probability distribution of the training data [30]. In machine learning, other generative methods encompass auto-encoders, normalizing flows, energy-based models, and reinforcement learning techniques. An autoencoder consists of an encoder and a decoder, where the encoder first encodes the input data into a smaller representation (i.e., latent variables), and then, the decoder decodes the smaller representation to regenerate a new data that is similar to the input data [30, 31]. Auto-encoders are the simplest and yet most appealing approach among generative models [32]. Normalizing flows construct probability distributions by transforming a probability density via consecutive mappings, where the initial probability density flows through the sequence of mappings [33, 34]. While other generative models involve approximate sampling, normalizing flows usually utilize exact inference and analytical computations [35]. Energy-based models incorporate the energy function with the learning model and do not deal with suitable normalization of probability distributions [36, 37]. Unlike other generative methods, energy-based models are more flexible to handle probability distributions [38]. Reinforcement learning techniques integrate a feedback loop between the learning model and the input data and can be implemented using recurrent neural networks [39, 40]. In contrast to other generative methods, reinforcement learning techniques do not just reply on a fixed input data set [30]. The promising GAN-based models have been a focus of attention since the first prototype has been invented [41, 42].

This chapter provides a comprehensive overview of how generative AI is applied in de novo protein and peptide design. This is one of the hot topics in the multidisciplinary fields of machine learning, deep learning, generative chemistry, and computer-aided drug design and discovery. Consequently, biological and/or clinical implications from this topic could then serve as a basis for future research in de novo peptide and protein design using GAN based models.

3.2 Deep Learning Model: Deep Neural Network (DNN), Convolutional Neural Network (CNN), Recurrent Neural Network (RNN), Generative Adversarial Network (GAN) Model

Various deep learning models has been developed that can handle analysis in many data types such as: time series, matrix, text and images dataset that can be seen in Fig. 3.1.

Mostly, DNN can be used for data matrix that consist of column and row dataset. CNN usually used for images dataset. RNN usually used for time series dataset and GAN usually used for generating dataset such as image, protein sequence and peptide sequence. However, in this book chapter, we will only focus on the GAN model.

GAN was proposed in 2014 by Goodfellow et al. to serve as a form of generative algorithms using an adversarial approach [43]. The GAN model provides the following merits. First, the GAN model can perform sampling tasks concurrently. As a result, it contributes to a significant speedup for generating samples [36]. Second, training the GAN model does not require real data distributions or mathematical conditions [43]. Third, in reference to the empirical experiments, the GAN model usually offers better performance than other generative approaches [36].

GAN model comprises two essential components such as a generative network module and a discriminative network module as depicted in Fig. 3.1. Moreover, for the example of research flow can be seen in Fig. 3.2 [43].

In the architecture of GAN, the generative and discriminative network modules are trained at the same time. These two modules are frequently implemented using two deep artificial neural networks with multiple layers. While the generative network

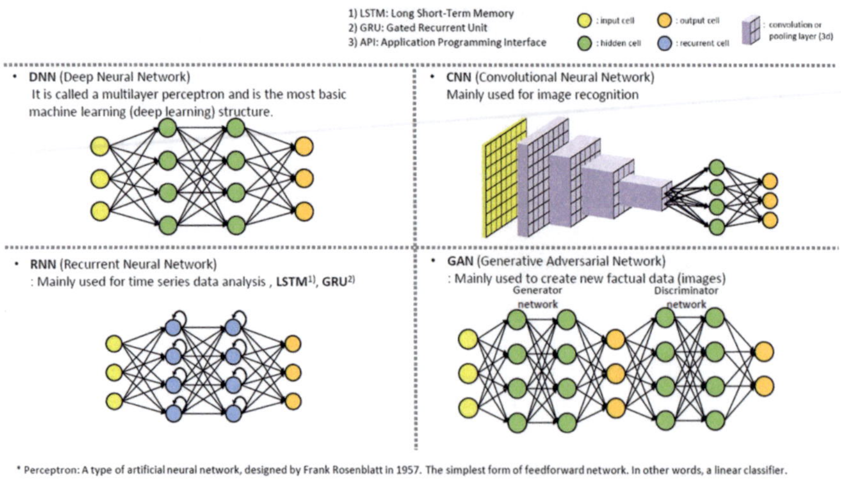

Fig. 3.1 Various deep learning models

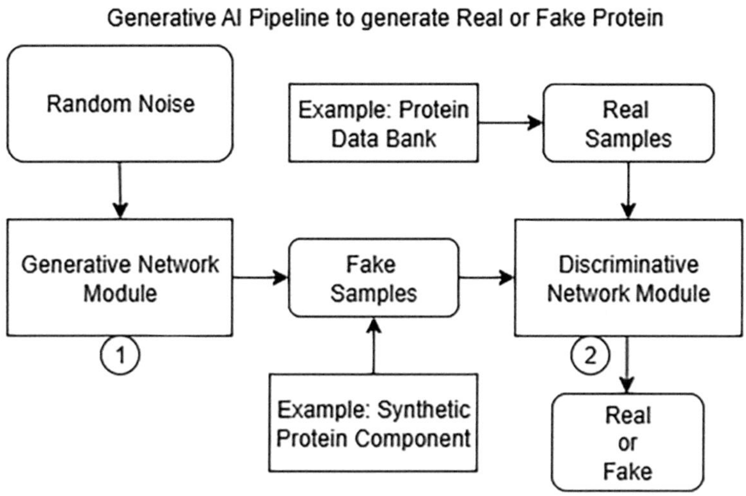

Fig. 3.2 GAN architecture to generate and predict real or fake protein

module is trained to generate fake instances in accordance with a latent variable, the discriminative network module obtains both real and fake instances to determine if its input is real or not. The discriminative network module estimates higher probability when it determines that an instance is more prone to be real. Simultaneously, the generative network module is trained to improve the chance of making a mistake in the discriminative network module. In other words, both the generative and discriminative networks compete concurrently against each other to achieve their objectives. Accordingly, the GAN model accomplishes an adversarial game between the generative and discriminative network modules.

3.3 Application of the GAN Model

The GAN model can be applied to a wide range of research areas, including image processing, video processing, speech processing, natural language processing, computer vision, medical imaging, and music, to name a few [36, 37]. Additionally, the GAN model has been implemented in chemistry (e.g., generative chemistry, molecular informatics, chemo-informatics) and biology (e.g., bioinformatics, computational biology, computer-aided drug design and discovery, multi-omics, and medical informatics).

There are a wide variety of GAN-based models (i.e., variants of the GAN model). In this subsection, we introduce the following five variants: Deep Convolutional Generative Adversarial Network (DCGAN), Wasserstein GAN, Long Short-Term Memory Generative Adversarial Network (LSTM-GAN), LeakGAN, and adversarial auto-encoder models. In addition, the reader can refer to recent reviews by Alqahtani et al. [36] and Lan et al. [37] for other variants of the GAN model.

Interestingly, the convolutional neural network module was initially applied to the tasks of object recognition and classification in the field of computer vision in the past. For instance, the DCGAN model has been implemented to acquire a hierarchy of image characterizations with an unsupervised learning technique in computer vision [39]. The DCGAN model is a hybrid approach based on the convolutional neural network module and the GAN model to combine the advantages of the convolutional neural network module in feature extraction with the robust generation ability of the GAN model. In the DCGAN model, both the generative and discriminative network modules are achieved by using the convolutional neural network module. While the DCGAN model is commonly applied to generate images in computer vision, it is lately utilized to create novel peptides and protein structures in generative chemistry and computer-aided drug design and discovery.

Wasserstein GAN model by employing a novel distance measurement approach called the Earth-Mover distance (i.e., Wasserstein distance) to deal with the instability of GAN training. In contrast, the initial GAN model utilizes the Jensen−Shannon divergence which is theoretically unsuitable for evaluating the distance between two distributions when the distributions do not overlap [40]. Moreover, the conditional Wasserstein GAN model, a variant of the Wasserstein GAN model, is primarily characterized by the Wasserstein GAN model with a gradient penalty term to further enhance performance [40, 41]. While the Wasserstein GAN or conditional Wasserstein GAN model is frequently implemented to construct images in the field of computer vision, it is newly leveraged to develop novel peptides and protein structures in generative chemistry and computer-aided drug design and discovery.

The LSTM-GAN model was implemented to synthesize sensory data in the fields of big data analytics and data encryption. The LSTM-GAN model [42] is a hybrid approach based on a long short-term memory unit and the GAN model to integrate the strengths of the long short-term memory unit in feature extraction with the persistent generation potential of the GAN model. The LSTM-GAN model utilizes a long short-term memory unit in the generative network module and another one in the discriminative network module, where the long short-term memory unit is frequently employed in the field of natural language processing. The LSTM-GAN model can be designed by using the SenseGen software package. While the LSTM-GAN model is typically exploited to compose sensory data in big data analytics and data encryption, it is currently repurposed to produce novel peptides and protein structures in generative chemistry and computer-aided drug design and discovery.

LeakGAN model was designed for the purpose of generating text which mimics human expression in the field of natural language processing. Similarly, the LeakGAN model is a hybrid approach based on a long short-term memory unit and the GAN model to incorporate the merits of the long short-term memory unit in feature extraction with the strong generation of the GAN model [44]. In the LeakGAN model, the generative network module consists of a high-level Manager module and a low-level Worker module. Both the Manager and Worker modules possess a long short-term memory unit, which is similar to the LSTM-GAN model. However, the overall architecture of the LeakGAN model is different from the LSTM-GAN model. During the adversarial training process, the discriminative network module provides the high-level feature information (i.e., the leaked information) to the Manager module in the generative network module. While the LeakGAN model is normally employed to construct text in natural language processing, it is recently used to generate novel peptides and protein structures in generative chemistry and computer-aided drug design and discovery.

The adversarial autoencoder model, a variant of the GAN model, is a probabilistic/generative model using an auto-encoder module [45]. Basically, the deep adversarial autoencoder structure comprises a conventional autoencoder module and an adversarial network module, where the autoencoder module consists of an encoder unit and a decoder unit. In brief, the training goal in the adversarial autoencoder model is to match the latent data generated by the generative network module with a particular prior latent distribution. On the contrary, the variational autoencoder model, a variant of the autoencoder-based algorithms, is not a variant of the GAN model. Essentially, the variational autoencoder model only consists of an encoder unit and a decoder unit, without an adversarial network module. The goal of the encoder unit is to generate the mean and covariance of the Gaussian distribution to represent the variational distribution in the variational auto-encoder model [46]. Lin et al. combined the variational autoencoder model with the adversarial autoencoder model to design the adversarial variational auto-encoder model [25]. A recent study also applied the variational autoencoder model to produce protein sequences [33]. On another note, the conditional variational autoencoder model exemplifies a variant of the variational autoencoder model. A recent study utilized the conditional variational autoencoder model to generate protein sequences [47].

3.4 De Novo Peptide and Protein Design

The usage of the GAN architecture is still incipient in the context of de novo peptide and protein design. Lately, there has been a revival of interest in developing GAN-based applications for de novo peptide and protein design in the cross-disciplinary fields of generative chemistry, machine learning, deep learning, and computer-aided drug design and discovery. De novo peptide and protein design using various GAN-based algorithms can be seen in Table 3.1 [48].

The aim of de novo peptide and protein design is to produce novel peptide and protein components based on prespecified protein folds and activities [2]. GAN-based algorithms are able to surpass other models only when the generative network module can generate continuous output values, for instance, a vector of numbers as in the task of image generation [11]. Using a vector of numbers, we then can train the generative network module and adjust its weights based on the gradient of the loss function from the discriminative network module. Nevertheless, peptide/protein structures are represented in text strings, not in continuous numbers. This is one key trick of GAN-based algorithms in de novo peptide and protein design.

Table 3.1 De novo peptide/protein design with various GAN-based algorithms

Study	Model	Main concept	Database (sample size)	Tasks
Rossetto et al. [49, 50]	GANDALF	DCGAN	THPdb (380) PDB (25,000)	Generate novel peptides
Gupta et al. [51]	Feedback-GAN	Conditional Wasserstein GAN	APD3 (2600)	Generate DNA (protein-coding) sequences for antimicrobial peptides with desired properties
Surana et al. [47]	PandoraGAN	LeakGAN	APD3 (423 + 130)	Generate bioactive antiviral peptides
Anand and Huang [52]	DCGAN	DCGAN	PDB (NA)	Generate novel protein structures
Li et al. [53]	DCGAN	DCGAN	PDB (250 – 617)	Perform the task of protein loop modeling
Karimi et al. [54]	gcWGAN	Conditional Wasserstein GAN	SCOPe (20,125)	Generate novel protein folds with high yields
Sabban et al. [55]	RamaNet	LSTM-GAN	PDB (303,500)	Generate protein backbones
Repecka et al. [56]	ProteinGAN	GAN	Uniprot (16,898)	Generate functional protein sequences
Yang et al. [57]	GANcon	GAN	SCOPe (7192 + 360)	Generate protein contact maps

Thus, we need to design an approach to facilitate gradients through peptide/protein structures as depicted in Fig. 3.2 [48]. The existing solutions in the literature are as follows. First, a pairwise distance matrix between alpha-carbons on the protein backbone is employed to represent protein structures. Second, a four-dimensional (4D) tensor is employed to represent the positions of active atoms in proteins. Third, DNA/gene sequences are used and then converted into protein sequences. Moreover, Ramachandran angles, the main chain torsion angles (i.e., phi and psi) in each amino acid, are used to represent a protein structure [58]. In addition, both the generative and discriminative network modules directly deal with a latent vector which is encoded by 20 canonical amino acids.

Several public databases, including Antimicrobial Peptide Database 3 (APD3) [59], Gene Ontology (GO) [60], Protein Data Bank (PDB) [61], Structural Classification of Proteins-extended (SCOPe) [62], Universal Protein (Uniprot) [63], Swiss-Prot [64], and Therapeutic Proteins database (THPdb) [65], are often utilized to facilitate de novo peptide and protein design. For instance, the PDB database is an online repository of experimentally determined protein structures [61].

On another note, AlphaFold1/2 [60, 61] and RoseTTAFold [62] can be considered as generative models for secondary/tertiary structures and contacts. AlphaFold1/2 incorporates deep neural network architectures such as a convolutional neural network, which is trained using PDB data sets. RoseTTAFold also utilizes deep learning techniques, where a three-track neural network is constructed to leverage three levels of information, including one-dimensional sequence data, two-dimensional data for the distance between pairs of amino acids, and three-dimensional data of backbone coordinates [62]. Taken altogether, AlphaFold1/2 and RoseTTAFold achieved similar accuracies for protein structure predictions [66].

Rossetto et al. proposed the Generative Adversarial Network Drug-target Ligand Fructifier (GANDALF) model to generate novel peptides that bind to specific proteins targeting cancer [49, 50]. The GANDALF model was trained using a total of 239 targets and 380 drugs from the THPdb [65] and 25,000 protein structures from the PDB database [61]. The GANDALF model was able to generate multiple peptides that bind to three specific cancer-related protein targets, where the best generated peptide for each protein target was comparable to and distinct from the existing FDA approved drugs [50]. Briefly, the GANDALF model leverages the concept of the DCGAN model where five-layer convolutional neural networks are applied to both the generative and discriminative network modules. To create novel peptides, a 4D tensor representing the peptide information serves as the "ground truth" to train the GANDALF model [49]. The GANDALF model also employs the information on active atoms, which are the atoms interacting with each other during the protein–peptide binding process [63]. The main drawback of the GANDALF model is that the active atoms were not properly aligned to have a tighter binding affinity as compared with the active atoms of the FDA approved drugs [49]. For example, the average error of binding affinities was near 10% [50]. Moreover, another drawback is that Rossetto et al. did not compare the GANDALF model with other existing deep learning or machine learning models to demonstrate its performance. It should be mentioned that Rossetto et al. compared the GANDALF model with PepComposer, a Monte Carlo approach [67]. On the other hand, the main merit of the GANDALF model is that the produced peptides were considerably comparable to the FDA-approved drugs (e.g., Avelumab, Durvalumab, and Pembrolizumab) [49, 50].

Surana et al. proposed the PandoraGAN model to computationally generate bioactive antiviral peptides in the context of de novo peptide design [47]. The PandoraGAN model was trained using the verified peptides with antiviral affinity (i.e., 423 active sequences and 130 highly active sequences), which were extracted from the existing experimentally validated peptide databases (e.g., the APD3 database) [59]. The PandoraGAN model was able to produce bioactive antiviral peptides with similar physicochemical properties to the training data of highly active peptide sequences [47]. In brief, the PandoraGAN model adopts the concept of the LeakGAN model which is a combination method involving the GAN architecture and long short-term memory units. Interestingly, the LeakGAN model has been broadly used in the text generation task. The main limitation of the PandoraGAN model is that it only focused on generating highly active peptides, not active peptides, in the current version [47]. Second, the distribution of residues in the generated peptides were needed to be improved by using both highly active and active peptides. Additionally, another limitation is that the sample size ($n = 423 + 130 = 553$) was small. On the other hand, the main benefit of the PandoraGAN model is that it outperformed several existing GAN-based models to generate novel antiviral peptides, including GSGAN [68], MaliGAN [69], RankGAN [70], SeqGAN [71], and Text-GAN [72]. These existing GAN-based models were developed for the text generation task and were adopted to generate bioactive antiviral peptides in the context of de novo peptide design.

Anand and Huang applied the DCGAN mode to generate novel protein structures [73]. As mentioned previously, the DCGAN model has been widely used to generate images in the field of computer vision. Anand and Huang adopted the concept of the DCGAN model to produce novel protein structures. The PDB database was used to train the DCGAN model [61]. Note that the number of sequences used (i.e., the sample size) was unavailable in their study. Anand and Huang showed that their method was able to create novel protein structures which were considerably different from biological macromolecular crystal structures in the PDB database. The methodology presented protein structures in terms of a pairwise distance matrix between alpha-carbons on the protein backbone. Then, the alternating direction method of multipliers (ADMM) algorithm was used to transform 2D pairwise distances into 3D Cartesian coordinates [73]. In other words, the ADMM algorithm was employed to fold 3D protein structures from the pairwise distance matrix generated by the DCGAN model. The main weakness of their method is that the ADMM algorithm was unable to guarantee correct secondary/local structures (e.g., deviations from helix or sheet structures) [52]. In addition, another weakness is that the sample size was unknown. On the other hand, the main strength of their method is that the DCGAN model was faster than the traditional approach such as the Rosetta modeling suite [52]. Moreover, another strength is that the DCGAN model exceeded other existing models such as TorusDBN, FB5-HMM, a multiscale torsion angle GAN [73], 3D-GAN, and a full-atom GAN [73].

Repecka et al. proposed the ProteinGAN model to computationally generate functional protein sequences in the context of de novo protein design by using the idea of the self-attention GAN model [74]. It was originally applied to accomplish high-resolution image generation in the field of computer vision. In the ProteinGAN model, the purpose of the self-attention layer was to target the functionally significant regions across the entire protein sequences (e.g., catalytic residues). Moreover, bacterial malate dehydrogenase sequences (including 16,898 sequences) from the Uniprot database [63] were used to train the ProteinGAN model. Repecka et al. indicated that 24% of the generated enzymes were soluble and also exhibited catalytic activities equivalent to natural enzymes. In essence, the framework of the ProteinGAN model derives from the self-attention GAN model, which is a variant of the GAN model with a self-attention layer. In the ProteinGAN model, both the generative and discriminative network modules contain residual blocks and a self-attention layer, where each residual block consists of convolutional neural network modules. In the ProteinGAN model, the generative and discriminative network modules communicate with each other using a latent vector, which is encoded with 20 canonical amino acids.

The main weakness of the ProteinGAN model is that it was laborious to train the ProteinGAN model as it had over 60 million trainable parameters with 45 layers of neural networks [56]. Moreover, another weakness is that Repecka et al. did not compare the ProteinGAN model with other existing deep learning or machine learning models to demonstrate its performance. However, they exploited two other models (e.g., transfer learning and recurrent neural networks) which did not show promising results. On the other hand, the main strength of the ProteinGAN model is that the highly diverse enzymes were produced and these generated enzymes may possess completely different catalytic properties from those of natural enzymes. Another strength of the ProteinGAN model is that the spectral normalization method was implemented to stabilize the training of the discriminative network module and thereby to reduce training instability.

References

1. Protein function | Learn science at scitable (cited 2025 Sep 15). Available from: https://www.nature.com/scitable/topicpage/protein-function-14123348/
2. Forbes J, Krishnamurthy K. Biochemistry, peptide. StatPearls. 2023 (cited 2025 Sep 15). Available from: https://www.ncbi.nlm.nih.gov/books/NBK562260/
3. Regan L, Caballero D, Hinrichsen MR, Virrueta A, Williams DM, O'hern CS. Protein design: past, present, and future. Biopolymers. 2015;104(4):334–50 (cited 2025 Oct 6). Available from: https://pubmed.ncbi.nlm.nih.gov/25784145/
4. Guo RC, Zhang XH, Ji L, Wei ZJ, Duan ZY, Qiao ZY, et al. Recent progress of therapeutic peptide based nanomaterials: from synthesis and self-assembly to cancer treatment. Biomater Sci. 2020;8(22):6175–89 (cited 2025 Oct 6). Available from: https://pubs.rsc.org/en/content/articlehtml/2020/bm/d0bm01358g
5. Craik DJ, Fairlie DP, Liras S, Price D. The future of peptide based drugs. Chem Biol Drug Des. 2013;81(1):136–47 (cited 2025 Oct 6) Available from: https://pubmed.ncbi.nlm.nih.gov/23253135/
6. Shehu A, Barbará D, Molloy K. A survey of computational methods for protein function prediction. Big Data Anal Genomics. 2016;225–98 (cited 2025 Oct 6). Available from: https://www.researchgate.net/publication/309432914_A_Survey_of_Computational_Methods_for_Protein_Function_Prediction
7. Chen H, Engkvist O, Wang Y, Olivecrona M, Blaschke T. The rise of deep learning in drug discovery. Drug Discov Today. 2018;23(6):1241–50 (cited 2025 Oct 6). Available from: https://pubmed.ncbi.nlm.nih.gov/29366762/
8. Vamathevan J, Clark D, Czodrowski P, Dunham I, Ferran E, Lee G, et al. Applications of machine learning in drug discovery and development. Nat Rev Drug Discov. 2019;18(6):463–77 (cited 2025 Oct 6). Available from: https://pubmed.ncbi.nlm.nih.gov/30976107/
9. Hessler G, Baringhaus KH. Artificial Intelligence in Drug Design Mol. 2018;23(10):2520 (cited 2025 Oct 6). Available from: https://www.mdpi.com/1420-3049/23/10/2520/htm
10. Butler KT, Davies DW, Cartwright H, Isayev O, Walsh A. Machine learning for molecular and materials science. Nature. 2018;559(7715):547–55 (cited 2025 Oct 6). Available from: https://www.researchgate.net/publication/326608140_Machine_learning_for_molecular_and_materials_science

11. Lin E, Lin CH, Lane HY. Relevant applications of generative adversarial networks in drug design and discovery: molecular de novo design, dimensionality reduction, and de novo peptide and protein design. Molecules. 2020;25(14):3250 (cited 2025 Oct 6). Available from: https://www.mdpi.com/1420-3049/25/14/3250/htm
12. (PDF) Generative chemistry: drug discovery with deep learning generative models (cited 2025 Oct 6). Available from: https://www.researchgate.net/publication/343786548_Generative_chemistry_drug_discovery_with_deep_learning_generative_models
13. Vanhaelen Q, Lin YC, Zhavoronkov A. The advent of generative chemistry. ACS Med Chem Lett. 2020;11(8):1496–505 (cited 2025 Oct 6). Available from: https://pubmed.ncbi.nlm.nih.gov/32832015/
14. Segler MHS, Preuss M, Waller MP. Planning chemical syntheses with deep neural networks and symbolic AI. Nature. 2018;555(7698):604–10 (cited 2025 Oct 6). Available from: https://pubmed.ncbi.nlm.nih.gov/29595767/
15. Baskin II, Winkler D, Tetko IV. A renaissance of neural networks in drug discovery. Expert Opin Drug Discov. 2016;11(8):785–95 (cited 2025 Oct 6). Available from: https://pubmed.ncbi.nlm.nih.gov/27295548/
16. Jing Y, Bian Y, Hu Z, Wang L, Xie XQS. Deep learning for drug design: an artificial intelligence paradigm for drug discovery in the big data era. AAPS J 2018;20(3) (cited 2025 Oct 6). Available from: https://pubmed.ncbi.nlm.nih.gov/29603063/
17. Rifaioglu AS, Atas H, Martin MJ, Cetin-Atalay R, Atalay V, Doğan T. Recent applications of deep learning and machine intelligence on in silico drug discovery: methods, tools and databases. Brief Bioinform. 2019;20(5):1878–912 (cited 2025 Oct 6). Available from: https://pubmed.ncbi.nlm.nih.gov/30084866/
18. Goh GB, Hodas NO, Vishnu A. Deep learning for computational chemistry. J Comput Chem. 2017;38(16):1291–307 (cited 2025 Oct 6). Available from: https://pubmed.ncbi.nlm.nih.gov/28272810/
19. Mamoshina P, Vieira A, Putin E, Zhavoronkov A. Applications of deep learning in biomedicine. Mol Pharm. 2016;13(5):1445–54 (cited 2025 Oct 6). Available from: https://pubmed.ncbi.nlm.nih.gov/27007977/
20. Ekins S. The next era: deep learning in pharmaceutical research. Pharm. 2016;33(11):2594–603 (cited 2025 Oct 6). Available from: https://pubmed.ncbi.nlm.nih.gov/27599991/
21. Ghasemi F, Mehridehnavi A, Pérez-Garrido A, Pérez-Sánchez H. Neural network and deep-learning algorithms used in QSAR studies: merits and drawbacks. Drug Discov Today. 2018;23(10):1784–90 (cited 2025 Oct 6). Available from: https://pubmed.ncbi.nlm.nih.gov/29936244/
22. Zhao H, Li H, Maurer-Stroh S, Cheng L. Synthesizing retinal and neuronal images with generative adversarial nets. Med Image Anal. 2018;49:14–26 (cited 2025 Oct 6). Available from: https://www.sciencedirect.com/science/article/pii/S1361841518304596
23. Hu B, Tang Y, Chang EIC, Fan Y, Lai M, Xu Y. Unsupervised learning for cell-level visual representation in histopathology images with generative adversarial networks. IEEE J Biomed Heal informatics. 2019;23(3):1316–28 (cited 2025 Oct 6). Available from: https://pubmed.ncbi.nlm.nih.gov/29994411/
24. Mardani M, Gong E, Cheng JY, Vasanawala SS, Zaharchuk G, Xing L, et al. Deep generative adversarial neural networks for compressive sensing MRI. IEEE Trans Med Imaging. 2019;38(1):167–79 (cited 2025 Oct 6). Available from: https://pubmed.ncbi.nlm.nih.gov/30040634/
25. Lin E, Mukherjee S, Kannan S. A deep adversarial variational autoencoder model for dimensionality reduction in single-cell RNA sequencing analysis. BMC Bioinf. 2020;21(1):1–11 (cited 2025 Oct 6). Available from: https://doi.org/10.1186/s12859-020-3401-5
26. Lin E, Kuo PH, Liu YL, Yu YWY, Yang AC, Tsai SJ. A deep learning approach for predicting antidepressant response in major depression using clinical and genetic biomarkers. Front Psychiatr. 2018;9(JUL) (cited 2025 Oct 6]. Available from: https://pubmed.ncbi.nlm.nih.gov/30034349/

27. Lin E, Tsai SJ. Machine learning in neural networks. Adv Exp Med Biol. 2019;1192:127–37 (cited 2025 Oct 6). Available from: https://pubmed.ncbi.nlm.nih.gov/31705493/
28. Ching T, Himmelstein DS, Beaulieu-Jones BK, Kalinin AA, Do BT, Way GP, et al. Opportunities and obstacles for deep learning in biology and medicine. J R Soc Interface. 2018;15(141) (cited 2025 Oct 6). Available from: https://pubmed.ncbi.nlm.nih.gov/29618526/
29. Dana D, Gadhiya SV., Surin LGS, Li D, Naaz F, Ali Q, et al. Deep learning in drug discovery and medicine; scratching the surface. Molecules. 2018;23(9). (cited 2025 Oct 6). Available from: https://pubmed.ncbi.nlm.nih.gov/30231499/
30. Deep Learning (cited 2025 Oct 6). Available from: https://www.deeplearningbook.org/
31. Sinai S, Kelsic E, Church GM, Nowak MA. Variational auto-encoding of protein sequences. 2017 (cited 2025 Oct 6). Available from: https://arxiv.org/pdf/1712.03346
32. Oussidi A, Elhassouny A. Deep generative models: survey. 2018 Int Conf Intell Syst Comput Vis ISCV. 2018; 2018-May:1–8
33. Kobyzev I, Prince SJD, Brubaker MA. Normalizing flows: an introduction and review of current methods. IEEE Trans Pattern Anal Mach Intell. 2019;43(11):3964–79 (cited 2025 Oct 6). Available from: https://arxiv.org/pdf/1908.09257
34. Luo Y, Yan K, Ji S. GraphDF: A discrete flow model for molecular graph generation. Proc Mach Learn Res. 2021;139:7192–203 (cited 2025 Oct 6). Available from: https://arxiv.org/pdf/2102.01189
35. Papamakarios G, Nalisnick E, Rezende DJ, Mohamed S, Lakshminarayanan B. Normalizing flows for probabilistic modeling and inference. J Mach Learn Res. 2019;22:1–64 (cited 2025 Oct 6). Available from: https://arxiv.org/pdf/1912.02762
36. Alqahtani H, Kavakli-Thorne M, Kumar G. Applications of generative adversarial networks (GANs): an updated review. Arch Comput Methods Eng. 2021;28(2):525–52 (cited 2025 Oct 6). Available from: https://www.researchgate.net/publication/338050169_Applications_of_Generative_Adversarial_Networks_GANs_An_Updated_Review
37. Lan L, You L, Zhang Z, Fan Z, Zhao W, Zeng N, et al. Generative adversarial networks and its applications in biomedical informatics. Front Public Heal. 2020;8:523500 (cited 2025 Oct 6). Available from: www.frontiersin.org
38. Song Y, Kingma DP. How to train your energy-based models. 2021 (cited 2025 Oct 6). Available from: https://arxiv.org/pdf/2101.03288
39. Radford A, Metz L, Chintala S. Unsupervised representation learning with deep convolutional generative adversarial networks. In: 4th International conference on learn represent ICLR 2016– conference track process. 2015 (cited 2025 Oct 6). Available from: https://arxiv.org/pdf/1511.06434
40. Arjovsky M, Chintala S, Bottou L. Wasserstein GAN. 2017 (cited 2025 Oct 6). Available from: https://arxiv.org/pdf/1701.07875
41. Gulrajani I, Ahmed F, Arjovsky M, Dumoulin V, Courville A. Improved training of Wasserstein GANs. Adv Neural Inf Process Syst. 2017;2017-December:5768–78 (cited 2025 Oct 6). Available from: https://arxiv.org/pdf/1704.00028
42. Palangi H, Deng L, Shen Y, Gao J, He X, Chen J, et al. Deep sentence embedding using long short-term memory networks: analysis and application to information retrieval. IEEE/ACM Trans Audio Speech Lang Process. 2015;24(4):694–707 (cited 2025 Oct 6). Available from: https://arxiv.org/pdf/1502.06922
43. Goodfellow IJ, Pouget-Abadie J, Mirza M, Xu B, Warde-Farley D, Ozair S, et al. Generative adversarial nets. Adv Neural Inf Process Syst. 2014;27 (cited 2025 Oct 6). Available from: http://www.github.com/goodfeli/adversarial
44. Guo J, Lu S, Cai H, Zhang W, Yu Y, Wang J. Long text generation via adversarial training with leaked information. Proc AAAI Conf Artif Intell. 2018;32(1):5141–8 (cited 2025 Oct 6). Available from: https://ojs.aaai.org/index.php/AAAI/article/view/11957
45. Makhzani A, Shlens J, Jaitly N, Brain G, Openai IG, Frey B. Adversarial autoencoders. Elem Dimens Reduct Manifold Learn. 2015;577–96 (cited 2025 Oct 6). Available from: https://arxiv.org/pdf/1511.05644

46. Kingma DP, Welling M. Auto-encoding variational bayes. In: 2nd International conference on learning representations ICLR 2014—conference track proceedings. 2013 (cited 2025 Oct 6). Available from: https://arxiv.org/pdf/1312.6114
47. Surana S, Arora P, Singh D, Sahasrabuddhe D, Valadi J. PandoraGAN: generating antiviral peptides using generative adversarial network. bioRxiv. 2022 (cited 2025 Oct 6) (2021.02.15.431193). Available from: https://doi.org/10.1101/2021.02.15.431193v2
48. Lin E, Lin CH, Lane HY. De novo peptide and protein design using generative adversarial networks: an update. J Chem Inf Model. 2022;62(4):761–74 (cited 2025 Oct 6). Available from: https://doi.org/10.1021/acs.jcim.1c01361
49. Rossetto AM, Zhou W. Gandalf: a prototype of a GAN-based peptide design method. In: ACM-BCB 2019—Proceedings of 10th ACM international conference on bioinformatics (Comput Biol Heal Inf. 2019;61–6)
50. Rossetto A, Zhou W. GANDALF: peptide generation for drug design using sequential and structural generative adversarial networks. In: Proceedings of 11th ACM international conference on bioinformatics (Comput Biol Heal Inf BCB. 2020 (cited 2025 Oct 6). Available from: https://www.researchgate.net/publication/347841282_GANDALF_Peptide_Generation_for_Drug_Design_using_Sequential_and_Structural_Generative_Adversarial_Networks
51. Gupta A, Zou J. Feedback GAN (FBGAN) for DNA: a novel feedback-loop architecture for optimizing protein functions. 2018 (cited 2025 Oct 6). Available from: https://arxiv.org/pdf/1804.01694
52. Anand N, Huang PS. Generative modeling for protein structures. Adv Neural Inf Process Syst. 2018;31
53. Li Z, Nguyen SP, Xu D, Shang Y. Protein loop modeling using deep generative adversarial network. In: Proceedings of international IEEE conference on tools for artificial intelligence ICTAI. 2017;2017-November:1085–91 (cited 2025 Oct 6). Available from: https://www.researchgate.net/publication/325635270_Protein_Loop_Modeling_Using_Deep_Generative_Adversarial_Network
54. Karimi M, Zhu S, Cao Y, Shen Y. De novo protein design for novel folds using guided conditional wasserstein generative adversarial networks. J Chem Inf Model. 2020;60(12):5667–81 (cited 2025 Oct 6). Available from: https://pubmed.ncbi.nlm.nih.gov/32945673/
55. Sabban S, Markovsky M. RamaNet: Computational *de novo* helical protein backbone design using a long short-term memory generative neural network. F1000Research 2020 9298. 2020;9:298 (cited 2025 Oct 6). Available from: https://f1000research.com/articles/9-298
56. Repecka D, Jauniskis V, Karpus L, Rembeza E, Rokaitis I, Zrimec J, et al. Expanding functional protein sequence spaces using generative adversarial networks. Nat Mach Intell. 2021;3(4):324–33.
57. Yang H, Wang M, Yu Z, Zhao XM, Li A. GANcon: protein contact map prediction with deep generative adversarial network. IEEE Access. 2020;8:80899–907.
58. Gopalakrishnan K, Sowmiya G, Sheik SS, Sekar K, Ramachandran plot on the web (2.0). Protein Pept Lett. 2007;14(7):669–71 (cited 2025 Oct 6). PMID: 17897092. https://doi.org/10.2174/092986607781483912
59. Wang G, Li X, Wang Z. APD3: the antimicrobial peptide database as a tool for research and education. Nucleic Acids Res. 2016;44(D1):D1087–93 (cited 2025 Oct 6). Available from: https://pubmed.ncbi.nlm.nih.gov/26602694/
60. Blake JA, Christie KR, Dolan ME, Drabkin HJ, Hill DP, Ni L, et al. Gene ontology consortium: going forward. Nucleic Acids Res. 2015;43(Database issue):D1049–56 (cited 2025 Oct 6). Available from: https://pubmed.ncbi.nlm.nih.gov/25428369/
61. Berman HM, Westbrook J, Feng Z, Gilliland G, Bhat TN, Weissig H, et al. The protein data bank. nucleic acids res. 2000;28(1):235–42 (cited 2025 Oct 6). Available from: https://pubmed.ncbi.nlm.nih.gov/10592235/
62. Kolodny R, Pereyaslavets L, Samson AO, Levitt M. On the universe of protein folds. Annu Rev Biophys. 2013;42(1):559–82 (cited 2025 Oct 6). Available from: https://pubmed.ncbi.nlm.nih.gov/23527781/

63. Bateman A. UniProt: a worldwide hub of protein knowledge. Nucleic Acids Res. 2019;47(D1):D506–15 (cited 2025 Oct 6). Available from: https://pubmed.ncbi.nlm.nih.gov/30395287/
64. Boutet E, Lieberherr D, Tognolli M, Schneider M, Bansal P, Bridge AJ, et al. UniProtKB/Swiss-Prot, the MANUALLY ANNOTATED SECTION OF the UniProt knowledge base: how to use the entry view. Methods Mol Biol. 2016;1374:23–54 (cited 2025 Oct 6). Available from: https://pubmed.ncbi.nlm.nih.gov/26519399/
65. Usmani SS, Bedi G, Samuel JS, Singh S, Kalra S, Kumar P, et al. THPdb: database of FDA-approved peptide and protein therapeutics. PLoS One. 2017;12(7) (cited 2025 Oct 6). Available from: https://pubmed.ncbi.nlm.nih.gov/28759605/
66. Baek M, DiMaio F, Anishchenko I, Dauparas J, Ovchinnikov S, Lee GR, et al. Accurate prediction of protein structures and interactions using a three-track neural network. Science. 2021;373(6557):871–6 (cited 2025 Oct 6). Available from: https://pubmed.ncbi.nlm.nih.gov/34282049/
67. Obarska-Kosinska A, Iacoangeli A, Lepore R, Tramontano A. PepComposer: computational design of peptides binding to a given protein surface. Nucleic Acids Res. 2016;44(W1):W522–8 (cited 2025 Oct 6). Available from: https://doi.org/10.1093/nar/gkw366
68. Kusner MJ, Hernández-Lobato JM. GANS for Sequences of discrete elements with the Gumbel-softmax distribution. 2016 (cited 2025 Oct 6). Available from: https://arxiv.org/pdf/1611.04051
69. Che T, Li Y, Zhang R, Hjelm RD, Li W, Song Y, et al. Maximum-likelihood augmented discrete generative adversarial networks. 2017 (cited 2025 Oct 6). Available from: https://arxiv.org/pdf/1702.07983
70. Lin K, Li D, He X, Zhang Z, Sun MT. Adversarial ranking for language generation. Adv Neural Inf Process Syst. 2017;2017-December:3156–66 (cited 2025 Oct 6). Available from: https://arxiv.org/pdf/1705.11001
71. Yu L, Zhang W, Wang J, Yu Y. SeqGAN: sequence generative adversarial nets with policy gradient. Proc AAAI Conf Artif Intell. 2017;31(1):2852–8 (cited 2025 Oct 6). Available from: https://ojs.aaai.org/index.php/AAAI/article/view/10804
72. Zhang Y, Gan Z, Fan K, Chen Z, Henao R, Shen D, et al. Adversarial feature matching for text generation. PMLR. 2017;4006–15 (cited 2025 Oct 6). Available from: https://proceedings.mlr.press/v70/zhang17b.html
73. Boyd S, Parikh N, Chu E, Peleato B, Eckstein J. Distributed optimization and statistical learning via the alternating direction method of multipliers. Found Trends Mach Learn. 2010;3(1):1–122 (cited 2025 Oct 6). Available from: https://www.researchgate.net/publication/220416607_Distributed_Optimization_and_Statistical_Learning_via_the_Alternating_Direction_Method_of_Multipliers
74. Zhang H, Goodfellow I, Metaxas D, Odena A. Self-attention generative adversarial networks. PMLR. 2019;7354–63 (cited 2025 Oct 6). Available from: https://proceedings.mlr.press/v97/zhang19d.html

Chapter 4
Exploring Chemical Space with Generative Models

Bharghavi Balaji and Sinosh Skariyachan

Abstract Chemical space captures the whole spectrum of possible small molecules and their physicochemical, biological and structural characteristics. Latest advancements in computational chemistry, generative artificial intelligence and cheminformatics have transformed ways in which vast molecular universe is visualised, prioritised and navigated for drug discovery. Molecular descriptors act as a quantitative base for these studies that captures features vital for property prediction, molecular similarity and activity modeling assessment. The incorporation of cheminformatics pipelines, machine learning frameworks and structure-based docking simulations has made rapid identification of biologically relevant subspaces possible within immense compound libraries. Generative AI methods, encompassing latent-space optimisation, reinforcement learning, diffusion models, ADMET profiles and selectivity. Databases put together from fragment-based libraries and natural products remain vital for mapping chemical diversity and improving novelty of scaffold. Together, these AI-driven and computational techniques connect data-rich chemical landscape with mechanistic and predictive modeling, speeding up the discovery of therapeutically synthetic and viably accessible molecular entities.

Keywords Chemical space · Artificial intelligence · Cheminformatics · Molecular descriptors · Predictive modeling

4.1 Introduction

Investigation of chemical space—encompasses a vast diversity of potential drug-like molecules, that have been changed fundamentally by the exploration of deep generative models, which now avails the design of novel compounds with customised

B. Balaji
Department of Medical Genetics, Kasturba Medical College, Manipal Academy of Higher Education, Manipal, Karnataka, India

S. Skariyachan (✉)
Department of Microbiology, St. Pius X College, Rajapuram, Kasaragod, Kerala, India
e-mail: sinoshmicro@stpius.ac.in

A. A. Parikesit and A. N. M. Ansori (eds.), *Recent Computational Techniques in De Novo Drug Design*, Bio-IT and AI, https://doi.org/10.1007/978-3-032-20428-8_4

properties beyond the reach of conventional screening and fragment-based methods [170]. Drug discovery historically relied on the gradual extension of well-established chemical scaffolds, restricted by experimental throughput and human intuition [66]. Yet, the application of artificial intelligence (AI), particularly generative neural networks, has revolutionised the field of producing structurally diverse, synthetically accessible, and biologically significant molecules by learning intricate molecular representations from vast chemical datasets [80]. However, variational autoencoders (VAEs) and generative adversarial networks (GANs) encode particles into continuous latent spaces, which permits smooth interpolation, scaffold hopping, and optimisation of multiple objectives, including potency, selectivity, and synthetic feasibility [17]. Recurrent Neural network (RNNs) and transformer-based models are two such models that has shown notable proficiency in handling sequential molecular representations, like SMILES strings, that collects nuanced syntax and semantics of chemical structures to produce valid and novel compounds [207]. In recent times, generative models have made significant progress in integrating three-dimensional (3D) structural information [122]. By combining geometric constraints and molecular conformations, these models have facilitated design compounds with favourable binding poses and pharmacophoric features that are vital for targeting complex protein–protein interactions and allosteric sites, domains where conventional 2D models often fail [11]. Adding on, the development of multimodal generative frameworks that include written information, chemical structure data and biological assay results, have made it less difficult to design context-aware molecules, combining produced compounds to therapeutic goals and accelerating the discovery of promising leads in intricate chemical systems [66]. The evaluation and interpretability of generative models have been challenging despite remarkable advancements as standard metrics for novelty, diversity and drug-likeliness occasionally capture biological relevance [188]. Therefore, hybrid evaluation frameworks that incorporates experimental validation and predictive bioactivity models are constructed to ensure the practical utility of generated compounds [148]. In the future, it is anticipated that the combination of generative AI with advanced simulations methods like molecular dynamics and quantum chemistry, as well as the creation of benchmarking datasets and standardised molecular representations [111]. These will improve the cross-model comparability, synthetic accessibility and physical realism of generated molecules [168]. This would be a game-changing tool to establish deep generative models for chemical space exploration and drug discovery process [135, 136].

4.2 Chemical Space—Introduction (Definition, Concepts, Basic Principles)

Chemical space contains all imaginable molecular structures and configurations representing a vast, multifaceted conceptual structure ranging from fundamental arrangements to extremely complicated molecular arrangements [231]. This conceptual framework can be understood as theoretical construct where molecules are placed across intricate, interrelated dimensions obtained by their physical, structural and chemical properties [231]. The sheer magnitude of chemical space is profound, with researchers predicting around 10^{60} possible structures to different molecular structures, far beyond our understanding and exceeding conventional exploratory approaches. Chemical space fundamentally offers an advanced, conceptually grounded method of comprehending molecular classifications and possibly groundbreaking interactions, beyond standard molecular classification schemes [112]. Latest computational approaches, especially advanced generative models and artificial intelligence algorithms, have transformed our understanding to navigate and learn this enormous, mostly uncharted molecular cosmos [18]. These revolutionising methods allow scientists to methodically traverse chemical space, producing unique molecular structures that upend existing theories and molecular discovery and design [163].

The basic principle of investigating chemical space includes advanced computational strategies that incorporates deep learning, neural networks, and unpredictable modelling techniques to generate, collect and analyse essential molecular arrangements [234]. Scientists use multiple identifiable tactics, encompassing molecular string depiction like SMILES (Simplified Molecular-Input Line-Entry Systems) and sophisticated graph-based models, to transform and change molecular structures [18]. These allow researchers to build generative models that are able to sample vast molecular databases, replicating notable portions of chemical space with outstanding efficiency [235]. For example, research have demonstrated that training repeated neural networks with only 1 million molecular structures can generate up to 68.9% of the whole database containing about 975 million molecules [70]. The involvement of chemical space discoveries is complicated by difficulty of exploring chemical space, especially with numerous rings and diverse atomic compositions [18]. Cutting edge methodological techniques, like quantum generative models, variational autoencoders and generative adversarial networks (GANs) are pushing boundaries of our knowledge, providing prior unheard-of insights into molecular configurations possibility that can exist within this vast, mainly unexplored domain [234].

4.3 Molecular Descriptors

Molecular descriptors are quantitative in nature that captures physicochemical, biological and also structural characteristics of chemical compounds, establishing a foundation of contemporary cheminformatics, virtual screening and QSAR modelling in drug discovery [39] (Table 4.1). Descriptors have historically included topological indices, physicochemical properties (like logP, polar surface areas and molecular weight) and simple counts (like bond types or atoms), in order to map and compare chemical space and to predict bioactivities and molecular properties [243]. The latest development of machine learning and deep learning techniques have revolutionized the domain by making it possible to automatically extract high-dimensional, non-linear characteristics in architecture from SMILES and molecular graphs like transformers and graph neural networks (GNNs) [108]. These architectures outperform traditionally created descriptors in predicting ADMET profiles, binding affinity and toxicity. Recent development has also highlighted the significance of three-dimensional (3D) descriptors, that accounts for pharmacophoric patterns, steric and electrostatic fields and flexible conformations [122]. This helps in further improvement of modelling protein–ligand interactions and estimating diverse biological target's activity [173]. These descriptors bridge the gap between computational design and experimental validations by offering a realistic and accurate understanding of molecular behaviour when coupled with hybrid machine learning frameworks and MD simulations [155]. In addition, the hybrid evaluation metrics that assure the therapeutic potential and relevance of generated molecules have been created because of combining data-driven descriptors with experimental bioassay outcomes [144]. Research into better-performing, interpretable, and biologically meaningful descriptor systems continues to be ongoing, because, despite these advancements, concerns such as interpretability, descriptor degeneracy, and alignment of computational outputs persist [56]. Artificial intelligence and high-throughput experimentations have poised to speed up the advancement of molecular descriptor engineering, and these tools have the capability to redefine drug-likeness criteria, fasten up the discovery of new therapeutics, and widen the scope of chemical space exploration [73].

4.4 Scope of Generative AI Within the Boundaries of Chemical Space

By means of systemic exploration and de novo design of compounds with specific properties, generative AI is becoming an influential factor in orienting chemical space, the enormous theoretical ensemble of any potential molecule [215] (Fig. 4.1). Large-scale molecular datasets are utilised by contemporary architectures such as generative adversarial networks (GANs), variational autoencoders (VAEs) and recurrent neural networks (RNNs) for decoding structural and functional patterns and to

Table 4.1 Major common descriptors used in chemical space analysis

Descriptor type	Description with example
Constitutional (1D)	1D is based on single counts which work based on molecular formula and composition, like molecular weight, atom types, functional group counts and number of atoms [134]
Topological (2D)	Obtained via representation from molecular graph, which includes Wiener index, connectivity indices, fingerprint bits and Balaban index [160]
Geometric (3D)	Apprehend 3D surface area, size, spatial attributes and shape like surface area, WHIM descriptors, radium of gyration and molecular volumes [123]
Electronic/ quantum	Represents electronic attributes computed based on quantum calculations, which include partial charges, dipole moment and HOMO/LUMO energy levels [94]
Physicochemical	Describes attributes affecting solubility and interaction, like LogP (lipophilicity), polar surface area and solubility [210]
Hybrid/descriptor sets	Integrates multiple descriptor domains, for instance MOE, PaDEL and Dragon descriptors [69]
Fingerprints	Presence or absence of substructures encoded with Binary or bit vectors like ECFP (circular fingerprints), Daylight fingerprints and MACCS keys [244]

generate novel scaffolds that can maximise synthetic accessibility, drug-likeliness and binding affinity [241, 242]. For instance, VAEs and GANs map molecules into continuous latent spaces, allowing scaffold hopping and interpolation amongst bioactive regions, whereas RNNs and transformer models are exceptional at processing sequential molecular representations like SMILES strings, learning syntactic rules to produce valid and diverse compounds. According to enzyme engineering, these models have proven effective [67]. Generative AI predicts the impact of mutations on catalytic activity and stability, which justifies the growth of de novo biocatalysts and enhances the identification of functional enzyme sequences [75]. In addition, 3D generative models utilizes conformational and spatial information to generate molecules with precise target complementarity, addressing issues with allosteric modulation and the inhibition of protein–protein interactions, generative AI exceeds traditional virtual screening techniques in hit identification and lead optimisation by utilising multi-objective optimization framework to custom properties like solubility, binding affinity and metabolic stability, simultaneously [240]. Given the vast size of chemical space (10^{60} molecules), such capacity is vital as generative models' productivity focuses on areas with therapeutic potential while avoiding hazardous or impractical compounds [171].

Despite these developments, it is still hard to match generative outputs with empirical support and real-world applications. Hybrid frameworks that join both AI-driven models with molecular dynamics simulations and retrosynthetic planning tools are essential as current models often fail to achieve balance between novelty and synthetic accessibility [143]. For example, generative models may recommend high-affinity ligands, yet their ability to be computationally achievable and stable in physiological settings [61]. Latest advancements in differentiable scoring functions, which

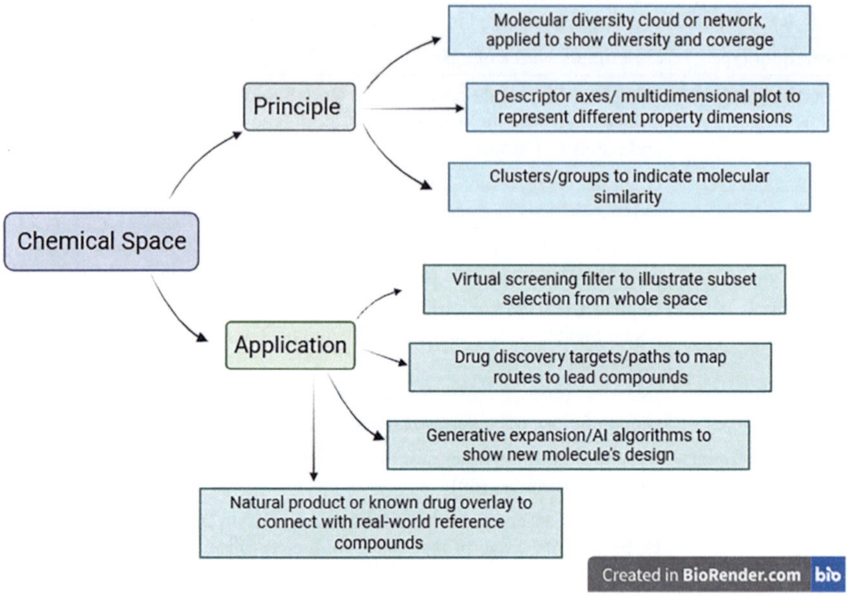

Fig. 4.1 Major principles and applications of chemical space (Created in https://BioRender.com)

guides latent space exploration towards synthesizable compounds, complete this gap. Additionally, interpretability remains a challenge as the connection between bioactivity and latent variables is often unclear, making rational design challenging [38]. With the objective to encourage context-aware generation that is associated with specific therapeutic targets, future directions will emphasize on multimodal integration, incorporating textual, chemical and biological data. To bridge the gap between empirical validation and computational design, emerging paradigms are centred on autonomous experimental platforms, where robotic synthesis and testing are guided as generative AI [208]. As technologies advances, generative AI can transform drug discovery by creating hitherto unreachable areas of chemical space and hastening the creation of next-generation treatments [106].

4.5 Major Types of Dimensions in Chemical Space and Molecular Descriptors

Many molecular descriptors that represent physicochemical, structural and biological properties of compounds characterize chemical space, a concept that is instinctively multidimensional and facilitates systemic exploration and analysis in drug discovery [142]. The principle dimensions arises form descriptor categories like constitutional (0D), that quantifies essential molecular characteristics such as atom/bond counts, molecular weight and functional groups; topological (1D/2D), which utilizes graph-based indices (such as Wiener index) or structural fingerprints to identify patterns of connectivity; and geometrical/3D descriptors, that uses spatial attributes such as surface area, molecular volume and steric/electrostatic fields to model target interactions [166]. Although 4D descriptors extend analysis to dynamic interaction fields, as demonstrated in CoMFA/GRID methods, to predict the binding affinities and pharmacokinetics, advanced quantum-chemical descriptors produced form computational simulations provide insights into electronic properties such as HOMO–LUMO energies and electron density distributions [95]. Collectively, these descriptors produce high-dimensional chemical spaces that are mostly reduced to 2D/3D visualization using methods like principal component analysis (PCA) and t-distributed stochastic neighbour embedding (t-SNE), that highlights structural outliners or groups of bioactive compounds [170]. The notion of the chemical multiverse emphasises the need for complementary descriptors sets, like integrating 3D pharmacophoric features with 2D fingerprints, with the objective to fully map regions of interest while avoiding the biases present in single-descriptor frameworks [40]. In the effort to improve predictive accuracy and automate descriptor selection, contemporary approaches include machine learning, guaranteeing alignment with therapeutic objectives like target specificity or ADMET optimisation. Multidimensional descriptor system proceeds to play a vital role in striking an equilibrium between the exploration of novel scaffolds and targeted lead optimisation as chemical libraries expand into ultra-large scales ($> 10^{12}$ compounds), improving the domains of material science and drug discovery [237].

4.5.1 Scope of Chemical Space in the Design of Molecules with Optimized Properties

Modern molecular designs are based on the ideology of chemical space, encompassing large theoretical ensemble of all possible molecules [107]. This is particularly valid in materials science and drug discovery [89]. Compounds can be explored systematically and optimised for specific features such as solubility, binding affinity, synthetic feasibility and metabolic stability. Through the multifaceted framework, these features can be defined by biological, physicochemical and structural descriptors [136]. Generative AI models that utilises huge datasets to efficiently traverse and develop chemical space are outperformed conventional techniques that are restrained by experimental throughput and rely on well-known scaffolds [226]. For instance, by indicating chemical reactions as changes in a Markov Decision Process, reinforcement learning (RL) frameworks now enhance molecules through synthetic accessible pathways, promising that the compounds created are both synthesizable and novel [116]. By incorporating retrosynthetic planning tools with productive bioactivity models, these models hit a balance between multi-objective enhancement tasks, like improving selectivity and activity while conserving drug-likeness [128,

130]. By incorporating both conformational and spatial data, advanced architectures like 3D generative models also enhance precision while also rendering design molecules with the best target complementarity for challenging protein–protein interactions or allosteric sites [138]. By integrating molecular dynamics simulations with quantum-chemical descriptors, physical realism is introduced, while stability and synthetic accessibility can be predicted and experimental attrition rates are lowered [78]. Notwithstanding these advancements, there are still challenges in maintaining the equilibrium between experimental validation with computational predictions, particularly with complex endpoints that account for multi-modal integrations or dynamic interaction fields [252]. To bridge this gap, hybrid frameworks that incorporate active learning protocols with generative AI are being developed [110]. These frameworks repeatedly refine models using experimental feedback, assuring that generated molecules fulfil both synthetic and therapeutic criteria.

The range of chemical space involves holistic molecular design paradigms along with single-property optimisation. Nowadays, ultra-large virtual libraries (> 10^{12} compounds) and AI-driven prioritisation algorithms are utilised to identify potential area of chemical space with high therapeutic value [20]. For example, context-aware generation is rendered achievable by transformer-based models that are trained on reaction databases [161]. These models recommend molecules with predetermined synthetic pathways and low retrosynthetic complexity [146]. By ensuring smooth interpolation in latent chemical space, this capability is vital for avoiding "activity cliffs" I which minor structural changes result in abrupt changes in bioactivity [230]. Furthermore, the ideology of chemical multiverse addresses the biases present in single-presentation frameworks by linking diverse regions through complementary descriptor sets, enabling an extensive examination of scaffold diversity and functional group substitutions [175]. Enzyme technology case studies suggest how generative models predict the effect of genetic changes on catalytic activity, accelerating the discovery of de novo biocatalysts with specific stability and functionality [51]. Yet, data quality and computational costs continue to limit scalability, resulting in innovations in transfer learning and federated learning to limit redundant and align disparate datasets. With the goal to overcome the gap between computational design and empirical validation, autonomous platforms will be the future focus where AI will guide robotic synthesis and testing [190]. Chemical space manoeuvring holds promise for both speeding up drug discovery and improving drug-likeness paradigms, facilitating the exploration of prior unreachable domains with a wealth of therapeutic potential [93].

4.5.2 *Major Types of Molecular Descriptors*

Molecular descriptors are quantitative depictions of molecular attributes that can be widely classified according to their application and dimensions [153]. Basic molecular attributes such as molecular weight, functional groups and atom/bond counts are represented by 0D descriptors (constitutional descriptors), which gives a basic overview on molecular composition but often lacks structural specificity [167]. By reserving structural fragments or substructures like fingerprints or SMILES patterns, 1D descriptors are above this, enabling fragment-based drug design and similarity comparisons [167]. 2D descriptors are based on graph theory, which measures the connectivity patterns using metrics like Wiener index or Zagreb index that are correlated with bioactivity and physicochemical properties [167]. To model protein–ligand interactions and conformational changes, 3D descriptors account to spatial attributes such as steric/electrostatic fields, molecular volume, and surface area. To be able to predict pharmacokinetics and binding affinities under different conditions, intricate 4D descriptors include dynamic interaction fields (e.g., CoMFA/GRID methods) [29]. Adding on to dimensional classification, descriptors are divided by the type of property: physicochemical descriptors (logP, Polar surface area) enhance drug-likeness and ADMET profiles, while electronic descriptors (e.g., HOMO–LUMO energies) estimate reactivity [238]. To bridge the gap for neglected compounds, emerging bioactivity descriptors utilise machine learning to derive biological properties (like target engagement and cellular responses) from structural data. To minimize degeneracy and enhance predictive accuracy in QSAR/QSPR models, contemporary frameworks gradually integrate these descriptors, like hybridising 2D fingerprints with 3D pharmacophoric features. As AI-driven descriptors selection evolves, its usefulness is further increased by giving priority to features that are interpretable and biologically significant for drug discovery [224].

4.6 Approaches for Navigating and Prioritizing Regions of Chemical Space

Computational techniques that maintain an equilibrium between the investigation of new scaffolds and targeted optimisation of therapeutic potential are essential to navigate and rank areas of chemical space, a vast, diverse framework of molecular diversity. By studying latent representations of chemical structures from enormous datasets, generative AI models-like variational autoencoders (VAE) and generative adversarial networks (GANs)-facilitate systemic exploration and generate molecules with optimal characteristics such as drug-likeness, synthetic accessibility, and binding affinity [251] (Fig. 4.2). To ensure visibility, these models incorporate predictive bioactivity models and retrosynthetic planning tools, arranging domains of interest utilising multi-objective optimisation frameworks. By connecting chemical space into comprehensive dimensions utilising quantitative metrics (like topological indices and 3D pharmacophoric features), molecular descriptors-based navigation making it easier to identify structural outliers and cluster bioactive compounds [198]. Principal components analysis (PCA) and uniform manifold approximation (UMAP) are two methods that efficiently determine "islands" of bioactivity or regions enriched for target-specific interactions by minimising high-dimensional descriptors data into 2D/3D visualisations [81]. Fragment-based methods, which grows or links computationally accessible fragments withing predefined reactions spaces, are a prime instance of how ultra-large virtual libraries (> 10^{12} compounds) utilise similarity searching and docking algorithms to identify molecules with high predicted target engagement [6]. To enhance generative models and reduce false positives, hybrid frameworks further enhance organisation by incorporating both computational predictions with experimental validation loops, like iterative screening rounds. To guarantee whether resources are prioritised towards the most promising domains, active learning protocols that modifies dynamically exploring strategies based on real-time bioassay feedback are developed in response to ongoing challenges in reconciling the vastness of chemical space with practical constraints. To facilitate context-aware generation and correspond molecular design with specific therapeutic aim, upcoming paradigms also include multimodal data (like assay, structural and textual result) [140]. When put together, these approaches alter chemical space navigation from the library-centric, static process to an AI-driven, dynamic discovery, that accelerates the search for novel treatment while also decreasing the possibility of missing important regions [99].

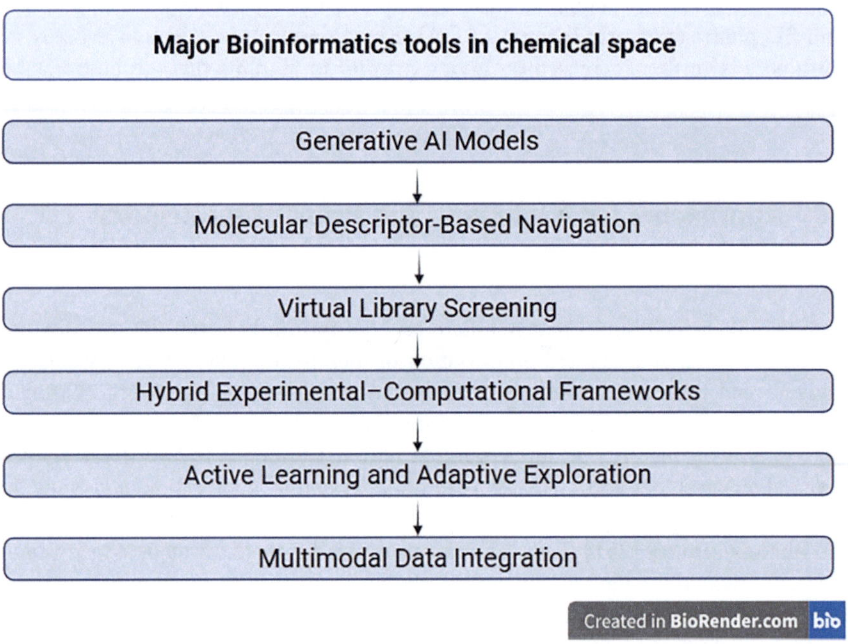

Fig. 4.2 Schematic representation of the major approaches for navigating and prioritizing chemical space. The flowchart outlines six interconnected approaches, generative AI, modeling, descriptor-based mapping, hybrid experimental-computational workflows, active learning-driven exploration, ultra-large virtual screening, and multimodal data integration, which together enable efficient discovery, prioritization and optimisation of new bioactive molecules (Created in https://BioRender.com)

4.6.1 Cheminformatics

Cheminformatics is a multidisciplinary discipline which incorporates both computer-based and information science techniques, commonly referred to as "in silico approaches" with the foundation of physical chemistry to solve a variety of descriptive and prescriptive chemistry problems, like usage in drug discovery, biology and other associated molecular sectors [103]. Scientists can utilise vast compound databases and the meticulous layout of molecular information for computational exploration due to this field, which comprises storage, analysis, retrieval and comprehending chemical data [53]. Primarily, cheminformatics utilises structure–activity relationships (SARs) and molecular descriptors, like 2D fingerprints and topological indices, to unlock opportunities for small-molecule drug discovery by quantitative structure-activity relationship (QSAR) modelling [91]. With algorithms such as support vector machines and regression analysis indicating an outstanding capacity to foresee and comprehend relationships amidst biological activity and molecular structures, the incorporation of machine learning techniques has been recognised as transformative [28]. Computational methods can be utilised in natural product-based drug discovery, in which they aid with organisation of data, filtering large chemical datasets in advance of experimental screening and design. While novel techniques for visualising chemical space are still in progress, the field consists of profiling natural product datasets based on complexity, diversity, ADME/Tox characteristics, fragment analysis and acid/base properties [32, 33]. Through improved compound activity predictions, reaction modelling and de novo designing, latest enhancements have highlighted deep learning and machine learning as vital components, addressing hitherto unsolved problems in pharmaceutical research. To enhance the compound quality and likelihood of identifying novel hit chemicals, virtual chemical libraries and virtual screening methods are becoming popular [52]. The precise prediction of intricate biological interactions, the prerequisite for comprehensive and varied datasets, adequate baseline comparisons, powerful validation method, and the development of novel applicability domains are few of the ongoing problems faced by cheminformatics despite tremendous advancements. Cheminformatics exceeds greater limits with the incorporation of artificial intelligence and advanced computational resources, which holds the potential to speed up the discovery of new therapeutic agents through predictive molecular analysis and design [151].

4.6.2 Molecular Docking

In generative AI-driven chemical space investigation, molecular docking—a computational approach that predicts ligand binding affinity and the orientation across target proteins—is becoming vital for designing molecules with ideal target complementarity [2]. Docking scores are utilised as reward functions in current generative models, such as reinforcement learning (RL), frameworks, to continuously enhance molecular structures. The MORLD (Molecule Optimisation by Reinforcement Learning and Docking) technique, for instance, produces unfamiliar DDR1 kinase inhibitors by directly enhancing docking-predicted binding affinity, surpassing size-matching controls by 10–20% and avoiding the necessity for time-consuming virtual screening [100]. As demonstrated in the design of DDR1 inhibitors, where 3.5% of the generated compounds met hybrid architecture, activity criteria like Bayesian docking approximations paired with RL, additionally enhances efficiency by diversifying chemotypes above the constrained experimental data [54]. By integrating spatial constraints and atomic coordinates, 3D generative models (like 3D-scaffold) manage conformational dynamics and produce molecules that have improved binding poses for targets such as NSP15 [49]. There persist problems with data quality and computational complexity, but AI-accelerated docking facilitates quick navigation of ultra-large libraries ($> 10^{12}$ compounds) by reducing screening times by $130 \times$ when compared to traditional techniques.

By resolving prior constraints in affinity estimation and pose prediction, AI has entirely altered scalability and docking accuracy. With resources such as AI-Bind predict binding sites from amino acid sequences without 3D structures, consequently mitigating data scarcity, neural networks (like IGModel) and geometric graph capture spatial atomic interactions to refine descriptions of binding pockets [164]. With the support of conformal prediction docking and machine learning, billion-compound libraries can now be reviewed in a few hours, demonstrating dual-target ligands with high specificity, like A2A adenosine/D2 dopamine receptors [179]. The bridge between experimental validation and computational design has been bridged by generative active learning frameworks: molecular dynamics simulations (such as GROMACS) confirm stability, docking-guided RL implies synthesizable compounds and bioassay feedback enhances models [181]. AI-docking serves as an organising principle for drug discovery in undiscovered chemical domains as future development is centred on differentiable scoring functions for multi-objective optimisation and synthetic accessibility (such as ADMET properties for balancing affinity) [10].

4.6.3 Machine Learning

Beside traditional similarity-driven and fragment-based techniques, machine learning, particularly through deep generative models, has revolutionized the exploration of chemical space through facilitating de novo design of molecules with optimised properties [3] (Table 4.2). Key structures like generative adversarial networks (GANs), recurrent neural networks (RNNs), transformers and variational autoencoders (VAEs), utilise wide molecular datasets (like GDB-13) to acquire complicated structural patterns and generate novel compounds with a particular drug-like characteristic [68, 222]. While reinforcement learning (RL) frameworks include predictive scoring functions (like docking affinity) to continuously refine designs towards multi-objective goals like metabolic stability and target specificity such as SMILES or SELFIES to generate synthetically accessible scaffolds [119]. For instance, RL-driven models ensure feasibility and novelty by enhancing chemical reaction pathways as Markov Decision Processes [5]. Despite their potential, there are still problems associated with interpretability as black-box architecture make it hard to see how latent variables relate to bioactivity, and with validation as there are high chances that insufficient experimental confirmation could result in unstable or unpatentable compounds. These shortcomings are filled by hybrid approaches that incorporate generative AI and active learning loops. They focus on domains with high therapeutic assurance by utilising real-time bioassay feedback. To enhance context-aware design, upcoming developments will focus on interactive exploration tools (such as ChemSpacE) for multimodal integration of biological, structural and textual data and for human-guided generation [218]. By establishing biologically essential domains of chemical space that were earlier inaccessible by conventional methods, these models hold potential to accelerate drug discovery as they develop [130].

4.7 Major Libraries of Chemical Space Exploration

Computationally accessible libraries which permit generative models to generate novel compounds with optimised properties are vital to systematically examine chemical space, which includes vast ensemble of theoretically possible molecules [70]. With around 2.8 trillion virtual product molecules, the eXplore chemical space is the biggest enumerated library to date. Effective similarity screening algorithm like SpaceLight, SpaceMACS and FTrees make it simpler to discover ligands [176]. Additionally, "SARS-by-Space" method that mine bioactive domains around targets such as BRD4 in the span of weeks accelerate the elucidation of structure–activity relationships (SARs) utilising REAL Space (11 billion readily available compounds [185]. Software platforms like SECSE (Systemic Evolutionary Chemical Space Explorer) integrate deep learning with fragment-based drug design to generate novel inhibitors for targets by "Lego-building" them within target binding pocket with phosphoglycerate dehydrogenase (PHGDH), being a classic example [138]. Through the application of combinational chemistry to evaluate biological targets at a reduced cost conventional high-throughput screening, DNA-encoded chemical libraries (DELs) additionally expand the diversity of chemicals that can be accessible [79]. Experienced vendor collections that highlight synthetic accessibility and high purity for targeted screening, like Chemspace's fragment sets (Rule-of-3 compliant molecules comprising 164,000 compounds) and pe-plated libraries (comprising 917,000 compounds), enhance these libraries. To effectively prioritise bioactive domains within ultra-large chemical spaces ($> 10^{12}$ compounds), generative models, such as variational autoencoders and reinforcement learning, pave way to these libraries by enhancing the multi-objective reward functions (like synthetic feasibility and docking scores) [115].

Table 4.2 Major bioinformatics tools/computational software available for cheminformatics, molecular docking and machine learning studies in chemical space

Tool/software	Category	Description/application
RDKit	Cheminformatics	Open-source Python toolkit for molecular descriptors, ML integration and chemical similarity [197]
CDK	Cheminformatics	Java-based library for descriptors, data processing and molecular structures [60]
Molinspiration	Cheminformatics	Web based tools to calculate property and QSAR or fragment-based screening [60]
ChEMBL/myChEMBL	Cheminformatics	Integrated tools with bioactivity databases for activity and compound analysis [60]
Indigo	Cheminformatics	Open-source tool that supports multiple languages for depiction and substructure searches [172]
AutoDock/Vina	Molecular docking	Open-source docking platform for protein–ligand interaction and virtual screening [169]
Glide (Schrödinger)	Molecular docking	Commercial docking software for high-accuracy, flexible docking for ligands [169]
GOLD	Molecular docking	Docking based on genetic algorithm with function customizable scoring [169]
FlexX	Molecular docking	Protein–ligand docking based on fast, incremental construction [169]
MOE-Dock	Molecular docking	Flexible protein docking and modeling based on comprehensive suite [169]
SwissDock	Molecular docking	Small molecule docking platform based on web-based CHARMM [59]
ZDOCK	Molecular docking	Large molecule virtual screening based on FFT-based rigid-body docking [169]
Surflex	Molecular docking	Protomol based tool for fast docking and ligand screening [169]
DeepChem	ML/cheminformatics	QSAR, molecular design, and property prediction based on deep learning [232]
KNIME	ML/cheminformatics	Integrates ML, docking and cheminformatics analysis [156]
LegoGram	ML and Generation	Molecule generation and chemical space visualization based on AI framework
DeSMILES	ML/cheminformatics	Activity and property prediction for SMILES-based deep learning [162]
Parametric t-SNE	ML/cheminformatics	Tool for large scale chemical space visualization mapping [15]
Conformal predictors	ML and docking	ML based confidence tool for drug discovery and screening [21]

4.7.1 Natural Products Libraries

With generative AI models currently allowing previously novel expansion and exploration of this chemical space, organic product libraries serve as an essential resource for drug discovery, employing bioactivity of compounds obtained from biological substrates and structural complexity [242]. Recurrent neural networks (RNNs) and Variational autoencoders (VAEs) are two such examples of deep generative architectures that have revolutionised conventional libraries, that were constrained by small diversity of existing natural products (approximately around 400,000 documented structures) [189]. Ultra-large libraries like RNNs which produces 67 million natural products, are trained by databases like COCONUT [209]. These libraries enhance the accessible chemical space by around 165 times yet preserving the bioactivity relevant attributes like scaffold diversity and chirality [209]. By developing a latent chemical space which enables high-fidelity reconstruction and the development of novel structures with optimised functions, the NP-VAE framework particularly tackles challenges associated with analysing complex natural compounds [236]. By predicting target engagement and bioactivity (like discovery of antibiotics agonist *Acinetobacter baumannii*), AI-driven prioritisation further enhances these libraries, and generative models recommend synthetically viable analogues to accelerate development of drugs [200]. Identification of targets and dereplication are still challenging despite advancements, but these issues can be minimised by integrating AI-powered databases with multimodal data (like genomic and spectroscopic) to connect structure–property relationships and identify bioactive domains [184]. Generative AI and biological product libraries together accelerate the process of identifying therapeutic candidates, turning resource-intensive screening into focused, computer-driven procedures [22].

4.7.2 Fragment-Based Libraries

In chemical space exploration, fragment-based libraries are key component building blocks for generative AI models that permit designing novel compounds by methodically assembling synthetically accessible fragments with predetermined bioactivity profiles [46]. As proved by commercial collections like FDSL-DD fragments and Chemspace's Rule-of-3-compliant sets (164,000 fragments) that utilises in-silico screening to identify fragments with target-specific binding characteristics, these libraries, which often involve small, diverse molecular fragments($<$ 300 Da), are carefully selected to enhance coverage of chemical space while ensuring synthetic feasibility [46]. These libraries are utilised by generative models, such as fragment-based VAEs and reinforcement learning (RL) architectures, to continually place fragments together into drug-like molecules according to the framework of multi-objective reward functions that increase target affinity, synthetic accessibility and metabolic stability (Table 4.3). One such instance is FragGen model that triumphs over the synthesizability limitations of atom-wise method by incorporating a hybrid strategy along with Deep Geometry Handling to produce geometrically reliable inhibitors (like nM-level kinase inhibitors). "Lego-building" approaches by medicinal chemistry have been emulated by AI-driven forums like SECSE (Systemic Evolutionary Chemical Space Explorer) [129], which additionally enhances the efficiency by evolving fragments within target binding pockets to identify new scaffolds for challenging targets such as PHGDH [182]. Extensive exploration is made achievable by integrating fragment-based generative designs with ultra-large virtual libraries (> 10^{12} compounds), and also for active learning protocols emphasizing importance to high-value domains by enhancing the output using experimental feedback [183].

4.8 Systematic Exploration of Chemical Spaces

Scientists can now efficiently identify and alter bioactive molecules while conserving their chemical diversity and novelty [128]. This is gaining attention due to the integration of sophisticated sampling methods, active learning protocol and deep generative models [133]. This has evolved the systematic exploration of chemical space, the huge, multidimensional landscape that comprises all possible chemical compounds. As virtual libraries advance to billions or even trillions of compounds, such as those in GDB-17 database, conventional exhaustive screening techniques-in which every molecule is identified and assessed separately (for instance, through docking or free energy calculations)—has become unfeasible computationally [47]. Active learning frameworks that incrementally update predictive models on novelly acquired data and utilise uncertainty or diversity criteria to choose novel candidates for evaluation indicate the way contemporary workflows utilise machine learning-enhanced selection protocol that equalises exploitation (such as determining the highest scoring compounds) and exploration (diverse region sampling) [76]. One such example, by incorporating first-principles alchemical free energy calculations, automated redocking and active learning, Khalak et al. [110] were able to recover over 80% of the experimental confirmed hits while retaining hit diversity and cutting computational costs by approximately 14 times. With various trade-offs in coverage and complexity, deep generative models, such as molecular transformers, additionally promote nearly exhaustive local exploration through the generation of new SMILES strings via deterministic beam search or multinomial sampling [42]. These models can be created by property-based constraints or reward functions, ensuring that the compounds established are not only unique but also biologically and synthetically accessible [45]. The determination of active clusters and unexplored domains are made easier by visualising methods like t-distributed stochastic neighbour embedding (t-SNE) and principal component analysis (PCA), which aids in locating high-dimensional chemical space into incorporating 2D or 3D projections [174]. Chemical space navigation has evolved to a dynamic, data-driven paradigm from being static

Table 4.3 Major examples of databases act as natural and fragment-based libraries

Database	Type	Description
BIOFACQUIM	Natural product	Used for Latin Americans for expanding and visualising natural products chemical space in generative model training [25]
PeruNPDB	Natural product	Peruvian Natural Product database for generative molecular design and comprehensive chemical space mapping [34]
NPASS	Natural product	Global Natural product database containing bioactivity data, widely used in AI-driven NP space generation and exploration [248]
ZINC/ZINC fragments	Fragment-based	Used in fragment and AI-assisted molecular design which contains commercially available scaffolds [43]
SP3, 3D, DIV fragment libraries	Fragment-based	Diverse fragment 3D-rich for generative synthesis and benchmarking [96]
eMolFrag	Fragment-based	Produces diverse building blocks for generative model by AI-enabled molecular fragmentation [102]
Generated NP-like libraries	Natural product	To expand natural product chemical space by in-silico NP-like libraries by deep generative models [50]
PDBbind fragments	Fragment-based	Used in generative synthesis validation for already fragments approved from biomolecular complexes [74]
ChEMBL	Mixed (natural + fragment)	Used to train generative AI models by curating bioactive molecules like NP-like and fragment subsets [245]

and library-centric process. This data is driven through the incorporation of systemic exploration strategies with active learning and generative models. This facilitates the creation of powerful, novel scaffolds while effectively allocating experimental and computational resources [178].

4.8.1 Need and Importance

Conventional drug discovery techniques, which depends on trial-and-error, serendipity and restricted experimental throughput, resulting in high costs and low success rates, fundamentally limit the exploration of chemical space, a vast theoretical ensemble exceeding 10^{60} drug-like compounds [132]. This is solved by generative AI models, which transforms the efficacy and reach of molecular discovery by enabling data-driven, systematic navigation and de novo design of new compounds with personalised attributes [231]. Without the need for diligent screening, these models—which consists of generative adversarial networks, variational autoencoders (VAEs) and transformer architectures—learn complex patterns from huge datasets to generate synthetically accessible molecules that are be personised for drug-likeness, metabolic stability and target affinity [113]. Reinforcement frameworks (RL), for instance, integrate synthetic feasibility metrics and docking scores to constantly improve designs, that accelerate hit identification by over 100 times when compared to high-throughput methods [63]. Additionally, generative models assure synthetic tractability via retrosynthetic planning tools while creating domains of chemical space that are not accessible by conventional techniques, like novel scaffold for challenging protein–protein interaction or allosteric sites [117]. By incorporating active learning and experimental feedback loops, model outputs have been advanced to be improved while resource waste and product positives are reduced. Apart from drug discovery, these also incorporate material design, as experimented by text-guided frameworks such as Chemeleon, which produces stable crystal structures for applications such as energy storage [121]. By coordinating diversity (exploration) with property optimisation (exploitation), generative AI characterizes access to ultra-large chemical libraries (over 10^{12} compounds), placing itself as an indispensable tool for fast, cost-effective discovery across catalysis, pharmaceuticals and material science [231].

4.8.2 Major Methods for the Systematic Exploration of Chemical Spaces

The integration of active learning, deep generative models and complex computational strategies targeted at efficient navigation and optimising this enormous landscape for material science and drug discovery have altered the systematic exploration of chemical space—the holistic universe of every possible small molecule [221] (Fig. 4.3). Because of computational cost and sheer volume, conventional exhaustive screening of specified virtual libraries—which may sometimes include thousands or even millions of compounds—has rapidly grown impractical. Rather, contemporary workflows employ deep learning frameworks, encompassing generative adversarial networks (GANs), transformers and variational autoencoders (VAEs), that are trained based on existing molecular datasets to create and optimise new structures that fulfil established property constraints like drug-likeliness, synthetic feasibility or target affinity. As suggested in studies where focused sampling and iterative retraining recover high-affinity ligands while significantly lowering the essential number of molecular simulations, the integration of active learning protocols—in which additional potential molecules are selected in each iteration on the basis of model's disagreement or uncertainty—significantly accelerating lead identification [195]. These approaches make sure that both relevance and uniqueness are methodically addressed by establishing an equilibrium between exploitation of high potential "islands" of bioactivity and the exploration of chemically diverse domains. Dimensional reduction approaches, like uniform manifold approximation (UMAP) [223]*and* principal component analysis (PCA) [137], are vital for this process as they convert high-dimensional descriptor spaces into understandable 2D or 3D projections, which aids in finding uncharted territory or active-compound clusters. Along with utilising differentiable scoring functions, such as docking scores or synthetic accessibility, to guide and restrict molecular generation in an essential, property-rich direction, latest advancements enable the "local" broad generation of compounds in targeted domains using beam search throughout latent spaces or multinomial sampling techniques [36]. Nowadays, hybrid models integrate physics-based free energy calculations and experimental feedback, wrapping up the cycle from computational design to empirical validation and enhancing candidate selection and enhancing model accuracy [228]. Notwithstanding these advancements, there are still recurring issues, like representing multi-target profiles productively, incorporating multimodal biological, textual and chemical data for context-aware designs, and to retain diversity despite creating redundancies [69]. In navigating chemical space, comprehensibility along with knowledge transfer are also being promoted by the development of interactive tools which facilitates human-in-the-loop molecule manipulation. Systematic exploration is anticipated to open formerly unexplored domains of chemical space and accelerate the process from in-silico design to assured bioactive compounds as it becomes more interpretable, context-driven and sample-efficient [202]. This symbolises a paradigm shift in how molecular innovations are approached by life sciences in the era of generative AI.

4.8.2.1 Virtual Screening

Hit identification in drug discovery has been highly aided by virtual screening (VS), that are developed into a vital technology for methodically traversing an enormous multidimensional expanse of chemical space [57]. Employing artificial intelligence innovations like molecular docking, generative models and deep learning, modern Virtual Screening (VS) enables a quick computational analysis of millions of compounds, giving priority to those that are more probable to have the desired drug-like qualities and biological activity [154]. Techniques such as ligand-based virtual screening (LBVS) [186], which utilises molecular similarity, quantitative structure–activity relationship (QSAR) [26, 27] and machine learning models to determine the

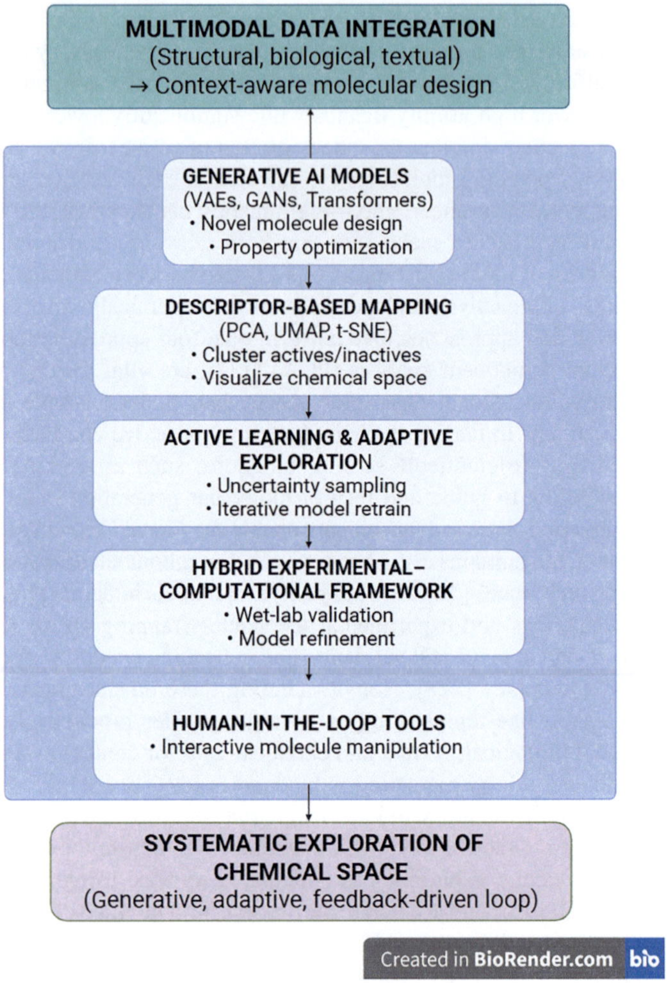

Fig. 4.3 Identification of small-molecule candidates in drug discovery by integrating drug-likeness assessment, ADME prediction, and toxicity evaluation following medicinal chemistry screening workflows (Created in https://BioRender.com)

activity from the features of known active compounds, and structure-based virtual screening (SBVS), that utilises 3D structure information in predicting interactions between protein targets and small molecules [87]. By docking systematically only for a portion of a large library and iteratively enhancing hit lists with ligand-based predictions, breakthroughs such as Deep Docking substantially decrease computational requirements [77]. This permits screening speeds up to 100 times faster results when compared to conventional full library docking, rendering multi-billion scale chemical libraries suitable for in silico campaigns [101]. To control prediction error rates and oversee naturally imbalanced collections where only a minor percentage of molecules are true binders, new workflows integrate the power of ensemble docking, machine learning classifiers and conformal prediction frameworks to guide searches towards high-confidence bioactive domains [19]. By demonstrating novel, synthetically accessible scaffolds that can be customised for pharmacokinetic attributes and target affinity, generative AI—encompassing reinforcement learning models and variational autoencoders—further enhances virtual screening pipelines [53]. This is often done as a part of an iterative cycle that integrates experimental feedback, synthesis and prediction. The scope is further extended by fragment-based virtual screening and divide-and-conquer algorithms, which facilitates quick identification of guaranteeing chemotypes from shallow screens of scaffold fragments that later expands to leadlike molecules [250]. Notwithstanding these advancements, there are still essential obstacles to overcome: computational models can produce false positives because of erroneous docking scores or absence of training data, and compound synthesizability and feed-forward experimental validation consideration are presently necessary for the practical success of virtual screening [37]. Drug and materials discovery is evolving, however, because of the current paradigm, that involves AI-driven scoring, ultra-large virtual libraries and prioritisation, and adaptive, data-driven cycles. These attributes have reduced attrition rates significantly, diminished hit-to-lead timelines, and have rendered it achievable to investigate formerly unobtainable domains of chemical space [219].

4.8.2.2 Quantitative Structure–Activity Relationship (QSAR)

The structured relationship between the molecule's biological activity and its chemical structure is made achievable by Quantitative Structure–Activity Relationship (QSAR) models, which are the basic computational strategy for contemporary drug discovery [35]. The primary objective of QSAR is to quantify the possibilities in which molecular attributes—like steric, hydrophobic, topological and electronic features—impact a compounds efficacy, affinity or toxicity for specific biological targets [145]. Classification or regression algorithms are performed to mathematically relate the predicator variables (X), which are often molecular descriptors obtained from chemical structures to a biological response variable (Y), like receptor binding affinity or inhibitory concentration [13]. Before performing a high-priced experimental validation, this predictive framework aids to design and establish importance to novel candidates by enabling virtual screening of vast chemical libraries [35]. The diversity and richness of QSAR models have enhanced drastically in the recent times mainly due to advancements in machine learning algorithms that can enhance variable selection and curated molecular databases, high-throughput screening and model performance [88]. As suggested by advanced methodologies such as comparative molecular field analysis (CoMFA) or fragment-based QSAR (FB-QSAR), integrating multidimensional descriptors—fragment-based, quantum mechanical, 2D and 3D—aid in modelling complex nonlinear interactions [64]. Consequently, QSAR has since been used to model multi-omics data, physico-chemical outcome pathways linked to organ-specific toxicity, compared to predicting straightforward end points such as toxicity or activity.

QSAR has major practical consequences in the larger context of generative models and chemical space, particularly as QSAR pipelines majorly incorporates AI-driven and deep learning approaches [35]. To improve predictive accuracy for endpoints and complex targets, novel techniques utilise architectures such as convolutional neural networks and graph neural networks that can absorb raw molecular graphs and obtain intricate characteristics that conventional descriptors might neglect [8]. These advancements have elevated QSAR from conventional linear or nonlinear regression into versatile models that can record multifaceted, complex relationships. Therefore, QSAR is currently considered a valuable tool to generate novel molecules in uncharted chemical space [227]. Machine learning-augmented QSAR approaches, for instance, are currently used frequently for ADMET profiling, risk assessment and toxicity prediction in regulatory context, as well as for establishing priorities of lead compounds in oncology and anti-infective pipelines [229]. The interpretability of models has additionally been brought by the utilisation of explainable artificial intelligence (XAI) approaches, which have assisted in giving clarity to as which molecular substructures impact toxicity or activity predictions [180]. This is a vital factor for transparency and trust in drug development workflows. Building robust, generalisable QSAR models from ever-larger chemical spaces is currently possible due to cloud computing, big data analytics, and active learning protocols, emphasising their crucial role as a connector between experimental innovation and computational design in the domains of toxicology, medicinal chemistry, environmental health research and material science [35].

4.8.2.3 De Novo Molecular Design

De novo molecular design is the technique of creating novel chemical structures via computational techniques from beginning to end, without any pre-existing compound libraries, in order to fulfil specific property specifications within the vast discipline of chemistry [213] (Table 4.4). Deep generative models—like variational autoencoders (VAEs), transformer-based architectures, recurrent neural networks (RNNs) and generative adversarial networks (GANs)—have been the inspiration behind major advances from 2020 and 2025 [246]. These models are programmed on vast molecular datasets to learn complicated, high-dimensional chemical distributions, and subsequently generate molecules that meets user-defined constraints such as selectivity, synthetic accessibility and binding affinity [213]. With the aim to perform targeted exploration and effectiveness through reinforcement learning or conditional generation strategies, where molecular structures are recorded by these generative models—expressed as molecular graphs or SMILES strings—into latent spaces [82]. Identification of target-specific ligands are made more straightforward by latest frameworks that integrates ligand–protein interaction fingerprints to generate molecules that are both uniquely structured and intended to retain specific binding modes. Transformer architectures and large language models have also made engaging de novo design achievable, automating molecular evaluation and reward function construction, while assisting chemists to manage property optimisation via natural language interfaces. The identification and experimental validation of novel PPARγ modulators with superior activity and specificity, along with the implementation of hybrid models that incorporate data-driven, deep learning and rule-based methodologies to address issues such as multi-target selectivity and synthetic tractability [12]. Regardless of renowned development, challenges of striking an equilibrium with validity, practical synthesizability and chemical diversity persists, which favours continued work on more accurate scoring functions and more understandable methodological generative workflows. Despite this, de novo molecular design using generative models is currently the basic component of contemporary drug discovery, facilitating effective and guided examination of chemical domains that were once unreachable. Real-world usage ranges to advanced materials starting from targeted therapies [213].

4.9 Scope of Chemical Space Analysis in Medicinal Chemistry

In medical chemistry, chemical space analysis is a basic technique employed for investigating, designing and mapping vast collection of potential drug-like compounds. This concept permits scientists to represent and compare the relationship and diversity among molecules [107] (Fig. 4.4). It is defined mathematically as a multidimensional space where every point symbolises an individual molecule encoded by physicochemical and structural descriptors [18]. Chemical space analysis has several practical applications, which includes supporting identification of uncharted molecular domains, aiding in the creation of novel compounds with intended biological activities, and providing data necessary for organising and picking candidates for experimental screening campaigns. Recent advancements, particularly from 2020 to 2025, have modernised the field with sophisticated cheminformatics and computational tools that utilise artificial intelligence, ultra-large molecular databases and data-driven molecular generators to facilitate rapid, theory-driven compound discovery [107]. Understanding clusters of bioactive molecules and structure–activity relationships relevant to medicinal chemistry has improvised by utilising contemporary visual representations such as t-SNE, generative topographic mapping and principal component analysis, which aid to decrease this high-dimensional space into maps that can be understood [198]. The significance of scalable and effective technique for chemical space analysis for early-phase lead optimisation and drug discovery highlighted by the steady rise in the variety and quantity of chemical entities readily accessible through DNA-encoded libraries, combinatorial synthesis and make-on-demand forums [107].

Table 4.4 Major bioinformatics tools/computational software available for virtual screening, QSAR and de novo design

Software/tool	Category	Feature and implementation
Glide (Schrödinger)	Virtual screening	Prediction dicking and ligand pose scoring [58]
AutoDock Vina	Virtual screening	Quick, open-source docking with enhanced accuracy scoring [58]
GOLD	Virtual screening	Adaptable docking that utilizes genetic algorithms scoring [58]
LigandScout	Pharmacophore QSAR	Screening and 3D pharmacophore modeling [105]
Schrödinger suite	QSAR/virtual screening	Docking workflows and QSAR model development [7]
MOE (molecular operating environment)	QSAR & de novo design	QSAR, pharmacophore molecular design and docking [109]
QSAR toolbox	QSAR	Toxicity and hazard prediction QSAR modeling [83]
DTC-QSAR	QSAR	Graphics User Interphase based regression and classification modeling [65]
AlvaBuilder	De novo design	Personalized platform for molecule generation [147]
RosettaVS	Virtual screening	AI-improved scoring and docking [249]
DRAGONFLY	De novo design	Generative design of ligands based on structure [23]
LEGEND	De novo design	Ligand generation in receptor sites based on atoms [159]
Deep learning frameworks (like DeepChem)	De novo design	Molecule generation for deep learning models [141]

Chemical space analysis has an influence that goes outside of enumeration and compound navigation. By providing structure–activity relationship (SAR) and structure–property analysis at a unique scale and depth, it actively impacts the trajectory of medicinal chemistry [206]. By observing and tracking changes in cluster density, target preferences, or common chemical scaffolds among medications authorised after 2020, researchers have discovered novel patterns and potential therapeutic targets [18]. For instance, the continued usage of kinase inhibitors or the recent advancements of epigenetic writer proteins might offer insights into potential future treatment approaches [14]. Sophisticated chemical space models are crucial for computational drug repurposing, AI-guided hit-to-lead optimisation and network pharmacology to enhance innovation and decrease resource waste [177]. Furthermore, by identifying pharmacological coverage gaps and promoting logical enquiries into previously overlooked domains, chemical space analysis aids lessen the gap

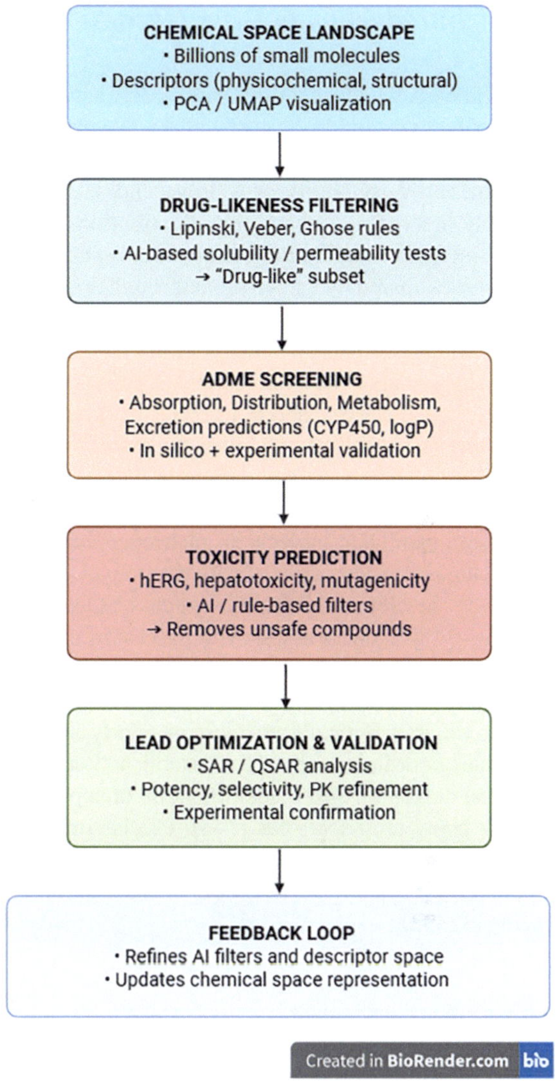

Fig. 4.4 Overview illustrating the mechanism of medicinal chemistry screening of small molecules, containing drug-likeness evaluation, ADME profiling and toxicity prediction. Compounds from AI-generated or virtual chemical spaces undergo sequential filtering depending on physicochemical attributes, pharmacokinetic parameters and safety assessment to identify the optimised lead candidates which has high efficacy and low toxicity (Created in https://BioRender.com)

between clinical practice and drug design. Chemical space analysis is revolutionising from a theoretical framework to a vital tool that is encouraging the advancement of genuinely new and patient-centred therapies, owing to the convergence of machine learning, high-throughput screening and cheminformatics [51]. Subsequent functional insights, translational success and molecular diversity in drug discovery will be stimulated by the expansion of chemical space analysis in medicinal chemistry as a visual representation tool and descriptor techniques improve, and datasets expand to billions of compounds [230].

4.9.1 Introduction to Drug Likeness

Drug-likeness, which quantifies how closely a molecule's physicochemical and structural attributes resemble those of well-known pharmaceuticals, is an essential thought in drug discovery and medicinal chemistry [253]. It predicts a likelihood of success and molecule's suitability as a drug candidate. This thorough assessment, which primarily incorporates absorption, distribution, metabolism and excretion (ADME) profile, integrates qualitative and quantitative analysis [32, 33]. Based on an analysis of properties shared by known breakthrough drugs, the broadly recognised Lipinski's "Rule of Five" serves as a foundation for evaluating drug-likeness [150]. It indicates that an oral active drug would generally have a molecular weight of less than 500 Da, a logP (octanol–water partition coefficient) of less than 5, not greater than five hydrogen bond donors, and not greater than ten hydrogen bond acceptors [97]. By addressing permeability, solubility, polar surface area and the function of molecular flexibility, and additional regulations and guidelines like the Ghose, Egan and Veber rules, further improves these standards [124]. Using computational analysis tools that assess candidate molecules' pharmacophores or structural motifs against vast drug databases—occasionally with the help of quantitative scoring frameworks like the newly developed AI-driven models or Quantitative Estimate of Drug-likeness (QED)—drug-likeness assessment passes over fundamental property checks [120]. With an effort to accelerate the drug development pipeline and enhance the likelihood of clinical success, scientists can recognise compounds that are most likely to be toxic, metabolically unstable or poorly absorbed early on by evaluating how molecules coordinate with these established attributes, whether cluster analysis in chemical descriptor space, identification of superior building blocks or similarity to already know drug scaffolds [143]. Precise drug-likeness predictors have become increasingly critical as virtual screening and generative chemical models achieve recognition in order focus resources on the development of truly effective therapeutic agents [32, 33].

4.9.1.1 Major Molecular Filters Used for Drug Likeness

Due to effective screening and prioritising compounds with attributes that aligns with those of clinically successful pharmaceuticals, essential molecular filters for drug-likeness plays a vital role in early-stage drug discovery [204]. To be able to predict ADME-related features and oral bioavailability [32], the basic physicochemical filters—like Lipinski's Rule of Five, Veber, Muegge, Ghose and Egan rules—assess molecular features such as hydrogen bons donors and acceptors, polar surface area, molecular weight, number of rotatable bonds and lipophilicity [118]. These requirements are further refined across four main dimensions by modern day filtering pipelines, such as deep learning-driven druglikeFilter: (1) toxicity alerts by sophisticated AI toxicity prediction and toxicophoric rules; (2) examination of physicochemical characteristics (with incorporates classical to conformity rules); (3)

synthetic accessibility determined using retrosynthetic analysis to estimating the ease of synthesis of a molecule in lab setting; (4) binding affinity assessment leveraging both structure-based and sequence-based models for probable drug-target interactions [158]. Groundbreaking quantitative metrics like Quantitative Estimate of Drug-likeness (QED) and robust combined ML models (like DrugMetric), enable extensive scoring and ordering of candidate molecules based on their compatibility to empirical pharmacokinetic data and standards of industry [124]. By integrating biomedical expertise with property and structural-based metrics, latest advancements—like model-free fragment analysis, boundary-optimised deep learning frameworks and fragment-based filters further advance the ability to classify and rank huge chemical libraries, removing possible non-drug like and toxic candidates while focusing efforts on promising possibilities [32]. When placed together, these multi-dimensional filters largely enhance the likelihood of downstream success by identifying compounds with the balance of safety, synthetic feasibility and bioactivity. They also improve generative frameworks for de novo molecular design and virtual screening [118].

4.9.1.2 Major Bioinformatics Software/Web-Based Tools in Predicting the Drug Likelihood

The exact prediction of drug-likeness is presently supported by rapidly developing arrays of web-based applications and bioinformatics software's, which are crucial for ranking candidate compounds in the initial stages of drug discovery [211]. A biologically based framework for AI-generated molecules is offered by advanced deep learning models such as BounDr.E, which establishes drug-likeness as a tight domain around approved drugs in chemical space employing one-class boundary optimisation and biomedical knowledge placement [33]. These models reach robust, dataset efficiency and succeed in zero-shot toxic compound filtering. By deploying both attention mechanisms and graphical convolution, tools like D-GCAN [212] are now able to determine drug-likeness directly from outperforming conventional models, exhibiting high-throughput screening capacity and molecular graphs such as SARS-CoV-2 protease on vast datasets such as GDB-13 [212]. By incorporating sequential ADME property prediction with drug-likeness classification, the ADME-DL framework enriches molecular foundation models by multi-task learning and facilitates thorough compound evaluation across a range of benchmark datasets (DrugMAP, ZINC, PubChem, ChEMBL) [32]. Adding to medicinal chemistry and pharmacokinetics properties, widely recognised web-based tools such as SwissADME offers free, intuitive user interfaces that access drug-likeness, facilitating quick, visualisation and property-based filtering [187]. Designed structural descriptors or extended connectivity fingerprints are utilised by Support Vector Machine (SVM)-based systems (such as DrugMint) and random forest models (such as PrOCTOR) to enhance interpretability and accuracy. SVMs are most effective when coupled with carefully designed features. Moreover, thorough in silico ADMET and property assessment have been rendered possible by online drug-likeness toolboxes and platforms integrated in recent interviews, which accelerates molecular triage along the pipeline [196]. When coupled together, these software programs, which includes conventional machine learning, hybrid ADMET modeling and deep learning, provide precise, comprehensive and scalable forecasts, accelerating up and decreasing the risk of lead selection because chemical space discoveries expand [32].

4.9.2 ADME Prediction

A critical part of drug discovery continues to be ADME (Absorption, Distribution, Metabolism and Excretion) prediction, which guides the selection of early-stage candidates by anticipating pharmacokinetic behaviour and possible toxicity [1]. Though essential, conventional in vitro and in vivo techniques are commonly pricy, time-consuming and restricted by throughput requirements. The adaptability and precision of ADME predictions have been significantly enhanced by the latest advancements in machine learning (ML) [51], specifically in graph neural networks (GNNs), which renders quick and systematic evaluation such as permeability, toxicity, solubility and metabolic stability. These machine learning models far exceed existing QSAR techniques in several criteria by exploiting an array of molecular descriptors and significant, meticulously chosen datasets to accurately reflect the intricate, nonlinear relationships between biological fate and chemical structure [1]. Scientists minimize development cost and clinical trial period by optimising compound prioritisation and lowering attrition rates through the integration of predictive ADME modeling at the beginning of the discovery pipeline. Despite, these developments, hurdles still persist, like those related to data quality, regulatory approval and model interpretability [51]. However, ADME predictions may develop into faster, more economical, and more reliable tool for coordinating medicinal chemistry efforts if experimental pharmacology and ML are incorporated constructively [26, 27].

To enhance the drug-likeness predictions in addition to other ADME features, further enhancements focus on directly incorporating pharmacokinetic knowledge into molecular representation frameworks [165]. Pharmacokinetics-related multi-task instruction, which acknowledges the natural sequential flow of drug metabolism—absorption affecting distribution, impacting metabolism and finally excretion—is utilised in the ADME-DL framework as a representation of this methodology. Sequential modeling removes competing signals that are prominent in prior models and precisely depicts interdependencies among ADME endpoints [32]. By integrating these encapsulations into the drug-likeness predictions process, ADME-DL is capable to precisely and biologically separate authorised medications from chemical libraries [32]. Changes in performance across various molecular foundations models highlights how robust and generalisable this approach is. ADME-DL links the differentiation between structural-based virtual screening and physiology informed evaluation by integrating computational predictions will well-established pharmacokinetic principle, facilitating for enhanced early-stage candidate filtering.

It anticipates to enhance molecular priority and decrease expensive late-stage failures, and its successful implementation in clinical drug case studies highlights the effectiveness of encoding PK information for precise drug discovery scenarios [32, 33].

4.9.2.1 Major Statistical Models

Generative and predictive competencies that accelerates early-stage research are made achievable by various kinds of deep learning and machine learning frameworks that are employed in major statistical models in drug discovery [139]. Toxicity prediction and quantitative structure–activity relationship (QSAR) modeling retain to depend on conventional techniques which includes random forests, decision trees, support vector machine (SVM) and regression models. However, advanced models such as convolutional neural networks (CNNs), graph neural networks (GNNs), deep neural networks (DNNs) and transformer-based designs have acquired importance as the recent spike in complexity and data volume. These models enhance hit discovery, ADME/Tox property prediction, and multi-objective optimisation by properly expressing the high-dimensional, non-linear interaction present in bioactivity profiles and chemical space [247]. By achieving molecule developing as a sequential decision-making process and weighing trade-offs such as synthetic accessibility and potency, reinforcement learning (RL) models additionally facilitate the optimisation of molecular characteristics in molecular workflows [100]. For optimal chemical library discovery and fine-tuning, integrative platforms like Pharma. AI integrate generative adversarial networks (GANs) along with policy-gradient-based reinforcement learning (RL) [98]. In addition, phenotype, chemical, textual data and omics are merged into comprehensive predictive frameworks utilising multimodal data integration and systems biology approaches. Natural language processing (NLP) and knowledge graphs are utilised to enhance compound prioritisation and target identification [55]. Furthermore, novel automated processes utilising statistical modeling methodologies, like generalised additive models (GAMs) and Bayesian inference, have been developed for dynamic assessment of drug development transition probabilities across clinical development phases and real-time reaction analysis [217]. In combination, these growing statistical methodologies facilitate the transition from empirical to data-driven drug discovery, resulting in quicker turnarounds and enhanced success rates in pharmaceutical research and development.

4.9.2.2 Major Bioinformatics Software/Web-Based Tools for Predicting ADME

Modern ADME predictions is made possible by a range of comprehensive web-based tools and bioinformatics software, each of which offering an independent set of capabilities for high-throughput, accessible pharmacokinetic and precise profiling [214].

With the potential to foresee over 30 ADME endpoints, about 20 medicinal chemical properties (which involves drug-likeness filters), toxicophoric alerts, toxicity endpoints, API access, unique uncertainty estimation and batch evaluation for every outcome, ADMETlab 3.0 represents one of the most complete web platforms offered [247]. It also additionally provides, user-friendly interface and thorough outcome and clarification that assist in decision-making for both large and small datasets. Preliminary rational assessment of chemical libraries has been rendered simpler by the renowned SwissADME, an inexpensive and easy-to-use software designed by the Swiss Institute of Bioinformatics that permits quick computation of vital pharmacokinetic attributes, medicinal chemistry friendliness [187]. ADMET Predictor® supplies machine learning-driven estimation of over 175 ADMET and safety properties, like permeability, hepatic risk and solubility, in addition to mechanistic PBPK simulations and sophisticated data analytics for AI-integrated, personalised workflows that serves both industrial and regulatory contexts [247]. To forecast ADMET endpoint, MOE (Molecular Operating Environment) incorporates QSAR, integrative 3D visualisation and molecular modeling, providing molecular workflows for both seasoned and experienced users [239]. The ecosystem is further improved by tools such as pkCSM, admetSAR 2.0, FAF-Drugs4, and novel AI-powered platforms such as Interpretable-ADME and ADMETboost, which facilitates batch processing, board endpoint coverage and interpretability modules [84]. Deep learning-driven techniques, like ADME-DL, incorporates both molecular and clinical data for higher biological relevance and utilises multi-task learning in linear with pharmacokinetic sequence for more complex ADME and drug-likeness predictions [32]. When paired together, these platforms form an essential toolkit that enables scientists to integrate predictive ADME analytics into early drug development, minimizing risks, increasing efficacy and accelerating the process of identifying clinical candidates from the chemical design [239].

4.9.3 Toxicity Prediction

A vital element of modern drug discovery is toxicity prediction, which utilises machine learning (ML) for predicting adverse consequences of chemical compounds early in the development pipeline [92]. This enhances safety profiles and minimises expensive late-stage failures. By evaluating fingerprints and molecular structures, recent developments in machine learning models, like graph convolutional networks, ensemble approaches and deep neural networks, make it possible to foresee various toxicity endpoints, particularly genotoxicity, nephrotoxicity, cardiotoxicity and hepatotoxicity [194]. These models utilise multi-task learning, data integration and transfer learning techniques that incorporates clinical, chemical and genomic datasets to enhance robustness and accuracy, overcoming substantial problems by the heterogeneity and lack of in vivo toxicity data [92]. To facilitate the analysis and regulatory approval, multi-task deep learning frameworks, for instance, model with clinical toxicity, in vivo and in vitro endpoints [201]. They also utilise explainability ways to find significant toxicophores. To obtain outstanding prediction performance across a vast variety of datasets, tools like MolToxPred utilise stacked ensemble classifiers enhanced via Bayesian techniques, demonstrating higher precision in comparison to base learners and standard approaches [191]. Regardless of these advancements, predictive validity of machine learning toxicity models is still highly dependent on precise dataset selection, algorithm selection, thorough model validation and chemical structure representation to prevent bias and overfitting, which might end up in poor decision-making and incorrect interpretation [194]. The integration of curated toxicology databases and AI is boosting prediction confidence by recognising essential molecular substructures that pose toxicity hazards. By encouraging early identification and exclusion of potentially toxic compounds, machine learning-driven toxicity prediction platforms are transforming drug development [26]. This has culminated in selecting safe candidates and minimizing dependence on animal models and on resource-intensive experimental testing [247].

4.9.3.1 Major Statistical Models

Significant statistical models that currently accelerate advances in drug discovery are based on artificial intelligence and machine learning, employing extensive biological, clinical and chemical datasets for guiding crucial processes like compound prioritization, ADMET property estimation, virtual screening and target prediction [85]. The technology encompasses standard techniques like support vector machines (SVM), random forests for QSAR (quantitative structure–activity relationship) and regression modelling; still, the last five years of deep learning frameworks—particularly graph neural networks, transformer-based models and deep neural network—that effectively captures high-dimensional and nonlinear relationships vital for efficient prediction of molecular activity and properties [35]. By implementing patient-level phenotypic and genomic data and often integration attention mechanism for interpretability, multi-task deep learning architectures have enhanced classification transparency and accuracy, addressing requirements in toxicity and binding affinity assessment prediction [32]. The in-silico creation of novel compounds is made possible by generative models, like reinforcement learning algorithms and variational autoencoders, which personalises molecular structures across a variety of goals, comprising selectivity, potency and synthetic accessibility [100]. Meta-learning strategies are another further development that enables models like MAML to rapidly adjust to novel prediction tasks, even in the lack of data for novel compounds or potential therapeutic targets [9]. This significantly accelerates hit discovery and candidate optimisation. A new era of data-driven, dynamic and reliable drug development workflow has been ushered in by these AI-powered statistical models, which not only accelerates prediction power and efficiency but also increasingly perform as specialised, integrated pipelines with adaption capabilities and real-time feedback [85].

4.9.3.2 Major Bioinformatics Software/Web-Based Tools for Predicting ADME Features

Drug development has become increasingly dependent on cutting-edge web-based tools and bioinformatics software for ADME feature predictions, which provide precise, interpretable and high-throughput pharmacokinetic profiling [214] (Table 4.5). ADMET Predictor® incorporates metabolic predictions, physiologically based pharmacokinetics (PBPK) simulations and data analysis into a single platform to predict over 175 attributes, like permeability, mechanistic liver safety risk, solubility and metabolic stability, utilising complex AL/ML algorithms [247]. To optimise lead compounds and to screen for clinical success, Schrödinger QikProp rapidly predicts a vast range of ADME and physicochemical parameters from 3D molecular parameters including solubility, cell permeability, serum protein binding, HERG liability, logP and CNS activity [152]. With the incorporation of machine learning and robust visualisation tools, Molecular Operating Environment (MOE) frameworks provide a unified framework for structure based molecular docking, drug design, providing accurate ADME prediction [32]. Rapid prioritisation and screening are made possible by ADMETlab 2.0/3.0, which provides an intuitive web-based interface for structured, batch assessment of upto 31 ADMET endpoints, toxicity predictions, similarity searches and drug-likeness analysis [72]. SwissADME is an invaluable resource for both academic and early-stage researchers as its accessible and fast, as well as the fact that it produces accurate and free predictions for drug-likeness, physicochemical and pharmacokinetics parameters [187]. Moreover, iterative hit-to-lead development and to promote instructed compound selection, methods like deepmirror and StarDrop integrate generative AI and data analytics with predictive ADME modelling [239]. When employed in a pair, these resources permit researchers to analyse more accurately with early filters, rank and enhance candidate compounds, minimising expensive failures brought on by subpar pharmacokinetics and improving drug discovery processes [32].

4.10 Approaches to Enhance the Drug's Likeliness, ADME and Toxicity

Efforts to enhance ADME, drug-likeness, and toxicity together concentrate on increasing molecular attributes to boost the possibility of clinical success while minimising side effects and pharmacokinetic liabilities [247] (Fig. 4.5). As emphasized in drug half-life optimisation, where selectivity, metabolic clearance and adjusting potency is of importance, an essential approach of balancing physicochemical factors, especially molecular weight, polar surface area, lipophilicity, and hydrogen bond donors/acceptors, to enhance metabolic stability and oral bioavailability [193]. Furthermore, structural clarity is vital as it reduces molecular

Table 4.5 Major bioinformatics tools/computational software/web-based servers available for the prediction of drug likeliness, ADME and toxicity features

Tool/software	Category	Description and feature
SwissADME	Drug-likeness, ADME	Free web tool for prediction of pharmacokinetics, drug-likeness, physicochemical properties, medicinal chemistry [247]
ADMETlab 3.0	ADME and toxicity	Extensive online tool with over 100 ADMET endpoints, that also supports batch processing and estimates uncertainty [247]
Molinspiration	Drug-likeness	Platform supports cheminformatics that provides molecular property calculations, that included bioactivity and drug-likeness [24]
ADMET Predictor	ADME and toxicity	Solubility, metabolism and permeability supporting prediction of over 175 ADMET attributes being a AI/ML-driven software [229]
ProTox 3.0	Toxicity	Webserver that predicts organ toxicity and molecular initiating events of over 61 toxicity endpoints [30]
pkCSM	ADME and toxicity	Checks toxicity, permeability and solubility being a web tool for pharmacokinetic and toxicity property prediction [247]
admetSAR 3.0	ADME and toxicity	ADMET endpoint predictive platform [247]
PreADMET	ADME and toxicity	In silico approach for web application with ADME prediction and building drug-like libraries [114]
QikProp (Schrödinger)	ADME prediction	Proprietary software predicting physicochemical and biological ADME properties using 3D molecular structures [4]
miDruglikeness	Drug-likeness	Machine learning tool where in vivo testing is done from market approval predicting likelihood of progression [44]
FAF-Drugs4	ADME and toxicity	Toxicity evaluation, medicinal chemistry rules and ADMET filtering webserver [247]
CSM-Toxin	Protein toxicity	Webserver that uses machine learning models to predict protein toxicity [157]

complexity, or "molecular obesity", which can improve efficacy and pharmacokinetics efficacy while minimizing the risk of toxicity through scaffold rigidification, the reduction of non-essential groups and bioisosteric replacement [86]. The estimation and optimization of ADME features and drug-likeness are enhanced by advanced machine learning techniques, like ensemble models that integrate Graph Neural Networks with standard algorithms (e.g., SVM, Random Forest) [32]. This minimizes false possibility in toxicity screening and permits accurate multi-parameter assessment. Medicinal chemists may develop molecules with more comprehensive profiles by utilising tools such as DrugMetric, which utilises balanced combination learning to quantitatively quantify drug-likeness [124]. Moreover, the advancements of candidate libraries with beneficial safety profiles and pharmacokinetic is encouraged by the usage of multi-objective optimization algorithms, like Monte Carlo tree and genetic algorithms searches, which permits the exploration of chemical space for compounds that effectively meet various ADME/toxicity constraints [247]. By proactively addressing toxicity and ADME liabilities while improving selectivity and potency, the application such as medicinal chemistry and integrated computational strategies, which include predictive modeling, iterative optimisation and molecular designs that simplifies early lead development [64].

4.10.1 Latent Space Exploration

By designing molecules as points in a continuous condensed mathematical space, study with latent space is a potent technique in generative models that permits effective discovery and navigation inside large and complex chemical domains [225]. Essential functional and structural attributes of molecules are captured by this latent space, facilitating meaningful modifications and seamless interpolation to generate novel compounds with ideal attributes [104]. As shown in the design of anticancer compounds, variational autoencoders (VAEs) like NP-VAE, for example, generate chemical latent space that includes both functional annotations and structural information such as bioactivity and chirality [225]. This facilitates the generation of target-optimised molecules by investigating the neighbourhoods around known drugs. By utilising manifold traversal controlled by heuristic scoring functions that includes drug-likeness, binding affinity, synthetic affinity and Tanimoto similarity indices, advanced latent space exploration methods outperform linear interpolation. The ChemoVerse platform's systematic manifold traversal strategies demonstrate that this reduces randomness in molecule production and highly enhance the yield of new chemicals that are both therapeutically relevant and chemically valid [205].

Adaptive sampling techniques adequately explore molecular subspaces and protein conformation by utilising latent representations, which further enhances chemical space exploration [220]. By incorporating adaptive selection of low-density domains with latent space embedding, the LAST method enhances the exploration of vital protein conformations and linked chemical states with more computing efficiency than conventional molecular dynamics simulations [220]. Latent spaces are vital low-dimensional encodings in the large context of machine learning that converse vital data attributes and allow generative models like VAEs and GANs to sample, interpolate and optimise inside structural molecular landscapes [225]. Therefore, it makes it simpler to balance properties such as selectivity, synthetic accessibility and potency through multi-objective optimisation in drug discovery [68]. Essentially, deep learning techniques that implicitly learn higher-order associations without the need for explicit binding information, that enables latent space exploration to identify detailed interactions across target proteins, phenotypic responses and molecular populations. Investigators can utilise interactive visualisation tools like ChemNav to visually analyse latent space neighbourhoods and identify potential

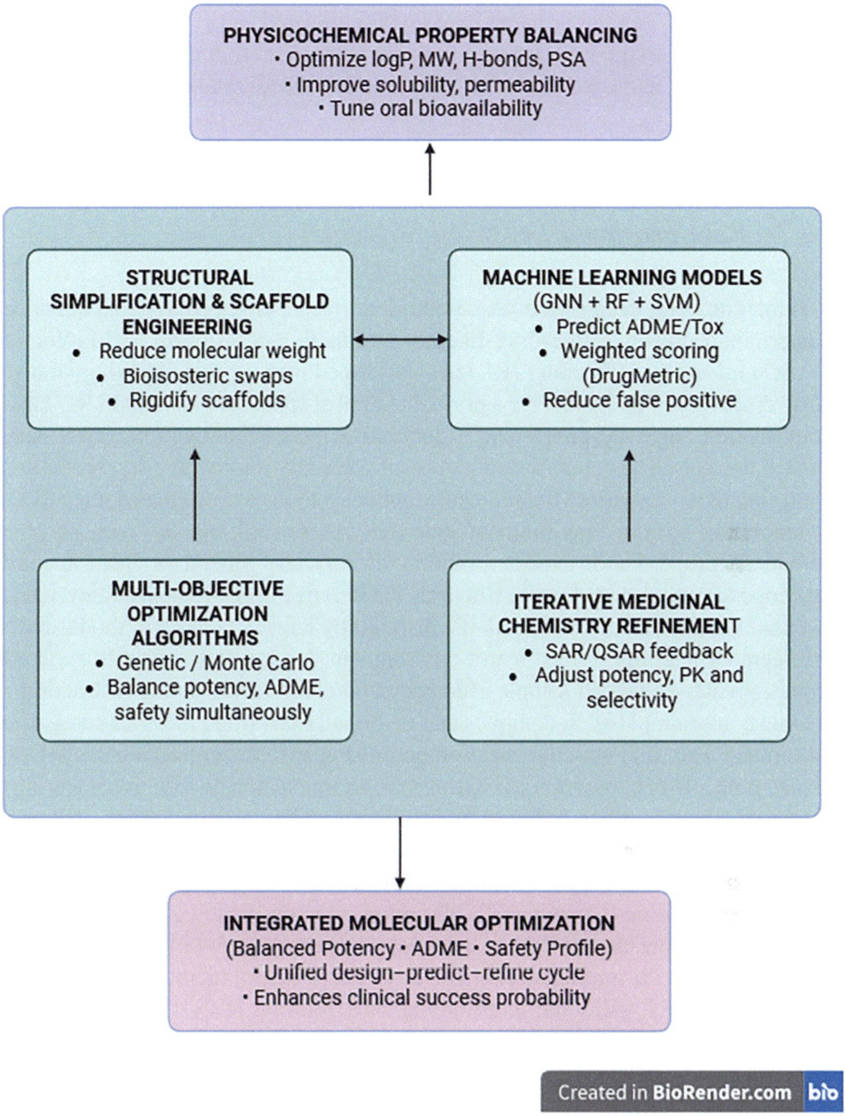

Fig. 4.5 Major approaches to enhance drug-likeness, ADME and toxicity profiles by balanced physicochemical optimisation, machine learning-based predictive modeling and structural refinement. The figure highlights how ensemble AI tool and multi-objective optimisations guide the design of compounds with therapeutic potential, improved bioavailability and safety (Created in https://BioRender.com)

chemicals [192]. Wherever is, latent space exploration enhances chemical innovation and de novo drug design by offering a scalable, interpretable and methodical framework for refining and creating novel compounds with pharmacological attributes [225].

4.10.2 Reinforcement Learning Strategies

By simulating drug design as a sequential decision-making process that enhances predetermined objective like drug-likeness, synthetic accessibility and biological activity, reinforcement learning (RL) has developed into a game-changing strategy in drug discovery [71]. To receive a normal chemical space representation RL-based methods often begin by pretraining a generative model incorporating preexisting chemical datasets. The generative model is subsequently improved using specialised reward functions that direct the molecule synthesis toward desired attributes [233]. By interacting with an environment that analyses reward signals, such as pharmacokinetic properties or expected binding affinity, the generative agent acquires knowledge to optimise molecule structures via iterative improvement achievable by this framework. Dyna-style algorithms, inspired by latest advances in model-based reinforcement learning, utilise learnt environment dynamics to generate synthetic samples, greatly enhanced sample efficiency and accelerating policy instruction in a chemical domain [216]. Techniques can be broadly classified into two categories: goal-directed learning, which focuses on optimising specific targeted characteristics or exact molecule in computational optimisation context, and distributional learning, which attempts to produce different molecule sets satisfying requirement. By overcoming the limitations of brute-force iteration and effectively choosing high-quality candidates, these RL strategies enable the exploration of large chemical spaces [216].

Unique applications show hoe RL algorithms, such as policy gradient and Proximal Policy Optimization (PPO) techniques, when integrated with deep generative models, such as recurrent neural networks or transformers, facilitate targeted drug design that is novel, valid and more biologically relevant [125, 126]. Recent findings, for instance, used RL to make compounds that are optimised for specific protein targets while also balancing a couple of goals, such as lipophilicity, drug-likeness and molecular weight scores. These studies have produced synthetically acceptable candidates with high success rates [216]. Adaptive reward functions, like composite metrics that include strength of the drug-target interaction and limitations of chemical property, enhance the creation of molecules and direct the agent towards domains of chemical space that are therapeutically relevant [48]. Moreover, integrating reinforcement learning with comprehensive language models include enhanced chemical representations and exploration capacities, offering a variety of therapeutic possibilities [71]. RL is also applied in practical uses, in personalised medicine to tailor medication formulation as per patient-specific information, enhancing the effectiveness and accuracy of therapeutic interventions. By facilitating effective, goal-directed, scalable exploration of chemical space, reinforcement learning methods, taken as a whole, present a data-driven, dynamic paradigm that revolutionaries de novo drug design [48].

4.10.3 *Multi-objective Optimization Techniques*

To handle the potential contradictory and sophisticated goals involved in the development and design of novel drug discovery, multi-objective optimisation (MOO) strategies have become essential in the process of drug discovery. MOO simultaneously takes many pharmaceutically relevant objectives into consideration, such as selectivity, toxicity, synthetic accessibility, and potency, which often trades off against one another, in comparison to single-objective optimisation, which increases or decreases a single feature [126]. Rather than generating a "best" answer, this method aims to identify a set of optimal compromises, or the Pareto front, that equilibrates these conflicting attributes [203]. Medicinal chemists might pick candidates based on project-specific priorities and acquire a better understanding of the trade-offs by examining these Pareto-optimal solutions. Incorporating evolutionary algorithms, particular swarm optimisation and genetic algorithms with multi-objective frameworks has significantly improved de novo design, quantitative structure–activity relationship (QSAR), compound library creation and molecular docking modeling [199].

De novo drug discovery workflows that integrate evolutionary computations and machine learning, are typical as latent advancements have further expanded multi-objective standardisation into sectors with many objective problems, where over three objectives are standardised at a single time. For instance, by considering pharmacokinetic attributes, multiple binding affinities across targets and consolidated toxicity endpoints, in parallel, this increase fosters deeper understanding and insights into behaviour of drug but also brings challenges related to enhanced solution interpretability and computational complexities [16]. By successfully narrowing the chemical space to individuals with both high potency and beneficial ADMET attributes, new approaches like Pareto Multi-Objective Molecular Generation (PMMG) utilises Pareto ranking and expertise in the field to generate molecules with outstanding compromises across a spectrum of objectives, accelerating real-world drug discovery projects [132]. Discovery of novel drug-like molecules appropriate for precision therapies can be assisted by the dynamic enhancement of complex molecular attributes, like scaffold diversity, multi-target functionality and blood–brain barrier permeability, which can be made feasible by the incorporation of conditional generative models and deep reinforcement learning within multi-objective frameworks. When implemented, multi-objective optimisation is a significant guiding principle that combines computational chemistry and medicinal chemistry, facilitating logical drug design that takes complicated properties of therapeutic objectives and biological systems into consideration [16].

4.11 Application of Chemical Space in Modern Drug Discovery

To efficiently detect novel bioactive molecules from an enormously large pool of intriguing compounds, chemical space analysis has become an ideal tool in modern drug discovery [107]. Chemical space mapping facilitates scientists to understand, organise, and meticulously study the structural diversity of chemical libraries by representing molecules as boundaries in a multidimensional descriptor space [149]. This facilitates them to focus on domains that are enriched with drug-like attributes and biological relevance [225]. To enhance predictive modeling, virtual screening and de novo design, latest advancements integrate artificial intelligence with machine learning [3]. This facilitates focused exploration of chemical subspaces specific to unique therapeutic domains or protein families, enhancing hit rates and reducing experimental costs [107]. Generative models, high-throughput combinatorial chemistry and DNA-encoded libraries, work together to generate synthetically acceptable and pharmacologically intriguing compounds that define novel chemical space that were unreachable earlier through traditional screening methods [18]. Adding on, visualisation tools and cheminformatics platforms enable the assessment of molecular diversity, structure–activity relationships (SAR) and scaffold novelty, aiding medicinal chemists in accelerating lead optimisation in drug development process [8]. An even more straightforward, effective and economical workflow is aided by the incorporation of chemical space concepts with scientific and computational techniques, which is revolutionising discovery paradigms towards personised therapies and precision medicine [216].

4.12 Potential Applications/Case Studies

With promising outcomes in an array of therapeutic domains, reinforcement learning (RL) has been applied progressively in drug discovery to improve the design of novel molecules with pharmacological attributes. For instance, of this is a model of ligands that target the adenosine A2A receptor utilising the DrugEx framework [131]. In this, an RNN-based generator optimised alongside reinforcement learning and supervised by the predictor model manufactured candidate molecule that provided better coverage of the already known ligand chemical space and enhances chemical diversity than earlier methods like ORGANIC and REINVENT [111]. This approach highlights how predictive models and RL can be integrated to achieve an equilibrium between exploitation and exploration, yielding compounds that have both structural novelty and biological relevance while preserving synthetic accessibility [71]. An additional significant application scenario is outlined in the ACEGEN toolkit, which renders flexible and user-friendly RL platform for optimisation and molecular generation. ACEGEN has been evaluated in many drug discovery case studies, displaying enhanced performance and flexibility than existing algorithms. Practical applications in practical initiatives centered on attributes such as toxicity reduction and protein target affinity is made achievable by its reproducibility and scalability [41].

Additionally, recent investigations targeting highly competitive proteins like SARS-CoV-2 main protease and dopamine receptor D2, have shown a variety and precision associated with RL-driven drug design platforms [127]. These studies have generated molecules with target specificity and tailored multi-parametric profiles which includes drug-likeness, synthetic feasibility and binding affinity [31]. It has been established that advancements utilising diversity-aware reward shaping can be associated to frequent issues like mode collapse, producing structurally diverse libraries to address drug resistance issues [127]. By facilitating a probability to create compounds with balanced drug-like qualities that are particular to specific protein targets, RL incorporates language models like MolT5 to further enhance targeted molecule generation [62]. Recent advances further broaden the application of RL to structure-guided discovery by integrating it with Monte Carlo Tree Search (MCTS) to further develop structure-based drug design by enhancing ligand-receptor complementarity in 3D [132]. All considered, these case studies can highlight how essential RL is to revolutionalise drug discovery into a versatile and constantly evolving procedure that manages complicated multi-objective issues while speeding up lead optimisation, building personalised treatments and hit identification [41].

4.13 Future Perspectives

To substantially accelerate and improve the pipeline of drug development, future projections on exploring chemical space with generative models place an important priority on incorporating cutting-edge machine learning and artificial intelligence techniques [185]. By 2025, artificial intelligence has already revolutionised from an exciting concept to a foundational R&D capability, contributing significant development such as enhanced hit enrichment rates, higher mechanistic accessibility, and higher predictable in silico screening platforms that decreases the expenses and challenges associated with experiments [66]. To extract additional information from a vast variety of chemical and biological datasets and offer customised pipelines specialised to therapeutic domains with accelerated developments to clinical stages, emerging trends emphasize on integrating multimodal, synthetic and self-supervised data methods [130]. It is expected that generative models integrated with multi-objective planning, reinforcement learning and latent space exploration will generate new compounds with optimised pharmacokinetic, efficacy profiles and safety, offering novel opportunities for precision medicine and peptide-based therapies [39]. Legal, data-governed, ethical mechanisms will be created in tandem with these methods advancements to guarantee the ethical, trustworthy and secure implementation of AI-driven platforms [90]. In the end, breaking long-standing drug discovery bottlenecks while providing revolutionary medicines to patients more frequently and at a cheaper price will require integrating AI-powered chemical space exploration with experimental feedback loops, real-time data incorporation and interactive ecosystems [38].

4.14 Conclusion

To sum up, the application of generative models to investigate chemical space is an advancement within modern drug development that makes it achievable to quickly build, prioritise and optimise novel compounds with desired pharmacological attributes. Researchers can successfully overcome the enormous quantity of chemical space by integrating cutting-edge computational methods such as reinforcement learning, multi-objective and deep learning optimisation. This decreases the necessity for complex experimental screening and accelerated the process from discovery to clinical implementation. The precision and success rates of candidate selection are now significantly better due to the incorporation of bioinformatics, cheminformatics tools and AI-driven predictive models which encompasses ADME, drug-likeness and toxicity prediction. Drug formulation procedures will further be enhanced as these techniques advance through additional developments in data integration, real-time feedback and model interpretability from experimental analysis. In the end, generative chemical space exploration provides the potential to revolutionize medicinal chemistry by enabling the enhancement of safer, personalised treatments, meeting unmet medicinal needs, and expediting pharmaceutical scientific and clinical procedures around the world.

References

1. Abdelwahab AA, Elattar MA, Fawzi SA. Advancing ADMET prediction for major CYP450 isoforms: graph-based models, limitations, and future directions. BioMed Eng OnLine. 2025;24(1). https://doi.org/10.1186/s12938-025-01412-6
2. Agu PC, Afiukwa CA, Orji OU, Ezeh EM, Ofoke IH, Ogbu CO, Ugwuja EI, Aja PM. Molecular docking as a tool for the discovery of molecular targets of nutraceuticals in diseases management. Sci Rep. 2023;13(1). https://doi.org/10.1038/s41598-023-40160-2
3. Ahmed W, Wani MA, Pawel P, Souham M, Mahmoud A, Hammad M. Machine learning-based academic performance prediction with explainability for enhanced decision-making in educational institutions. Sci Rep. 2025;15(1). https://doi.org/10.1038/s41598-025-12353-4
4. AlAjmi M, Azhar A, Hasan S, Alshabr A, Hussain A, Rehman MT. Identification of natural compounds (Proanthocyanidin and Rhapontin) as high-affinity inhibitor of SARS-CoV-2 Mpro and PLpro using computational strategies. Arch Med Sci. 2021. https://doi.org/10.5114/aoms/133706.
5. Alhumaidi NH, Dermawan D, Kamaruzaman HF, Alotaiq N. The use of machine learning for analyzing real-world data in disease prediction and management: systematic review. JMIR Med Inform. 2025;13:e68898. https://doi.org/10.2196/68898.

6. AlKharboush DF, Kozielski F, Wells G, Porta EOJ. Fragment-based drug discovery: a graphical review. Curr Res Pharmacol Drug Discov. 2025;9:100233. https://doi.org/10.1016/j.crphar.2025.100233.
7. Almuqdadi HTA, Kifayat S, Anwer R, Alrehaili J, Abid M. Fragment-based virtual screening identifies novel leads against Plasmepsin IX (PlmIX) of *Plasmodium falciparum*: homology modeling, molecular docking, and simulation approaches. Front Pharmacol. 2024;15:1387629. https://doi.org/10.3389/fphar.2024.1387629.
8. Alqarni A, Hosmani J, Alassiri S, Alqahtani AMA, Assiri HA, AlHousami T, Zaki H, Patil S. Quantitative structure-activity relationship-based neural network model in screening potential inhibitors from brown algae against RelA in oral squamous cell carcinoma. Int Dent J. 2025;75(6):103890. https://doi.org/10.1016/j.identj.2025.103890.
9. Alsaleh AM, Albalawi E, Algosaibi A, Albakheet SS, Khan SB. Few-shot learning for medical image segmentation using 3D U-net and model-agnostic meta-learning (MAML). Diagnostics. 2024;14(12):1213–1213. https://doi.org/10.3390/diagnostics14121213.
10. Alruwaili M, Tilal E, Mohamed, MA, Elderdery AY, Mills J, Yasir A, Hamza A, Salma, Alotaibi HA, Alrowily MJ, Althobiti MM. Molecular docking, free energy calculations, ADMETox studies, DFT analysis, and dynamic simulations highlighting a chromene glycoside as a potential inhibitor of PknG in *Mycobacterium tuberculosis*. Front Chem. 2025;13. https://doi.org/10.3389/fchem.2025.1531152
11. Amabilino S, Pogány P, Pickett SD, Darren. Guidelines for recurrent neural network transfer learning-based molecular generation of focused libraries. 2020;60(12):5699–713. https://doi.org/10.1021/acs.jcim.0c00343
12. Amin SA, Sessa L, Gayen S, Piotto S. PPARγ modulator predictor (PGMP_v1): chemical space exploration and computational insights for enhanced type 2 diabetes mellitus management. Mol Diversity. 2025;29(4):3305–21. https://doi.org/10.1007/s11030-025-11118-5.
13. Ancuceanu R, Popovici PC, Drăgănescu D, Busnatu Ș, Lascu BE, Dinu M. QSAR regression models for predicting HMG-CoA reductase inhibition. Pharmaceuticals. 2024;17(11):1448–1448. https://doi.org/10.3390/ph17111448.
14. Andrianov GV, Ong G, Serebriiskii I, Karanicolas J. Efficient hit-to-lead searching of kinase inhibitor chemical space via computational fragment merging. J Chem Inf Model. 2021;61(12):5967–87. https://doi.org/10.1021/acs.jcim.1c00630.
15. Andronov M, Fedorov MV, Sosnin S. Exploring chemical reaction space with reaction difference fingerprints and parametric t-SNE. ACS Omega. 2021;6(45):30743–51. https://doi.org/10.1021/acsomega.1c04778.
16. Angelo JS, Guedes IA, Barbosa HJC, Dardenne LE. Multi-and many-objective optimization: present and future in de novo drug design. Front Chem. 2023;11:1288626. https://doi.org/10.3389/fchem.2023.1288626.
17. Arús-Pous J, Johansson, S., Prykhodko O, Bjerrum EJ, Tyrchan C, Reymond J-L, Chen H, Engkvist O. Randomized SMILES strings improve the quality of molecular generative models. 2019;11(1). https://doi.org/10.1186/s13321-019-0393-0
18. Arús-Pous J, Blaschke T, Ulander S, Reymond J-L, Chen H, Engkvist O. Exploring the GDB-13 chemical space using deep generative models. J Cheminf. 2019;11(1). https://doi.org/10.1186/s13321-019-0341-z
19. Arvidsson McShane S, Norinder U, Alvarsson J, Ahlberg E, Carlsson L, Spjuth O. CPSign: conformal prediction for cheminformatics modeling. J Cheminf. 2024;16(1). https://doi.org/10.1186/s13321-024-00870-9
20. Ashenden SK. The era of artificial intelligence, machine learning and data science in the pharmaceutical industry. Academic Press; 2021.
21. Astigarraga M, Sánchez-Ruiz A, Colmenarejo G. Conformal prediction-based machine learning in cheminformatics: current applications and new challenges. Artif Intell Life Sci. 2025;7:100127. https://doi.org/10.1016/j.ailsci.2025.100127.
22. Atanasov AG, Zotchev SB, Dirsch VM, Supuran CT. Natural products in drug discovery: advances and opportunities. Nat Rev Drug Discov. 2021;20(3):200–16. https://doi.org/10.1038/s41573-020-00114-z.

23. Atz K, Cotos L, Isert C, Håkansson M, Focht D, Hilleke M, Nippa DF, Iff M, Ledergerber J, Schiebroek CCG, Romeo V, Hiss JA, Merk D, Schneider P, Kuhn B, Grether U, Schneider G. Prospective de novo drug design with deep interactome learning. Nat Commun. 2024;15(1):3408. https://doi.org/10.1038/s41467-024-47613-w.
24. Alanazi A, Younas, S., Khan, M. U., Saleem, H., Alruwaili M, Abdalla, A. E., Abosalif BK, Ejaz H. A combined in silico and MD simulation approach to discover novel LpxC inhibitors targeting multiple drug resistant *Pseudomonas aeruginosa*. Sci Rep. 2025;15(1). https://doi.org/10.1038/s41598-025-99215-1
25. Angélica Pilón-Jiménez B, Saldívar-González FI, Díaz-Eufracio BI, Medina-Franco JL. BIOFACQUIM: a Mexican compound database of natural products. Biomolecules. 2019;9(1):31–31. https://doi.org/10.3390/biom9010031.
26. Bai C, Wu L, Li R, Cao Y, He S, Bo X. Machine learning-enabled drug-induced toxicity prediction. Adv Sci. 2025. https://doi.org/10.1002/advs.202413405.
27. Bai H, Cheng Y, Jia S, Wang X, Jin R, Guo H, Tang Y, Wang Y. Machine learning-based QSAR and structure-based virtual screening guided discovery of novel mIDH1 inhibitors from natural products. J Comput Aided Mol Des. 2025;39(1):44. https://doi.org/10.1007/s10822-025-00624-1.
28. Bajorath J. Milestones in chemoinformatics: global view of the field. J Cheminfo. 2024;16(1). https://doi.org/10.1186/s13321-024-00922-0
29. Bandini E, Kajtazi A, Szucs R, Lynen F. The role and choice of molecular descriptors for predicting retention times in HPLC: a comprehensive review. TrAC Trends Anal Chem. 2025;187:118207. https://doi.org/10.1016/j.trac.2025.118207.
30. Banerjee P, Kemmler E, Dunkel M, Preissner R. ProTox 3.0: a webserver for the prediction of toxicity of chemicals. Nucleic Acids Res. 2024;52(W1). https://doi.org/10.1093/nar/gkae303
31. Banerjee R, Perera L, Tillekeratne LMV. Potential SARS-CoV-2 main protease inhibitors. Drug Discov Today. 2021;26(3):804–16. https://doi.org/10.1016/j.drudis.2020.12.005.
32. Bang D, Kim J, Song H, Kim S. ADME-drug-likeness: enriching molecular foundation models via pharmacokinetics-guided multi-task learning for drug-likeness prediction. Bioinformatics. 2025;41(Supplement_1):i352–i361. https://doi.org/10.1093/bioinformatics/btaf259
33. Bang D, Sung I, Piao Y, Lee S, Kim S. BounDr.E: predicting drug-likeness through knowledge alignment and EM-like one-class boundary optimization. Openreview.net. 2025. https://openreview.net/forum?id=TnlLMYPfx5&noteId=oDnPrr4cid
34. Barazorda-Ccahuana HL, Ranilla LG, Candia-Puma MA, Cárcamo-Rodriguez EG, Centeno-Lopez AE, Davila-Del-Carpio G, Medina-Franco JL, Chávez-Fumagalli MA. PeruNPDB: the Peruvian natural products database for in silico drug screening. Sci Rep. 2023;13:7577. https://doi.org/10.1038/s41598-023-34729-0.
35. Bastikar VA, Alpana B, Gupta PP. Quantitative structure–activity relationship-based computational approaches. Elsevier EBooks; 2022. pp. 191–205. https://doi.org/10.1016/b978-0-323-91172-6.00001-7
36. Batt AL, Brunelle LD, Quinete NS, Stebel EK, Ng B, Gardinali P, Chao A, Huba AK, Glassmeyer ST, Alvarez DA, Kolpin DW, Furlong ET, Mills MA. Investigating the chemical space coverage of multiple chromatographic and ionization methods using non-targeted analysis on surface and drinking water collected using passive sampling. Sci Total Environ. 2024;955:176922. https://doi.org/10.1016/j.scitotenv.2024.176922.
37. Beroza P, Crawford JJ, Ganichkin O, Gendelev L, Harris SF, Klein R, Miu A, Steinbacher S, Klingler F-M, Lemmen C. Chemical space docking enables large-scale structure-based virtual screening to discover ROCK1 kinase inhibitors. Nat Commun. 2022;13(1). https://doi.org/10.1038/s41467-022-33981-8
38. Bhat AR, Ahmed S. Artificial intelligence (AI) in drug design and discovery: a comprehensive review. In Silico Res Biomed. 2025:100049. https://doi.org/10.1016/j.insi.2025.100049
39. Bian Y, Xie X-Q. Generative chemistry: drug discovery with deep learning generative models. J Mol Model. 2021;7(3). https://doi.org/10.1007/s00894-021-04674-8
40. Boldini D, Ballabio D, Consonni V, Todeschini R, Grisoni F, Sieber SA. Effectiveness of molecular fingerprints for exploring the chemical space of natural products. J Cheminf. 2024;16(1):1–16. https://doi.org/10.1186/s13321-024-00830-3.

41. Bou A, Thomas M, Dittert S, Navarro C, Majewski M, Wang Y, Patel S, Tresadern G, Ahmad M, Moens V, Sherman W, Sciabola S, De Fabritiis G. ACEGEN: reinforcement learning of generative chemical agents for drug discovery. J Chem Inf Model. 2024;64(15):5900–11. https://doi.org/10.1021/acs.jcim.4c00895.
42. Brinkmann H, Argante A, ter Steege H, Grisoni F. Going beyond SMILES enumeration for data augmentation in generative drug discovery. Digital Discov. 2025. https://doi.org/10.1039/d5dd00028a.
43. Buehler Y, Reymond J-L. Expanding bioactive fragment space with the generated database GDB-13s. J Chem Inf Model. 2023;63(20):6239–48. https://doi.org/10.1021/acs.jcim.3c01096.
44. Cai C, Lin H, Wang H, Xu Y, Ouyang Q, Lai L, Pei J. MiDruglikeness: subdivisional drug-likeness prediction models using active ensemble learning strategies. Biomolecules. 2022;13(1):29. https://doi.org/10.3390/biom13010029.
45. Cai T, Xie L, Chen M, Liu Y, He D, Zhang S, Mura C, Bourne P, Xie L. Exploration of dark chemical genomics space via portal learning: applied to targeting the undruggable genome and COVID-19 anti-infective polypharmacology. Res Square. 2021. https://doi.org/10.21203/rs.3.rs-1109318/v1
46. Carbery A, Skyner R, von Delft F, Deane CM. Fragment libraries designed to be functionally diverse recover protein binding information more efficiently than standard structurally diverse libraries. J Med Chem. 2022;65(16):11404–13. https://doi.org/10.1021/acs.jmedchem.2c01004.
47. Carrel A, Yiannakas A, Roukens J-J, Reynoso-Moreno I, Orsi M, Thakkar A, Arus-Pous J, Pellegata D, Gertsch J, Reymond J-L. Exploring simple drug scaffolds from the generated database chemical space reveals a chiral bicyclic azepane with potent neuropharmacology. J Med Chem. 2025;68(9):9176–201. https://doi.org/10.1021/acs.jmedchem.4c02549.
48. Chapman M, Xu L, Lapeyrolerie M, Boettiger C. Bridging adaptive management and reinforcement learning for more robust decisions. Philos Trans R Soc B. 2023;378(1881). https://doi.org/10.1098/rstb.2022.0195
49. Chatziefthymiou SD, Kuzikov M, Afandi S, Kovacs G, Srivastava S, Zaliani A, Gruzinov A, Pompidor G, Lunelli M, Ahmed GR, Labahn J, Hakanpää J, Windshügel B, Kolbe M. Identification, validation, and characterization of approved and investigational drugs interfering with the SARS-CoV-2 endoribonuclease Nsp15. Protein Sci Publ Protein Soc. 2025;34(6):e70156. https://doi.org/10.1002/pro.70156.
50. Chávez-Hernández AL, Sánchez-Cruz N, Medina-Franco JL. Fragment library of natural products and compound databases for drug discovery. Biomolecules. 2020;10(11):1518. https://doi.org/10.3390/biom10111518.
51. Chen A, Peng X, Shen T, Zheng L, Wu D, Wang S. Discovery, design, and engineering of enzymes based on molecular retrobiosynthesis. MLife. 2025;4(2):107–25. https://doi.org/10.1002/mlf2.70009.
52. Chen Y, Kirchmair J. Cheminformatics in natural product-based drug discovery. Mol Inf. 2020;39(12):2000171. https://doi.org/10.1002/minf.202000171.
53. Chetry AB, Ohto K. From molecules to data: the emerging impact of chemoinformatics in chemistry. J Cheminf. 2025;17(1). https://doi.org/10.1186/s13321-025-00978-6
54. Chi X, Chen R, Yang X, He X, Pan Z, Yao C, Peng H, Yang H, Huang W, Chen Z. Discovery of novel DDR1 inhibitors through a hybrid virtual screening pipeline, biological evaluation and molecular dynamics simulations. ACS Med Chem Lett. 2025;16(4):602–10. https://doi.org/10.1021/acsmedchemlett.4c00634.
55. Chamarthi B, Polu OR, Anumula SK, Ushmani A, Kasralikar P, Syed AA. Natural language processing (NLP)-and machine learning (ML)-enabled operating room optimization: a preferred reporting items for systematic reviews and meta-analyses (PRISMA) Systematic review anchored in project planning theory. Cureus. 2025. https://doi.org/10.7759/cureus.82796.
56. Chu X, Zhang B, Koeken VACM, Gupta MK, Li Y. Multi-omics approaches in immunological research. Front Immunol. 2021;12:668045. https://doi.org/10.3389/fimmu.2021.668045.

57. Corrêa Veríssimo G, Salgado Ferreira R, Gonçalves Maltarollo V. Ultra-large virtual screening: definition, recent advances, and challenges in drug design. Mol Inf. 2024;44(1). https://doi.org/10.1002/minf.202400305
58. Cox PB, Gupta R. Contemporary Computational Applications And Tools In Drug Discovery. ACS Med Chem Lett. 2022;13(7):1016–29. https://doi.org/10.1021/acsmedchemlett.1c00662.
59. Daina A, Zoete V. Application of the SwissDrugDesign online resources in virtual screening. Int J Mol Sci. 2019;20(18):4612. https://doi.org/10.3390/ijms20184612.
60. David L, Thakkar A, Mercado R, Engkvist O. Molecular representations in AI-driven drug discovery: a review and practical guide. J Cheminf. 2020;12(1). https://doi.org/10.1186/s13321-020-00460-5
61. Decardi-Nelson B, Alshehri AS, You F. Generative artificial intelligence in chemical engineering spans multiple scales. Front Chem Eng. 2024;6:1458156. https://doi.org/10.3389/fceng.2024.1458156.
62. Deng Y, Ericksen SS, Gitter A. Chemical language model linker: blending text and molecules with modular adapters. J Chem Inf Model. 2025. https://doi.org/10.1021/acs.jcim.5c00853.
63. Devarakonda VS, Sun W, Tang X, Tian Y. Recent advances in reinforcement learning for chemical process control. Processes. 2025;13(6):1791–1791. https://doi.org/10.3390/pr13061791.
64. Er-rajy M, Fadili ME, Hamada I, Saeed A, Rehman AU, Zarougui S, Abdullah SA, Alahdab A, Parvez MK, Menana E. 3D-QSAR studies, molecular docking, molecular dynamic simulation, and ADMET proprieties of novel pteridinone derivatives as PLK1 inhibitors for the treatment of prostate cancer. Life. 2023;13(1):127–127. https://doi.org/10.3390/life13010127
65. Evangelista M, Papa E. A review of quantitative structure-activity relationship (QSAR) models to predict thyroid hormone system disruption by chemical substances. Toxics. 2025;13(9):799–799. https://doi.org/10.3390/toxics13090799.
66. Filella-Merce I, Molina A, Lucía D, Orzechowski M, Berchiche YA, Zhu YM, Júlia V-M, Malo L, Yekkirala AS, Ray S, Guallar, V. Optimizing drug design by merging generative AI with a physics-based active learning framework. Commun Chem. 2025;8(1). https://doi.org/10.1038/s42004-025-01635-7
67. Flanagan AR, Dalal D, Glavin FG. Exploring generative artificial intelligence and data augmentation techniques for spectroscopy analysis. Chem Rev. 2025;125(13):6130–55. https://doi.org/10.1021/acs.chemrev.4c00815.
68. Flores-Hernandez H, Martinez-Ledesma E. A systematic review of deep learning chemical language models in recent era. J Cheminf. 2024;16(1). https://doi.org/10.1186/s13321-024-00916-y
69. Fourches D, Ash J. 4D-quantitative structure—activity relationship modeling: making a comeback. Expert Opin Drug Discov. 2019;14(12):1227–35. https://doi.org/10.1080/17460441.2019.1664467.
70. Franco LS, de Jesus BSM, Pinheiro PSM, Fraga CAM. Remapping the chemical space and the pharmacological space of drugs: what can we expect from the road ahead? Pharmaceuticals. 2024;17(6):742. https://doi.org/10.3390/ph17060742
71. Frommeyer TC, Gilbert MM, Fursmidt RM, Park Y, Khouzam JP, Brittain GV, Frommeyer DP, Bett ES, Bihl TJ. Reinforcement learning and its clinical applications within healthcare: a systematic review of precision medicine and dynamic treatment regimes. Healthcare. 2025;13(14):1752–1752. https://doi.org/10.3390/healthcare13141752.
72. Fu L, Shi S, Yi J, Wang N, He Y, Wu Z, Peng J, Deng Y, Wang W, Wu C, Lyu A, Zeng X, Zhao W, Hou T, Cao D. ADMETlab 3.0: an updated comprehensive online ADMET prediction platform enhanced with broader coverage, improved performance, API functionality and decision support. Nucleic Acids Res. 2024;52(W1). https://doi.org/10.1093/nar/gkae236
73. Fugger L, Jensen LT, Rossjohn J. Challenges, progress, and prospects of developing therapies to treat autoimmune diseases. Cell. 2020;181(1):63–80. https://doi.org/10.1016/j.cell.2020.03.007.

74. Gans JD, Miller JE, Esen S, Nileena V, Jha RK. The impact of curation errors in the PDBBind database on machine learning predictions of protein–protein binding affinity. Database. 2025;2025. https://doi.org/10.1093/database/baaf061
75. Gao W, Luo S, Coley CW. Generative AI for navigating synthesizable chemical space. Proc Natl Acad Sci USA. 2025;122(41):e2415665122. https://doi.org/10.1073/pnas.2415665122.
76. Gaur AS, John L, Kumar N, Ram Vivek M, Nagamani S, Mahanta HJ, Sastry GN. Towards systematic exploration of chemical space: building the fragment library module in molecular property diagnostic suite. Mol Diversity. 2022;27(3):1459–68. https://doi.org/10.1007/s11030-022-10506-5.
77. Gentile F, Agrawal V, Hsing M, Ton A-T, Ban F, Norinder U, Gleave ME, Cherkasov A. Deep docking: a deep learning platform for augmentation of structure based drug discovery. ACS Cent Sci. 2020;6(6):939–49. https://doi.org/10.1021/acscentsci.0c00229.
78. Gertig C, Leonhard K, Bardow A. Computer-aided molecular and processes design based on quantum chemistry: current status and future prospects. Curr Opin Chem Eng. 2020;27:89–97. https://doi.org/10.1016/j.coche.2019.11.007.
79. Gironda-Martínez A, Donckele EJ, Samain F, Neri D. DNA-encoded chemical libraries: a comprehensive review with succesful stories and future challenges. ACS Pharmacol Transl Sci. 2021;4(4):1265–79. https://doi.org/10.1021/acsptsci.1c00118.
80. Gómez-Bombarelli R, Wei JN, Duvenaud D, Hernández-Lobato JM, Sánchez-Lengeling B, Sheberla D, Aguilera-Iparraguirre J, Hirzel TD, Adams RP, Aspuru-Guzik A. Automatic chemical design using a data-driven continuous representation of molecules. ACS Cent Sci. 2018;4(2):268–76. https://doi.org/10.1021/acscentsci.7b00572.
81. Gómez-García A, Acuña DA, Zamora WJ, Barazorda-Ccahuana HL, Chávez-Fumagalli MÁ, Valli M, Andricopulo ADV, Olmedo, DA, Solís PN, Núñez MJ, Rodríguez Pérez JR, Sánchez V, Cortés Hernández HF, Medina-Franco JL. Navigating the chemical space and chemical multiverse of a unified Latin American natural product database: LANaPDB. Pharmaceuticals. 2023;16(10):1388–1388. https://doi.org/10.3390/ph16101388
82. Grant LL, Sit CS. De novo molecular drug design benchmarking. RSC Med Chem. 2021;12(8):1273–80. https://doi.org/10.1039/d1md00074h.
83. Greco S, Bossa C, Battistelli CL, Giuliani A. A statistical exploration of QSAR models in cancer risk assessment: a case study on pesticide-active substances and metabolites. Toxics. 2025;13(4):299. https://doi.org/10.3390/toxics13040299.
84. Gu Y, Yu Z, Wang Y, Chen L, Lou C, Yang C, Li W, Liu G, Tang Y. admetSAR3.0: a comprehensive platform for exploration, prediction and optimization of chemical ADMET properties. Nucleic Acids Res. 2024. https://doi.org/10.1093/nar/gkae298
85. Haas BC, Kalyani D, Sigman MS. Applying statistical modeling strategies to sparse datasets in synthetic chemistry. Sci Adv. 2025;11(1). https://doi.org/10.1126/sciadv.adt3013
86. Hann MM. Molecular obesity, potency and other addictions in drug discovery. MedChemComm. 2025;2(5):349–55. https://doi.org/10.1039/c1md00017a.
87. Hany MS, Ahmed NS, Boeckler FM, Ibrahim TM. Benchmarking the structure-based virtual screening performance of wild-type and resistant. PubMed. 2025;19:7045–58. https://doi.org/10.2147/dddt.s537065.
88. Hartmann M, Rauscher M, Robinson J, Welsh J, Roush D. Integration of QSAR models with high throughput screening to accelerate the development of polishing chromatography unit operations. J Chromatogr A. 2025;1747:465818. https://doi.org/10.1016/j.chroma.2025.465818.
89. Hemant Kumar S, Venkatachalapathy M, Sistla R, Poongavanam V. Advances in molecular glues: exploring chemical space and design principles for targeted protein degradation. Drug Discov Today. 2024;29(11):104205. https://doi.org/10.1016/j.drudis.2024.104205.
90. Hermann E, Hermann G, Tremblay J-C. Ethical artificial intelligence in chemical research and development: a dual advantage for sustainability. Sci Eng Ethics. 2021;27(4). https://doi.org/10.1007/s11948-021-00325-6
91. Hoffmann T, Gastreich M. The next level in chemical space navigation: going far beyond enumerable compound libraries. Drug Discov Today. 2019;24(5):1148–56. https://doi.org/10.1016/j.drudis.2019.02.013.

92. Hong J, Kwon H. Multimodal deep learning for chemical toxicity prediction and management. Sci Rep. 2025;15(1). https://doi.org/10.1038/s41598-025-95720-5
93. Hu F, Jia X, Liao W, Chen Z, Bi H, Ge H, Liu D, Zhang R, Hu Y, Mei W, Zhao Z, Zhang K, Zhu L, Diao Y, Li H. Exploring the macrocyclic chemical space for heuristic drug design with deep learning models. Commun Chem. 2025;8(1). https://doi.org/10.1038/s42004-025-01686-w
94. Huang B, von Lilienfeld. Ab initio machine learning in chemical compound space. ArXiv (Cornell University). 2020. https://doi.org/10.48550/arxiv.2012.07502
95. Huo X, Xu J, Xu M, Chen H. An improved 3D quantitative structure-activity relationships (QSAR) of molecules with CNN-based partial least squares model. Artif Intell Life Sci. 2023;3:100065. https://doi.org/10.1016/j.ailsci.2023.100065.
96. Imbernon JR, Jacquemard C, Bret G, Marcou G, Kellenberger E. Comprehensive analysis of commercial fragment libraries. RSC Med Chem. 2021;13(3):300–10. https://doi.org/10.1039/d1md00363a.
97. Işık M, Bergazin TD, Fox T, Rizzi A, Chodera JD, Mobley DL. Assessing the accuracy of octanol–water partition coefficient predictions in the SAMPL6 Part II log P challenge. J Comput Aided Mol Des. 2020;34(4):335–70. https://doi.org/10.1007/s10822-020-00295-0.
98. Ivanenkov YA, Polykovskiy D, Bezrukov D, Zagribelnyy B, Aladinskiy V, Kamya P, Aliper A, Ren F, Zhavoronkov A. Chemistry42: an AI-driven platform for molecular design and optimization. J Chem Inf Model. 2023;63(3):695–701. https://doi.org/10.1021/acs.jcim.2c01191.
99. Jeon H, Lee JG, Shin W, Ji H, Joung I, Lee HS. STELLA provides a drug design framework enabling extensive fragment-level chemical space exploration and balanced multi-parameter optimization. Sci Rep. 2025;15(1). https://doi.org/10.1038/s41598-025-12685-1
100. Jeon W, Kim D. Autonomous molecule generation using reinforcement learning and docking to develop potential novel inhibitors. Sci Rep. 2020;10(1). https://doi.org/10.1038/s41598-020-78537-2
101. Jeong C-H, Na YJ, Kim TY, Lee SY, Kim J, Ryou S. In silico screening and experimental validation identify riboflavin as an RNA-targeted antiviral against SARS-CoV-2. Sci Rep. 2025;15(1). https://doi.org/10.1038/s41598-025-16949-8
102. Jinsong S, Qifeng J, Xing C, Hao Y, Wang L. Molecular fragmentation as a crucial step in the AI-based drug development pathway. Commun Chem. 2024;7(1):1–9. https://doi.org/10.1038/s42004-024-01109-2.
103. Journal of Cheminformatics. BioMed Central. 2025. https://jcheminf.biomedcentral.com/articles
104. Kabelac A, Eulig E, Maier J, Hammermann M, Knaup M, Kachelrieß M. Latent space reconstruction for missing data problems in CT. Med Phys. 2025;52(7). https://doi.org/10.1002/mp.17910
105. Kainrad T, Hunold S, Seidel T, Langer T. LigandScout remote: a new user-friendly interface for HPC and cloud resources. J Chem Inf Model. 2018;59(1):31–7. https://doi.org/10.1021/acs.jcim.8b00716.
106. Karthikeyan A, Priyakumar UD. Artificial intelligence: machine learning for chemical sciences. J Chem Sci. 2021;134(1). https://doi.org/10.1007/s12039-021-01995-2
107. Kattuparambil AA, Chaurasia DK, Shekhar S, Srinivasan A, Mondal S, Aduri R, Jayaram B. Exploring chemical space for "druglike" small molecules in the age of AI. Front Mol Biosci. 2025;12:1553667. https://doi.org/10.3389/fmolb.2025.1553667.
108. Kell DB, Samanta S, Swainston N. Deep learning and generative methods in cheminformatics and chemical biology: navigating small molecule space intelligently. Biochem J. 2020;477(23):4559–80. https://doi.org/10.1042/bcj20200781.
109. Khan A, Umbreen S, Hameed A, Fatima R, Zahoor U, Babar Z, Waseem M, Hussain Z, Rizwan M, Zaman N, Ali S, Suleman M, Shah A, Ali L, Ali SS, Wei D-Q. In silico mutagenesis-based remodelling of SARS-CoV-1 peptide (ATLQAIAS) to inhibit SARS-CoV-2: structural-dynamics and free energy calculations. Interdisc Sci Comput Life Sci. 2021;13(3):521–34. https://doi.org/10.1007/s12539-021-00447-2.

110. Khalak Y, Tresadern G, Hahn DF, Bert, Gapsys V. Chemical space exploration with active learning and alchemical free energies. J Chem Theory Comput. 2022;18(10):6259–6270. https://doi.org/10.1021/acs.jctc.2c00752
111. Khater T, Alkhatib SA, AlShehhi A, Pitsalidis C, Pappa AM, Ngo ST, Chan V, Truong VK. Generative artificial intelligence based models optimization towards molecule design enhancement. J Cheminf. 2025;17(1). https://doi.org/10.1186/s13321-025-01059-4
112. Kim S, Noh J, Gu GH, Aspuru-Guzik A, Jung Y. Generative adversarial networks for crystal structure prediction. ACS Central Sci. 2020;6(8):1412–1420. https://doi.org/10.1021/acscentsci.0c00426
113. Kim T, Baek E, Kim J. Exploring macrocyclic chemical space: strategies and technologies for drug discovery. Pharmaceuticals. 2025;18(5):617. https://doi.org/10.3390/ph18050617.
114. Klimoszek D, Jeleń M, Dołowy M, Morak-Młodawska B. Study of the lipophilicity and ADMET parameters of new anticancer diquinothiazines with pharmacophore substituents. Pharmaceuticals. 2024;17(6):725–725. https://doi.org/10.3390/ph17060725.
115. Knehans T, Boyles NA, Bos PH. Enumerable libraries and accessible chemical space in drug discovery. 2024;315–336. https://doi.org/10.1002/9783527840748.ch14
116. Karandashev K, Weinreich J, Heinen S, Jose D, von Rudorff GF, Hermansson K, von Lilienfeld A. Evolutionary Monte Carlo of QM properties in chemical space: electrolyte design. J Chem Theory Comput. 2023;19(23):8861–70. https://doi.org/10.1021/acs.jctc.3c00822.
117. Konstantinidou M, Virta JM, Arkin MR. Stabilization of native protein-protein interactions with molecular glues: a 14-3-3 case study. Acc Chem Res. 2025. https://doi.org/10.1021/acs.accounts.5c00441.
118. Kralj S, Jukić M, Bren U. Molecular filters in medicinal chemistry. Encyclopedia. 2023;3(2):501–11. https://doi.org/10.3390/encyclopedia3020035.
119. Krenn M, Häse F, Nigam A, Friederich P, Aspuru-Guzik A. Self-referencing embedded strings (SELFIES): a 100% robust molecular string representation. Mach Learn Sci Technol. 2020;1(4):045024. https://doi.org/10.1088/2632-2153/aba947.
120. Laguionie-Marchais C, Allcock AL, Baker BJ, Conneely E-A, Dietrick SG, Kearns F, McKeever K, Young RM, Sierra CA, Soldatou S, Woodcock HL, Johnson MP. Not drug-like, but like drugs: Cnidaria natural products. Mar Drugs. 2021;20(1):42–42. https://doi.org/10.3390/md20010042.
121. Latosińska M, Latosińska JN. The Chameleon strategy—a recipe for effective ligand screening for viral targets based on four novel structure-binding strength indices. Viruses. 2024;16(7):1073–1073. https://doi.org/10.3390/v16071073.
122. Lavecchia A. Navigating the frontier of drug-like chemical space with cutting-edge generative AI models. Drug Discov Today. 2024;29(9):104133. https://doi.org/10.1016/j.drudis.2024.104133.
123. Lessigiarska I, Peng Y, Tsakovska I, Alov P, Lagarde N, Jereva D, Villoutreix BO, Nicot AB, Pajeva I, Pencheva T, Miteva MA. Computational analysis of chemical space of natural compounds interacting with sulfotransferases. Molecules. 2021;26(21):6360–6360. https://doi.org/10.3390/molecules26216360.
124. Li B, Wang Z, Liu Z, Tao Y, Sha C, He M, Li X. DrugMetric: quantitative drug-likeness scoring based on chemical space distance. Brief Bioinf. 2024;25(4). https://doi.org/10.1093/bib/bbae321
125. Li C, Dong W, He L, Cai M, Wang D. Intelligent decision for joint operations based on improved proximal policy optimization. Sci Rep. 2025;15(1). https://doi.org/10.1038/s41598-025-86229-y
126. Li J, Gurgenci H, Guan Z, Wang J, Chen J, Wang C, Huang Z. Multi objective optimization algorithm for hybrid quantum harmonic oscillator and its application in rotor system optimization. Sci Rep. 2025;15(1). https://doi.org/10.1038/s41598-025-92070-0
127. Limanaqi F, Zecchini S, Dino B, Strizzi S, Cappelletti G, Utyro O, Vanetti C, Garziano M, Saulle I, Clerici M, Biasin M. Dopamine reduces SARS-CoV-2 replication in vitro through downregulation of D2 receptors and upregulation of type-I interferons. Cells. 2022;11(10):1691–1691. https://doi.org/10.3390/cells11101691.

128. Liu C-S, Yan B-C, Sun H-D, Lu J-C, Puno P-T. Bridging chemical space and biological efficacy: advances and challenges in applying generative models in structural modification of natural products. Nat Prod Bioprosp. 2025;15(1). https://doi.org/10.1007/s13659-025-00521-y
129. Liu M, Quinn RJ. Fragment-based screening with natural products for novel anti-parasitic disease drug discovery. Expert Opin Drug Discov. 2019;1–13:1653849. https://doi.org/10.1080/17460441.2019.1653849.
130. Liu X, Li M, Liu X, Luo Y, Yang D, Ouyang H, He J, Xia J, Xiao F. Clinical validation and optimization of machine learning models for early prediction of sepsis. Front Med. 2025;12:1521660. https://doi.org/10.3389/fmed.2025.1521660.
131. Liu X, Ye K, van Vlijmen HWT, IJzerman AP, van Westen GJP. An exploration strategy improves the diversity of de novo ligands using deep reinforcement learning: a case for the adenosine A2A receptor. J Cheminf. 2019;11(1). https://doi.org/10.1186/s13321-019-0355-6
132. Liu Y, Mathis C, Bajczyk MD, Marshall SM, Wilbraham L, Cronin L. Exploring and mapping chemical space with molecular assembly trees. Sci Adv. 2021;7(39). https://doi.org/10.1126/sciadv.abj2465
133. Llanos EJ, Leal W, Luu DH, Jost J, Stadler PF, Restrepo G. Exploration of the chemical space and its three historical regimes. Proc Natl Acad Sci. 2019;116(26):12660–5. https://doi.org/10.1073/pnas.1816039116.
134. Lo Y-C, Rensi SE, Torng W, Altman RB. Machine learning in chemoinformatics and drug discovery. Drug Discovery Today. 2018;23(8):1538–46. https://doi.org/10.1016/j.drudis.2018.05.010.
135. Loeffler HH, He J, Tibo A, Janet JP, Voronov A, Mervin LH, Engkvist O. Reinvent 4: Modern AI-driven generative molecule design. J Cheminf. 2024;16(1). https://doi.org/10.1186/s13321-024-00812-5
136. Loeffler HH, Wan S, Klahn M, Bhati AP, Coveney PV. Optimal molecular design: generative active learning combining REINVENT with precise binding free energy ranking simulations. J Chem Theory Comput. 2024. https://doi.org/10.1021/acs.jctc.4c00576.
137. López-Pérez K, López-López E, Medina-Franco JL, Miranda-Quintana RA. Sampling and mapping chemical space with extended similarity indices. Molecules. 2023;28(17):6333. https://doi.org/10.3390/molecules28176333.
138. Lu C, Liu S, Shi W, Yu J, Zhou Z, Zhang X, Lu X, Cai F, Xia N, Wang Y. Systemic evolutionary chemical space exploration for drug discovery. J Cheminf. 2022;14(1). https://doi.org/10.1186/s13321-022-00598-4
139. Lusa L, Kappenberg F, Collins GS, Schmid M, Sauerbrei W, Rahnenführer J. Evaluation of changes in prediction modelling in biomedicine using systematic reviews. BMC Med Res Methodol. 2025;25(1). https://doi.org/10.1186/s12874-025-02605-2
140. Luttens A, Cabeza de Vaca I, Sparring L, Brea J, Martínez AL, Kahlous NA, Radchenko DS, Moroz YS, Loza MI, Norinder U, Carlsson J. Rapid traversal of vast chemical space using machine learning-guided docking screens. Nat Comput Sci. 2025;5(4):301–312. https://doi.org/10.1038/s43588-025-00777-x
141. Manan A, Baek E, Ilyas S, Lee D. Digital Alchemy: the rise of machine and deep learning in small-molecule drug discovery. Int J Mol Sci. 2025;26(14):6807–6807. https://doi.org/10.3390/ijms26146807.
142. Mapari S, Camarda KV. Use of three-dimensional descriptors in molecular design for biologically active compounds. Curr Opin Chem Eng. 2020;27:60–4. https://doi.org/10.1016/j.coche.2019.11.011.
143. Mariam Z, Niazi SK, Magoola M. Unlocking the future of drug development: generative AI, digital twins, and beyond. BioMedInformatics. 2024;4(2):1441–56. https://doi.org/10.3390/biomedinformatics4020079.
144. Martorell-Marugán J, López-Domínguez R, García-Moreno A, Toro-Domínguez D, Villatoro-García JA, Barturen G, Martín-Gómez A, Troule K, Gómez-López G, Al-Shahrour F, González-Rumayor V, Peña-Chilet M, Dopazo J, Sáez-Rodríguez J, Alarcón-Riquelme ME, Carmona-Sáez P. A comprehensive database for integrated analysis of omics data in autoimmune diseases. BMC Bioinf. 2021;22(1). https://doi.org/10.1186/s12859-021-04268-4

145. Matsuzaka Y, Uesawa Y. A deep learning-based quantitative structure-activity relationship system construct prediction model of agonist and antagonist with high performance. Int J Mol Sci. 2022;23(4):2141. https://doi.org/10.3390/ijms23042141.
146. Maurer TS, Edwards M, Hepworth D, Verhoest P, Allerton CMN. Designing small molecules for therapeutic success: a contemporary perspective. Drug Discov Today. 2022;27(2):538–46. https://doi.org/10.1016/j.drudis.2021.09.017.
147. Mauri A, Bertola M. AlvaBuilder: a software for de novo molecular design. J Chem Inf Model. 2023;64(7):2136–42. https://doi.org/10.1021/acs.jcim.3c00610.
148. McGrady M, Colby S, Nuñez JR, Renslow R, Metz TO. AI for chemical space gap filling and novel compound generation. 2022. *ArXiv (Cornell University).* https://doi.org/10.48550/arxiv.2201.12398
149. Medina-Franco JL, Chávez-Hernández AL, López-López E, Saldívar-González FI. Chemical multiverse: an expanded view of chemical space. Mol Inf. 2022;41:2200116. https://doi.org/10.1002/minf.202200116.
150. Miebs G, Mielniczuk A, Kadziński M, Bachorz RA. Beyond the arbitrariness of drug-likeness rules: rough set theory and decision rules in the service of drug design. Appl Sci. 2024;14(21):9966. https://doi.org/10.3390/app14219966.
151. Miljković F, Medina-Franco JL. Artificial intelligence-open science symbiosis in chemoinformatics. Artif Intell Life Sci. 2024;5:100096. https://doi.org/10.1016/j.ailsci.2024.100096.
152. Mohan A, Rendine N, Mohammed MKS, Jeeva A, Ji H-F, Talluri VR. Structure-based virtual screening, in silico docking, ADME properties prediction and molecular dynamics studies for the identification of potential inhibitors against SARS-CoV-2 Mpro. Mol Diversity. 2021. https://doi.org/10.1007/s11030-021-10298-0.
153. *Molecular Descriptor—an overview | ScienceDirect Topics.* (n.d.). https://www.sciencedirect.com/topics/medicine-and-dentistry/molecular-descriptor
154. Monteiro PQ, Schaeffer E, da Silva AJM, Alves CR, Souza-Silva F. A virtual screening approach to evaluate the multitarget potential of a chalcone library with binding properties to oligopeptidase B and cysteine proteinase B from *Leishmania (Viannia) braziliensis.* Int J Mol Sci. 2025;26(5):2025. https://doi.org/10.3390/ijms26052025.
155. Moingeon P. Artificial intelligence-driven drug development against autoimmune diseases. Trends Pharmacol Sci. 2023;44(7):411–24. https://doi.org/10.1016/j.tips.2023.04.005.
156. Moreira-Filho JT, Ranganath D, Conway M, Schmitt C, Kleinstreuer N, Mansouri K. Democratizing cheminformatics: interpretable chemical grouping using an automated KNIME workflow. J Cheminf. 2024;16(1). https://doi.org/10.1186/s13321-024-00894-1
157. Morozov V, Rodrigues CHM, Ascher DB. CSM-toxin: a web-server for predicting protein toxicity. Pharmaceutics. 2023;15(2):431. https://doi.org/10.3390/pharmaceutics15020431.
158. Mou M, Zhang Y, Qian Y, Zhou Z, Liao Y, Niu T, Hu W, Chen Y, Jiang R, Zhao H, Dai H, Zhang Y, Fu T. druglikeFilter 1.0: an AI powered filter for collectively measuring the drug-likeness of compounds. J Pharm Anal. 2025;15(6):101298. https://doi.org/10.1016/j.jpha.2025.101298
159. Mouchlis VD, Afantitis A, Serra A, Fratello M, Papadiamantis AG, Aidinis V, Lynch I, Greco D, Melagraki G. Advances in de novo drug design: from conventional to machine learning methods. Int J Mol Sci. 2021;22(4). https://doi.org/10.3390/ijms22041676
160. Niazi SK, Mariam Z. Recent advances in machine-learning-based chemoinformatics: a comprehensive review. Int J Mol Sci. 2023;24(14):11488–11488. https://doi.org/10.3390/ijms241411488.
161. Nafiz Abeer ANM, Urban NM, Ryan Weil M, Alexander FJ, Yoon B-J. Multi-objective latent space optimization of generative molecular design models. Patterns. 2024;101042–101042. https://doi.org/10.1016/j.patter.2024.101042
162. Nag S, Baidya ATK, Mandal A, Mathew AT, Das B, Devi B, Kumar R. Deep learning tools for advancing drug discovery and development. 3 Biotech. 2022;12(5). https://doi.org/10.1007/s13205-022-03165-8
163. Nigam A, Pollice R, Krenn M, Gabriel, Aspuru-Guzik A. Beyond generative models: superfast traversal, optimization, novelty, exploration and discovery (STONED) algorithm for molecules using SELFIES. 2021. https://doi.org/10.26434/chemrxiv.13383266.v2

164. Nivatya HK, Singh A, Kumar N, Sonam N, Sharma L, Singh V, Mishra R, Gaur N, Mishra AK. Assessing molecular docking tools: understanding drug discovery and design. Future J Pharm Sci. 2025;11(1). https://doi.org/10.1186/s43094-025-00862-y
165. Niu Z, Xiao X, Wu W, Cai Q, Jiang Y, Jin W, Wang M, Yang G, Kong L, Jin X, Yang G, Chen H. PharmaBench: enhancing ADMET benchmarks with large language models. Sci Data. 2024;11(1). https://doi.org/10.1038/s41597-024-03793-0
166. Orlov AA. Akhmetshin TN, Horvath D, Marcou G, Varnek, A. From high dimensions to human insight: exploring dimensionality reduction for chemical space visualization. Mol Inf. 2024;44(1). https://doi.org/10.1002/minf.202400265
167. Orosz Á, Héberger K, Rácz A. Comparison of descriptor- and fingerprint sets in machine learning models for ADME-Tox targets. Front Chem. 2022;10:852893. https://doi.org/10.3389/fchem.2022.852893.
168. Özçelik R, Brinkmann H, Criscuolo E, Grisoni F. Generative deep learning for de novo drug design—a chemical space odyssey. J Chem Inf Model. 2025;65(14):7352–72. https://doi.org/10.1021/acs.jcim.5c00641.
169. Pagadala NS, Syed K, Tuszynski J. Software for molecular docking: a review. Biophys Rev. 2017;9(2):91–102. https://doi.org/10.1007/s12551-016-0247-1.
170. Park H, Onwuli A, Walsh A. Exploration of crystal chemical space using text-guided generative artificial intelligence. Nat Commun. 2025;16(1):4379. https://doi.org/10.1038/s41467-025-59636-y.
171. Pathan I, Raza A, Sahu A, Joshi M, Sahu Y, Patil Y, Raza MA. Revolutionizing pharmacology: AI-powered approaches in molecular modeling and ADMET prediction. Med Drug Discov. 2025;28:100223. https://doi.org/10.1016/j.medidd.2025.100223.
172. Pernaa J, Takala A, Ciftci V, Hernández-Ramos J, Cáceres-Jensen L, Rodríguez-Becerra J. Open-source software development in cheminformatics: a qualitative analysis of rationales. Appl Sci. 2023;13(17):9516–9516. https://doi.org/10.3390/app13179516.
173. Patil S, Gs V, Sarode GS, Sarode SC, Khurayzi TA, Beshir M, Gadbail AR, Gondivkar S. Exploring the role of immunotherapeutic drugs in autoimmune diseases: a comprehensive review. J Oral Biol Craniof Res. 2021;11(2):291–6. https://doi.org/10.1016/j.jobcr.2021.02.009.
174. Pavlović N, Milošević Sopta N, Mitrović D, Zaklan D, Tomas Petrović A, Stilinović N, Vukmirović S. Principal component analysis (PCA) of molecular descriptors for improving permeation through the blood-brain barrier of quercetin analogues. Int J Mol Sci. 2023;25(1):192. https://doi.org/10.3390/ijms25010192.
175. Pikalyova R, Akhmetshin T, Horvath D, Varnek A. CoLiNN: a tool for fast chemical space visualization of combinatorial libraries without enumeration. Mol Inf. 2025;44(3). https://doi.org/10.1002/minf.202400263
176. Pikalyova R, Zabolotna Y, Volochnyuk DM, Horvath D, Marcou G, Varnek A. Exploration of the chemical space of DNA-encoded libraries. Mol Inf. 2022;41:2100289. https://doi.org/10.1002/minf.202100289.
177. Pillaiyar T, Meenakshisundaram S, Manickam M, Sankaranarayanan M. A medicinal chemistry perspective of drug repositioning: recent advances and challenges in drug discovery. Eur J Med Chem. 2020;195:112275. https://doi.org/10.1016/j.ejmech.2020.112275.
178. Piriou M, Lorenzo C, Raymond-Letron I, Coronas-Dupuis S, Pieruccioni L, Rouquette J, Guissard C, Chaumont J, Casteilla L, Planat-Benard V, Kemoun P, Monsarrat P. A spectral principal component analysis-based framework for composite hard/soft tissue fluorescence image investigation. Front Physiol. 2022;13:899626. https://doi.org/10.3389/fphys.2022.899626.
179. Prasad K, Erik, Elsinga PH, Dierckx R, van Waarde A. Allosteric interactions between adenosine A2A and dopamine D2 receptors in heteromeric complexes: biochemical and pharmacological characteristics, and opportunities for PET imaging. Int J Mol Sci. 2021;22(4):1719–1719. https://doi.org/10.3390/ijms22041719
180. Qadri YA, Shaikh S, Ahmad K, Choi I, Kim SW, Vasilakos AV. Explainable artificial intelligence: a perspective on drug discovery. Pharmaceutics. 2025;17(9):1119–1119. https://doi.org/10.3390/pharmaceutics17091119.

181. Rahimi A, Khibech O, Benabbou A, Merzouki M, Bouhrim M, Al-Zharani M, Nasr FA, Ahmed Qurtam A, Abadi S, Challioui A, Mimouni M, Elbekay M. ADMET-guided docking and GROMACS molecular dynamics of ziziphus lotus phytochemicals uncover mutation-agnostic allosteric stabilisers of the KRAS switch-I/II groove. Pharmaceuticals. 2025;18(8):1110. https://doi.org/10.3390/ph18081110.
182. Ramezani H, Dietz H. Building machines with DNA molecules. Nat Rev Genet. 2019. https://doi.org/10.1038/s41576-019-0175-6.
183. Ramírez-Cid V, Chávez-Hernández AL, Sánchez López O, Novais RM, Alegbejo Price TO, Moraes Alves K, Lima Silva WJ, da Silva Emery F, Andrade CH, Medina-Franco JL. Fragment libraries from large and novel synthetic compounds and natural products: a comparative chemoinformatic analysis. ACS Omega. 2025;10(16):16921–16937. https://doi.org/10.1021/acsomega.5c01420
184. Rao SJA, Shetty NP. Structure-based screening of natural product libraries in search of potential antiviral drug-leads as first-line treatment to COVID-19 infection. Microb Pathog. 2022;165:105497. https://doi.org/10.1016/j.micpath.2022.105497.
185. Reymond J-L. Chemical space as a unifying theme for chemistry. J Cheminf. 2025;17(1). https://doi.org/10.1186/s13321-025-00954-0
186. Syahdi RR, Jasial S, Maeda I, Miyao T. Bridging structure- and ligand-based virtual screening through fragmented interaction fingerprint. ACS Omega. 2024;9(37):38957–69. https://doi.org/10.1021/acsomega.4c05433.
187. Roman DC, Roman M, Som C, Schmutz M, Hernandez E, Wick P, Casalini T, Perale G, Ostafe V, Isvoran A. Computational assessment of the pharmacological profiles of degradation products of chitosan. Front Bioeng Biotechnol. 2019;7:214. https://doi.org/10.3389/fbioe.2019.00214.
188. Romanelli V, Cerchia C, Lavecchia A. Unlocking the potential of generative artificial intelligence in drug discovery. Springer EBooks; 2024. pp. 37–63. https://doi.org/10.1007/978-3-031-46238-2_3
189. Saldívar-González FI, Aldas-Bulos VD, Medina-Franco JL, Plisson F. Natural product drug discovery in the artificial intelligence era. Chem Sci. 2022;13(6):1526–46. https://doi.org/10.1039/D1SC04471K.
190. Saldívar-González FI, Medina-Franco JL. Approaches for enhancing the analysis of chemical space for drug discovery. Expert Opin Drug Discov. 2022;1–10. https://doi.org/10.1080/17460441.2022.2084608
191. Setiya A, Jani V, Sonavane U, Joshi R. MolToxPred: small molecule toxicity prediction using machine learning approach. RSC Adv. 2024;14(6):4201–20. https://doi.org/10.1039/d3ra07322j.
192. Samha MH, Karas LJ, Vogt DB, Odogwu EC, Elward J, Crawford JM, Steves JE, Sigman MS. Predicting success in Cu-catalyzed C–N coupling reactions using data science. Sci Adv. 2024;10(3). https://doi.org/10.1126/sciadv.adn3478
193. Scotti L, Scotti MT. ADME properties in drug delivery. Pharmaceutics. 2025;17(5):617. https://doi.org/10.3390/pharmaceutics17050617
194. Seal S, Mahale M, García-Ortegón M, Joshi CK, Hosseini-Gerami L, Beatson A, Greenig M, Shekhar M, Patra A, Weis C, Mehrjou A, Badré A, Paisley B, Lowe R, Singh S, Shah F, Johannesson B, Williams D, Rouquie D, Clevert D-A. Machine learning for toxicity prediction using chemical structures: pillars for success in the real world. Chem Res Toxicol. 2025. https://doi.org/10.1021/acs.chemrestox.5c00033.
195. Serrano-Morrás A, Bertran-Mostazo A, Miñarro-Lleonar M, Comajuncosa-Creus A, Cabello A, Labranya C, Escudero C, Tian TV, Khutorianska I, Radchenko DS, Moroz YS, Defelipe L, Ruiz-Carrillo D, Garcia-Alai M, Schmidt R, Rarey M, Aloy P, Galdeano C, Juárez-Jiménez J, Barril X. A bottom-up approach to find lead compounds in expansive chemical spaces. Commun Chem. 2025;8(1). https://doi.org/10.1038/s42004-025-01610-2
196. Sengupta A, Singh SK, Kumar R. Support vector machine-based prediction models for drug repurposing and designing novel drugs for colorectal cancer. ACS Omega. 2024;9(16):18584–92. https://doi.org/10.1021/acsomega.4c01195.

197. Sharma N, Ojha H, Raghav PK, Goyal RK. Chemoinformatics and bioinformatics in the pharmaceutical sciences. London: United Kingdom Academic Press, Elsevier; 2021.
198. Sosnin S. MolCompass: multi-tool for the navigation in chemical space and visual validation of QSAR/QSPR models. J Cheminf. 2024;16(1). https://doi.org/10.1186/s13321-024-00888-z
199. Serra A, Önlü S, Festa P, Fortino V, Greco D. MaNGA: a novel multi-niche multi-objective genetic algorithm for QSAR modelling. Bioinformatics. 2019;36(1):145–53. https://doi.org/10.1093/bioinformatics/btz521.
200. SeyedAlinaghi S, Mehraeen E, Mirzapour P, Yarmohammadi S, Dehghani S, Zare S, Gholami S, Attarian N, Abiri A, Farahani Rad F, Tabari A, Afroughi F, Gholipour A, Roozbahani MM, Jahanfar S. A systematic review on natural products with antimicrobial potential against WHO's priority pathogens. Eur J Med Res. 2025;30(1):525. https://doi.org/10.1186/s40001-025-02717-x.
201. Sharma B, Chenthamarakshan V, Dhurandhar A, Pereira S, Hendler JA, Dordick JS, Das P. Accurate clinical toxicity prediction using multi-task deep neural nets and contrastive molecular explanations. Sci Rep. 2023;13(1):4908. https://doi.org/10.1038/s41598-023-31169-8.
202. Sharma S, Arya A, Cruz R, Cleaves H II. Automated exploration of prebiotic chemical reaction space: progress and perspectives. Life. 2021;11(11):1140. https://doi.org/10.3390/life11111140.
203. Sharma S, Kumar V. A comprehensive review on multi-objective optimization techniques: past, present and future. Arch Comput Methods Eng. 2022. https://doi.org/10.1007/s11831-022-09778-9
204. Silva AS, Martins V, Soares EP, Francisco R, Guzzo CR. Antagonistic trends between binding affinity and drug-likeness in SARS-CoV-2 Mpro inhibitors revealed by machine learning. Viruses. 2025;17(7):935–935. https://doi.org/10.3390/v17070935
205. Singh H, McCarthy N, Ain QU, Hayes J. *ChemoVerse: manifold traversal of latent spaces for novel molecule discovery*. 2020. https://doi.org/10.48550/arXiv.2009.13946
206. Singh P, Singh VK, Gond C, Singh D, Tiwari AK. Current advances in the structure-activity relationship (SAR) analysis of the old/new 18-kDa translocator protein ligands. Mol Diversity. 2025;29(3):2639–89. https://doi.org/10.1007/s11030-024-10963-0.
207. Skinnider MA, Greg Stacey R, Wishart DS, Foster LJ. Chemical language models enable navigation in sparsely populated chemical space. Nat Mach Intell. 2021;3(9):759–70. https://doi.org/10.1038/s42256-021-00368-1.
208. Smith LB, Hadar K-R. Episodes of experience and generative intelligence. Trends Cogn Sci. 2022;26(12):1064–5. https://doi.org/10.1016/j.tics.2022.09.012.
209. Sorokina M, Merseburger P, Rajan K, Yirik MA, Steinbeck C. COCONUT online: collection of open natural products database. J Cheminf. 2021;13(1). https://doi.org/10.1186/s13321-020-00478-9
210. Sosnin S. Chemical space visual navigation in the era of deep learning and Big Data. Drug Discov Today. 2025;30(7):104392. https://doi.org/10.1016/j.drudis.2025.104392.
211. Spanakis M, Tzamali E, Tzedakis G, Koumpouzi C, Pediaditis M, Tsatsakis A, Sakkalis V. Artificial intelligence models and tools for the assessment of drug-herb interactions. Pharmaceuticals. 2025;18(3):282–282. https://doi.org/10.3390/ph18030282.
212. Sun J, Wen M, Wang H, Ruan Y, Yang Q, Kang X, Zhang Y, Zhang Z, Lü H. Prediction of drug-likeness using graph convolutional attention network. Bioinformatics. 2022;38(23):5262–9. https://doi.org/10.1093/bioinformatics/btac676.
213. Sundin I, Voronov A, Xiao H, Papadopoulos K, Bjerrum EJ, Heinonen M, Patronov A, Kaski S, Engkvist O. Human-in-the-loop assisted de novo molecular design. J Cheminf. 2022;14(1). https://doi.org/10.1186/s13321-022-00667-8
214. Swanson K, Walther P, Leitz J, Mukherjee S, Wu JC, Shivnaraine RV, Zou J. ADMET-AI: a machine learning ADMET platform for evaluation of large-scale chemical libraries. Bioinformatics. 2024;40(7). https://doi.org/10.1093/bioinformatics/btae416
215. Talevi A, Alberca LN, Bellera CL. Tackling the issue of confined chemical space with AI-based de novo drug design and molecular optimization. Expert Opin Drug Discov. 2025;1–14. https://doi.org/10.1080/17460441.2025.2555275

216. Tan RK, Liu Y, Xie L. Reinforcement learning for systems pharmacology-oriented and personalized drug design. Expert Opin Drug Discov. 2022;17(8):849–63. https://doi.org/10.1080/17460441.2022.2072288.
217. Teng Y, Jian Y, Chen X, Li Y, Han B, Wang L. Comparison of three prediction models for predicting chronic obstructive pulmonary disease in China. Int J Chron Obstruct Pulmon Dis. 2023;18:2961–9. https://doi.org/10.2147/copd.s431115.
218. Tetko IV, Clevert D-A. Advanced machine learning for innovative drug discovery. J Cheminf. 2025;17(1). https://doi.org/10.1186/s13321-025-01061-w
219. Thaingtamtanha T, Ravichandran R, Gentile F. On the application of artificial intelligence in virtual screening. Expert Opin Drug Discov. 2025;1–13. https://doi.org/10.1080/17460441.2025.2508866
220. Tian H, Jiang X, Xiao S, Force HL, Larson EC, Tao P. LAST: latent space-assisted adaptive sampling for protein trajectories. J Chem Inf Model. 2022;63(1):67–75. https://doi.org/10.1021/acs.jcim.2c01213.
221. Trachtenberg A, Akabayov B. From patterns to pills: how informatics is shaping medicinal chemistry. Pharmaceutics. 2025;17(5):612–612. https://doi.org/10.3390/pharmaceutics17050612.
222. Tripathi S, Augustin AI, Dunlop A, Sukumaran R, Dheer S, Zavalny A, Haslam O, Austin T, Donchez J, Tripathi PK, Kim E. Recent advances and application of generative adversarial networks in drug discovery, development, and targeting. Artif Intell Life Sci. 2022;2:100045. https://doi.org/10.1016/j.ailsci.2022.100045.
223. Trozzi F, Wang X, Tao P. UMAP as a dimensionality reduction tool for molecular dynamics simulations of biomacromolecules: a comparison study. J Phys Chem B. 2021;125(19):5022–34. https://doi.org/10.1021/acs.jpcb.1c02081.
224. Ugasini Preetha P, Suresh M, Tolasa FT, Bonyah E. QSPR/QSAR study of antiviral drugs modeled as multigraphs by using TI's and MLR method to treat COVID-19 disease. Sci Rep. 2024;14(1). https://doi.org/10.1038/s41598-024-63007-w
225. Ullah F, Xiao A, Ullah S, Yang N, Lei M, Chen L, Wang S. Synergizing attribute-guided latent space exploration (AGLSE) with classical molecular simulations to design potent pep-magnet peptide inhibitors to abrogate SARS-CoV-2 host cell entry. Viruses. 2025;17(6):828. https://doi.org/10.3390/v17060828.
226. van Patel S, Gerard, Jespers W, Sherman W. In search of beautiful molecules: a perspective on generative modeling for drug design. J Chem Inf Model. 2025. https://doi.org/10.1021/acs.jcim.5c01203
227. Vasilev B, Atanasova M. A (comprehensive) review of the application of quantitative structure-activity relationships (QSARs) in the prediction of new compounds with anti-breast cancer activity. Appl Sci. 2025;15(3):1206–1206. https://doi.org/10.3390/app15031206.
228. Vázquez J, López M, Gibert E, Herrero E, Luque FJ. Merging ligand-based and structure-based methods in drug discovery: an overview of combined virtual screening approaches. Molecules. 2020;25(20):4723. https://doi.org/10.3390/molecules25204723.
229. Venkataraman M, Chand Rao G, Madavareddi JK, Maddi SR. Leveraging machine learning models in evaluating ADMET properties for drug discovery and development. ADMET DMPK. 2025;13(3):2772. https://doi.org/10.5599/admet.2772.
230. Vogt M. How do we optimize chemical space navigation? Expert Opin Drug Discov. 2020;1–3. https://doi.org/10.1080/17460441.2020.1730324
231. Vogt M. Exploring chemical space—generative models and their evaluation. Artif Intell Life Sci. 2023;100064. https://doi.org/10.1016/j.ailsci.2023.100064
232. Volkamer A, Riniker S, Nittinger E, Lanini J, Grisoni F, Evertsson E, Rodríguez-Pérez R, Schneider N. Machine learning for small molecule drug discovery in academia and industry. Artif Intell Life Sci. 2023;3:100056. https://doi.org/10.1016/j.ailsci.2022.100056.
233. Weaver DT, King ES, Maltas J, Scott JG. Reinforcement learning informs optimal treatment strategies to limit antibiotic resistance. Proc Natl Acad Sci USA. 2024;121(16). https://doi.org/10.1073/pnas.2303165121

234. Whitehouse C, Corbett N, Brownlees J. 3D models of neurodegeneration: implementation in drug discovery. Trends Pharmacol Sci. 2023. https://doi.org/10.1016/j.tips.2023.01.005.
235. Whittingham H, Ashenden SK. Hit discovery. Elsevier EBooks; 2021. pp. 81–102. https://doi.org/10.1016/b978-0-12-820045-2.00006-4
236. Wilson BAP, Thornburg CC, Henrich CJ, Grkovic T, O'Keefe BR. Creating and screening natural product libraries. Nat Prod Rep. 2020;37(7):893–918. https://doi.org/10.1039/c9np00068b.
237. Winter R, Montanari F, Noé F, Clevert D-A. Learning continuous and data-driven molecular descriptors by translating equivalent chemical representations. Chem Sci. 2019;10(6):1692–701. https://doi.org/10.1039/C8SC04175J.
238. Woon KL, Chong ZX, Ariffin A, Chan CS. Relating molecular descriptors to frontier orbital energy levels, singlet and triplet excited states of fused tricyclics using machine learning. J Mol Graph Model. 2021;105:107891. https://doi.org/10.1016/j.jmgm.2021.107891.
239. Wu F, Zhou Y, Li L, Shen X, Chen G, Wang X, Liang X, Tan M, Huang Z. Computational approaches in preclinical studies on drug discovery and development. Front Chem. 2020;8:726. https://doi.org/10.3389/fchem.2020.00726.
240. Xiang Y-T, Huang G-Y, Shi X-X, Hao G-F, Yang G-F. 3D molecular generation models expand chemical space exploration in drug design. Drug Discov Today. 2024;104282–104282. https://doi.org/10.1016/j.drudis.2024.104282
241. Xie E, Wang X, Siepmann JI, Chen H, Snurr RQ. Generative AI for design of nanoporous materials: review and future prospects. Digital Discov. 2025. https://doi.org/10.1039/d5dd00221d.
242. Xie S, Zhan F, Zhu J, Xu S, Xu J. The latest advances with natural products in drug discovery and opportunities for the future: a 2025 update. Expert Opin Drug Discov. 2025;20(7):827–43. https://doi.org/10.1080/17460441.2025.2507382.
243. Xu Z, Wauchope OR, Frank AT. Navigating chemical space by interfacing generative artificial intelligence and molecular docking. 2021;61(11):5589–5600. https://doi.org/10.1021/acs.jcim.1c00746
244. Yang J, Cai Y, Zhao K, Xie H, Chen X. Concepts and applications of chemical fingerprint for hit and lead screening. Drug Discov Today. 2022;27(11):103356. https://doi.org/10.1016/j.drudis.2022.103356.
245. Zdrazil B. Fifteen years of ChEMBL and its role in cheminformatics and drug discovery. J Cheminf. 2025;17(1). https://doi.org/10.1186/s13321-025-00963-z
246. Zeng X, Wang F, Luo Y, Kang S, Tang J, Lightstone FC, Fang EF, Cornell W, Nussinov R, Cheng F. Deep generative molecular design reshapes drug discovery. Cell Reports Med. 2022;3(12):100794. https://doi.org/10.1016/j.xcrm.2022.100794.
247. Zhang J, Li H, Zhang Y, Huang J, Ren L, Zhang C, Zou Q, Zhang Y. Computational toxicology in drug discovery: applications of artificial intelligence in ADMET and toxicity prediction. Brief Bioinf. 2025;26(5). https://doi.org/10.1093/bib/bbaf533
248. Zhao H, Yang Y, Wang S, Yang X, Zhou K, Xu C, Zhang X, Fan J, Hou D, Li X, Lin H, Tan Y, Wang S, Chu X-Y, Zhuoma D, Zhang F, Ju D, Zeng X, Chen YZ. NPASS database update 2023: quantitative natural product activity and species source database for biomedical research. Nucleic Acids Res. 2022. https://doi.org/10.1093/nar/gkac1069.
249. Zhou G, Rusnac D-V, Park H, Canzani D, Nguyen HM, Stewart L, Bush MF, Nguyen PT, Wulff H, Yarov-Yarovoy V, Zheng N, DiMaio F. An artificial intelligence accelerated virtual screening platform for drug discovery. Nat Commun. 2024;15(1). https://doi.org/10.1038/s41467-024-52061-7
250. Zhou H, Cao H, Skolnick J. FRAGSITE: a fragment-based approach for virtual ligand screening. J Chem Inf Model. 2021;61(4):2074–89. https://doi.org/10.1021/acs.jcim.0c01160.
251. Zhou J, Mroz A, Jelfs KE. Deep generative design of porous organic cages *via* a variational autoencoder. Digital Discov. 2023;2(6):1925–36. https://doi.org/10.1039/d3dd00154g.

252. Zhou Z, Li Y, Hong P, Xu H. Multimodal fusion with relational learning for molecular property prediction. Commun Chem. 2025;*8*(1). https://doi.org/10.1038/s42004-025-01586-z
253. Zhu W, Wang Y, Niu Y, Zhang L, Liu Z. Current trends and challenges in drug-likeness prediction: are they generalizable and interpretable? Health Data Sci. 2023;3. https://doi.org/10.34133/hds.0098

Chapter 5
Prediction of Drug Likeness and Synthetic Accessibility Using AI: A Case Study on Potential Therapeutic Compounds from Plants for SARS-CoV-2 Treatment

Dibyendu Seth, Sandip Debnath, and Sourish Pramanik

Abstract As scientists learn more about SARS-CoV-2, they are exploring plant-based compounds as possible treatments. These compounds may block the proteins the virus needs to multiply and enter cells. Traditional remedies like green and black tea, *Terminalia chebula* (haritaki), and various spices have shown potential in stopping the virus and supporting the immune system. Computer modeling has identified plant chemicals such as taraxerol, friedelin, stigmasterol (from Clerodendrum species), and EBDGp (from *Phyllanthus emblica* and *Aegle marmelos*) as strong binders to important SARS-CoV-2 proteins. In this study, we use artificial intelligence to help find and improve these plant-based candidates. AI tools can help design these compounds, check if they could work as drugs, and see how easy they are to make. This method helps move promising compounds from computer models into real-world drug development.

Keywords AI-driven drug discovery · Phytochemicals · SARS-CoV-2 · Synthetic accessibility · Plant-derived antivirals · And drug-likeness prediction

D. Seth (✉)
Department of Genetics and Plant Breeding, Chaudhary Charan Singh Haryana Agricultural University (CCSHAU), Hisar, India
e-mail: deep032002@gmail.com

S. Debnath
Department of Genetics and Plant Breeding, Institute of Agriculture, Visva-Bharati University, Sriniketan, West Bengal, India
e-mail: sandip.debnath@visva-bharati.ac.in

S. Pramanik
Department of Plant Pathology, Punjab Agricultural University, Ludhiana, India

A. A. Parikesit and A. N. M. Ansori (eds.), *Recent Computational Techniques in De Novo Drug Design*, Bio-IT and AI, https://doi.org/10.1007/978-3-032-20428-8_5

5.1 Introduction

The outbreak of severe acute respiratory syndrome coronavirus 2 (SARS-CoV-2), a member of the Coronaviridae family within the order Nidovirales, has emerged as one of the most disruptive global health crises in modern times. With a genome size of approximately 27–32 kilobases, it is the largest known positive-sense single-stranded RNA virus and known to cause the COVID-19 pandemic declared by the World Health Organization (WHO) on March 11, 2020 [14]. Originating in Wuhan, China, in December 2019, the virus rapidly spread worldwide, transmitted initially from bats through an intermediate host to humans, and now persists in the global population through both symptomatic and asymptomatic carriers [20]. The SARS-CoV-2 virus utilizes the angiotensin-converting enzyme 2 (ACE2) receptor for cellular entry facilitated by the spike (S) glycoprotein and primed by transmembrane protease serine 2 (TMPRSS2) [18]. Given the widespread expression of ACE2 in pulmonary, cardiac, and renal tissues, the infection leads to systemic inflammation and complications such as myocarditis, respiratory distress, and neurological manifestations [13]. The virus's ability to infect via multiple structural and non-structural proteins, including spike (S), membrane (M), envelope (E), nucleocapsid (N), 3CLpro (Mpro), PLpro, helicase, and RdRp, thereby makes it a formidable target for antiviral research.

Many synthetic antiviral drugs are available, but their limited effectiveness, side effects, and the risk of resistance have led researchers to reconsider natural products as treatments. Plant-based compounds and traditional remedies have long been used against viral infections. Recent computer studies show that some compounds from Clerodendrum species and EBDGp from *Phyllanthus emblica* and *Aegle marmelos* can bind well to important SARS-CoV-2 proteins. However, many plant compounds are not widely used because they are hard to produce or may not have the right properties to become drugs. Artificial intelligence can help address these challenges. Modern AI models can predict how to make these compounds, check their complexity, and see if they are practical to manufacture, making it easier to turn promising plant molecules into real medicines. AI models trained on reaction databases like USPTO and ChEMBL can also suggest ways to synthesize complex plant-derived compounds. This approach helps create lead compounds that are easier to produce and more likely to be effective, speeding up the development of new antivirals from traditional medicine. In summary, combining natural product research with AI tools can help find effective antiviral treatments for SARS-CoV-2 and other new viruses more quickly.

5.2 SARS-CoV-2 Structure and Drug-Targeting Relevance in AI-Assisted Molecular Design

SARS-CoV-2 is a positive-sense single-stranded RNA virus with a genome size between 26 and 32 kilobases, making it the largest known viral RNA genome [27]. Understanding its structure and function is important for AI-based drug design, especially for tasks like structure-informed retrosynthesis and bioactivity prediction. The virus is part of the beta-coronavirus group and is closely related to bat coronaviruses, and to a lesser extent, to SARS-CoV and MERS-CoV [31]. Structurally, the virus has four main proteins: spike (S), envelope (E), membrane (M), and nucleocapsid (N), along with accessory and non-structural proteins such as 3CLpro (main protease), PLpro (papain-like protease), and RNA-dependent RNA polymerase (RdRp) (Fig. 5.1) [2]. These components are the main targets in AI-driven drug discovery.

Computational methods now use detailed information about how viral proteins interact with drugs to train AI models. For example, the spike protein's S1 subunit binds to the ACE2 receptor, allowing the virus to enter human cells [18, 41]. AI systems use this knowledge to predict which compounds might block this step or to design new molecules that could help. The virus also relies on host enzymes like TMPRSS2 and cathepsin L to enter cells, and AI tools can suggest improved versions of known inhibitors for these enzymes (Fig. 5.2) [20]. Once inside, the virus produces large proteins that are cut by viral proteases, which are important drug targets. AI models can help design new compounds that are easier to make and less likely to cause side effects by learning from existing drugs and their chemical properties.

AI-powered tools are especially useful for finding ways to make new antiviral compounds from ingredients that are already available. These tools can predict the best methods to create these molecules and check if they are likely to work as drugs

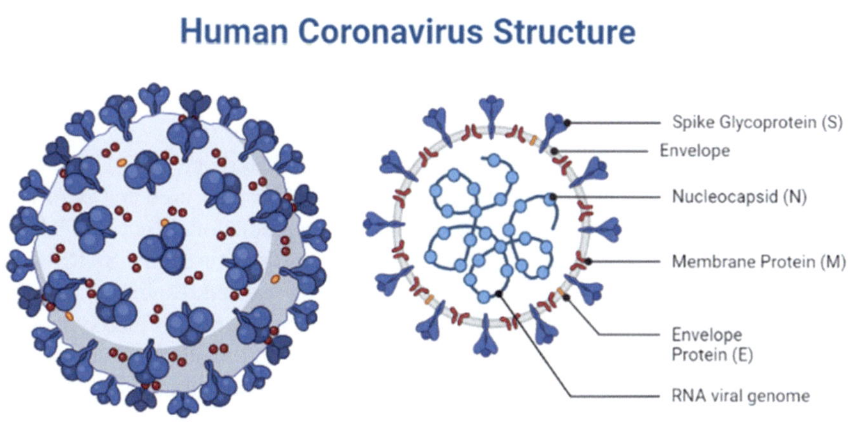

Fig. 5.1 Visualization of the SARS coronavirus and its structural components (Figure created by Biorender)

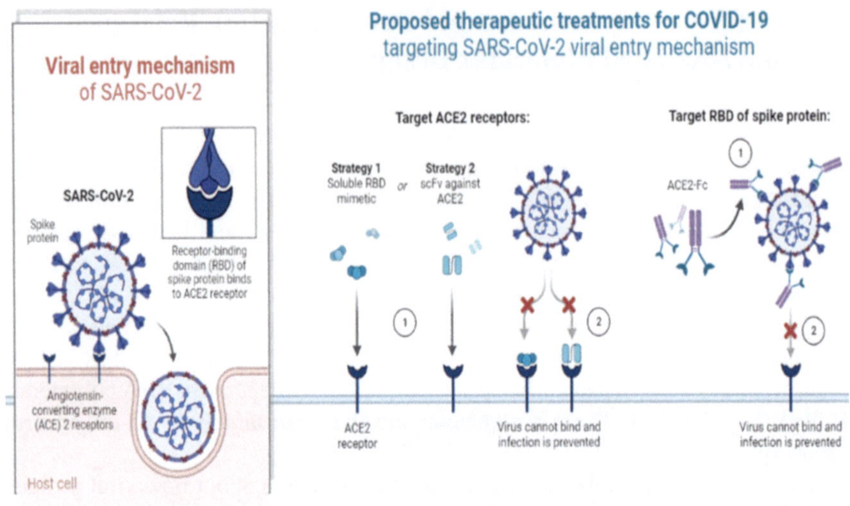

Fig. 5.2 Transmission of SARS-CoV-2 and its entrance points (Figure created by Biorender)

by looking at their chemical and biological properties. Early in the process, AI can also help predict how these compounds will behave in the body, such as how they are absorbed and processed. New deep learning models like AlphaFold2 can predict the shapes of viral proteins very accurately, which helps researchers design better drugs. By combining these structural insights with AI, scientists can discover new types of molecules that might block the virus at different stages. This is especially helpful for adapting natural products to target SARS-CoV-2, making drug discovery faster and more practical property forecasting [8, 28].

5.3 AI-Interpretable Epidemiological and Pathological Signatures of SARS-CoV-2 for Drug Design

Understanding the epidemiological and pathological landscape of SARS-CoV-2 infection provides crucial biological priors that fuel the data used for AI-driven retrosynthetic analysis and transformer-based reaction prediction. Clinically, SARS-CoV-2 infection mirrors many symptoms of earlier coronaviruses like SARS-CoV, including pyrexia (fever), dry cough, dyspnea, and generalized respiratory distress. Yet the variability in patient phenotypes, ranging from gastrointestinal symptoms (vomiting and diarrhoea) to neuromuscular (myalgia and headache), points to a complex systemic pathophysiology. This heterogeneity in symptomatology must be quantitatively encoded into AI models for accurate virtual screening and compound prioritization. From a mechanistic drug development standpoint, severe disease phenotypes such as acute respiratory distress syndrome (ARDS), multi-organ failure, and septic shock underscore the need for intervention strategies that address not only viral replication but also host immune dysregulation [45]. These complications are often characterized histologically by pulmonary alveolitis, bronchiolitis, squamous metaplasia, mucus hypersecretion, and interstitial fibrosis along with epithelial cell desquamation. Such biological features are increasingly being used as endpoints in AI models trained to predict host response to new chemical entities (NCEs) through multimodal datasets that combine transcriptomic, histopathologic, and pharmacokinetic information.

In AI-enabled retrosynthetic planning, this pathological diversity introduces constraints that must be modelled for effective hit-to-lead optimization. Molecules predicted to act on viral proteases or inflammatory mediators are further filtered against clinical manifestations of fibrosis, edema, and immune infiltration to predict potential adverse interactions or off-target effects. Transformer-based models are particularly well-suited to this task, as they can learn the conditional likelihoods of reaction outcomes in the context of such biological constraints. Moreover, immune-histochemical data and lung pathology are now being computationally mapped to guide the prioritization of scaffolds likely to exhibit therapeutic efficacy without exacerbating immune-mediated damage. Drug-like candidates designed via AI models can now be evaluated in silico against known tissue pathologies using neural chemical informatics, aligning molecule design with real-world biological fidelity. Hence, the pathological insights from SARS-CoV-2 not only inform clinical treatment strategies but also shape the training and validation datasets for AI systems. These models benefit from cross-disciplinary integration using features extracted from autopsy reports, histological slides, and immune-histochemical profiles to simulate realistic in vivo conditions. This represents a fundamental methodological advancement in making AI-designed molecules viable for translational application in pandemic drug development.

5.4 AI-Guided Approaches Are Helping Address Heart Complications Linked to SARS-CoV-2 Infection

These complications are more than just clinical observations; they provide important clues for AI-driven drug discovery. As artificial intelligence becomes more involved in predicting chemical reactions and designing new drugs, it is important to include outcomes like virus-related heart damage in model training. SARS-CoV-2, like SARS-CoV, uses the angiotensin-converting enzyme 2 (ACE2) to enter host cells Fig. 5.2) [33]. The high number of ACE2 receptors in heart tissues, such as heart muscle cells, blood vessel lining, and smooth muscle cells, makes the heart a main target for the virus [43]. Clinical reports have described various heart problems, including myocarditis, irregular heartbeats, low blood pressure, increased heart rate, and even cardiac arrest. Higher levels of biomarkers like troponin I and creatine kinase, along with signs of inflammation and viral mRNA in heart tissue, confirm direct infection of the heart by the virus [6].

From a computational chemistry perspective, this presents a dual challenge and opportunity. AI-based retrosynthetic models, especially those employing transformer architectures, must not only optimize chemical reactions for synthesis but also account for biological constraints such as cardiotoxicity and cytokine-mediated cardiac stress. Models must be trained to predict how drug candidates interact with ACE2 signalling and downstream inflammatory cascades. For instance, cytokine storm-induced systemic inflammation must be simulated within in silico pharmacodynamic frameworks to evaluate off-target risks associated with immunomodulators or kinase inhibitors. The development of small molecules, peptides, or antibody fragments intended to modulate ACE2-related pathways demands an integration of cardiac-specific datasets into virtual screening workflows. Transformer-based models can assist in designing ACE2-interacting compounds with reduced pro-arrhythmic potential by learning from cardiac-specific adverse event profiles, ECG datasets, and known cardiotoxic pharmacophores [32, 39].

Furthermore, molecular candidates proposed through AI retrosynthesis must undergo rigorous post hoc screening against cardiac safety panels using deep learning models trained on clinical pharmacology databases like FDA's FAERS or SIDER. In vitro cardiomyocyte infection models, such as those used in the study by Bojkova et al. [6], also provide benchmark data that can feed back into the fine-tuning of generative AI models for safer and context-aware molecule design. The elucidation of SARS-CoV-2's role in altering ACE2 and its downstream effects is not only vital for understanding COVID-19 pathogenesis but also for guiding the prioritization of AI-designed compounds with favourable efficacy-toxicity ratios. Ultimately, this reorients the AI pipeline from being reaction-centric to being mechanism-aware, enabling the generation of therapeutic candidates that are not only synthetically accessible but also clinically viable for high-risk complications such as virus-induced myocarditis.

5.5 Integrating AI Tools in Addressing SARS-Cov-2-Induced Thrombotic Complications

Blood clots are a serious issue in SARS-CoV-2 infection and have become a major focus in drug development. These clots can form in the lungs and other parts of the body, often due to problems with blood vessels, immune system overreactions, and changes in hormone systems. Because of this, AI models for drug design now need to include targets related to blood clotting and inflammation. When the virus enters cells through ACE2, it disrupts normal blood vessel function, leading to increased clotting and inflammation. Patients often have higher D-dimer levels, longer clotting times, and clots in multiple organs [32].

AI-assisted retrosynthetic tools, traditionally focused on optimizing molecular synthesis pathways, are now being retrofitted to prioritize compounds with antithrombotic or endothelium-protective effects. Transformer models trained on reaction libraries (e.g., USPTO, Reaxys) can be fine-tuned with pharmacological metadata pertaining to anticoagulants, PAI-1 inhibitors, and vWF-targeting biologics. These adaptations help the models not only design synthetically feasible molecules but also predict bioactivity against coagulopathic markers elevated in COVID-19, such as IL-6, TNF-α, and tissue factor. In this context, real-world drug development demands that AI-predicted compounds undergo in silico thrombosis risk profiling before moving into synthesis or biological evaluation. Integration of clinical data from COVID-19 patients with hypercoagulability—such as elevated vWF multimers, anticardiolipin IgA/IgG, and anti-β2-glycoprotein antibodies—can serve as predictive features in deep learning classification models for thrombotic toxicity [5, 7].

New AI technologies, like generative models and reinforcement learning, can help design molecules that support normal ACE2 function or block the processes that cause blood clots and inflammation. With these tools, researchers can develop new blood thinners that are more targeted and safer. Including blood clotting as a key factor in AI drug design is now essential, as it helps make sure new treatments address both the virus and its serious complications.

5.6 Platelets and Their Function in Covid-19

Thrombosis triggered by SARS-CoV-2 involves direct and indirect interactions with platelets, which are the second most abundant cell type in human blood after red blood cells. In a healthy adult, normal platelet counts range from 150,000 to 450,000 per microliter. A deficiency in platelet count, known as thrombocytopenia, can result from reduced production, increased destruction, or excessive consumption of platelets. Viruses, including coronaviruses, frequently target platelets due to their ability to bind to platelet surface receptors such as lectins, integrins, and Toll-like receptors (TLRs). Upon viral binding, platelets become activated and release the contents of their alpha granules and dense granules, which stimulate immune cells and contribute to the antiviral response. In SARS-CoV, thrombocytopenia, along with lymphopenia and leukopenia, is commonly observed. This is primarily due to viral presence in the lung tissues and pulmonary endothelial cells, which contributes to:

- Platelet consumption at sites of injury,
- Apoptosis or impaired maturation of megakaryocytes (platelet-producing cells),
- Disseminated intravascular coagulation (DIC),
- Elevated D-dimer levels.

These mechanisms have also been observed in MERS-CoV infections and are strongly suspected in SARS-CoV-2. A clinical study involving 1799 COVID-19 patients confirmed a strong link between platelet count and disease severity, with lower counts associated with higher mortality. Since the lungs are a major site of platelet production, damage to the pulmonary blood vessels in COVID-19 leads to increased platelet activation, aggregation, and clot formation, which further reduces circulating platelet levels [26]. In addition, type I interferons, while antiviral, can suppress megakaryocyte function and reduce platelet output. Lung injury caused by viral infection, oxygen toxicity, or mechanical ventilation makes this problem worse by promoting platelet trapping and microthrombi formation in the lungs and other organs. New evidence connects megakaryocytes to the formation of platelet–fibrin microthrombi in the heart and other tissues. Although platelets lack ACE2 receptors, viral RNA (including RdRp, E, and N1 genes) has been found inside platelets, suggesting that SARS-CoV-2 particles can be internalized. Several changes in platelet function have been noted in COVID-19 patients: release of extracellular vesicles (EVs) with phosphatidylethanolamine (PE) exposure, increased secretion of platelet granules such as 5-HT (a mild platelet agonist), PF4 (a regulator of T-cell activity), soluble CD40L, and IL-1 in response to thrombin, and elevated platelet hyperactivation and aggregation, which contribute to coagulopathy.

Hyperactivated platelets also interact with monocytes, forming platelet-monocyte aggregates, a process driven by P-selectin and integrin αIIbβ3. This interaction is critical for tissue factor (TF) expression on monocytes, promoting thrombosis and inflammation [19]. While the full mechanism of platelet involvement in SARS-CoV-2 remains under investigation, the current data suggest that:

- Platelets may internalize the virus even without ACE2.
- Platelets are hyper-activated and participate in thrombo-inflammation,
- SARS-CoV-2 induces profound changes in platelet signalling, secretion, and interaction with immune cells.

Further mechanistic insights into SARS-CoV-2-induced platelet dysfunction are urgently needed and expected to emerge from future research.

5.7 As New Variants of SARS-CoV-2 Appear, Researchers Are Focusing More on Plant-Based Compounds, Called Phytochemicals, as Possible Additional Treatments for COVID-19

These natural substances, such as alkaloids, phenolics, and tannins, can help reduce inflammation, act as antioxidants, and prevent blood clots. Some medicinal plants, including *Senna tora, Cibotium barometz, Gentiana scabra, Dioscorea polystachya,* and *Taxillus chinensis,* have been used in traditional medicine and may help stop the virus from multiplying (Table 5.1). These plant compounds could be especially useful in places where modern drugs are hard to get [29]. For example, withaferin A and withanone from ashwagandha have shown antiviral effects and may block a key enzyme the virus uses to enter cells, which could help reduce both infection and

Table 5.1 Plants and their compounds with known mechanism against COVID 19

Plant/compound	Mechanisms related to COVID-19 and platelets	Key activity
Senna tora, Cibotium barometz, Gentiana scabra, Dioscorea polystachya, Taxillus chinensis	Inhibit SARS-CoV genome replication	Antiviral
Ashwagandha (*Withania somnifera*) → Withaferin A, Withanone	Suppress TMPRSS2 synthesis (entry receptor), may block SARS-CoV-2 cell entry	Antiviral, TMPRSS2 inhibition
Curcuma longa (Turmeric) → Curcumin	Inhibits platelet aggregation via multiple pathways: ADP, PAF, collagen, AA; inhibits TMPRSS2; reduces apoptosis, fibrosis	Antiplatelet, anti-inflammatory, antiviral
Berberis vulgaris → Berberine	Prevents platelet aggregation by modulating calcium mobilization and AA metabolism; reduces cytokine storm	Antiplatelet, anti-inflammatory
Magnolia officinalis → Magnolol	Blocks intracellular calcium to suppress platelet activation and aggregation	Antiplatelet

blood clotting problems. Curcumin, found in turmeric, is another plant compound that has been widely studied. It can block platelet activation in several ways and may also interfere with the virus's ability to enter cells. This makes curcumin a promising option for both fighting the virus and reducing blood clots. Berberine, from *Berberis vulgaris*, is known for its antibacterial and anti-inflammatory effects and can also help prevent platelets from sticking together [40]. By lowering inflammation, berberine may help control the severe immune reactions seen in some COVID-19 cases [4, 24, 30]. *Magnolia officinalis* contains magnolol, which also blocks platelet activation by affecting calcium signals inside cells. Since too much platelet activity is a problem in severe COVID-19, these plant compounds deserve further study as part of a broader approach to managing the disease.

5.8 Phytochemicals that Have the Potential to Be Used in the Treatment of COVID-19

Plants are a rich source of chemicals that can be used as medicines, and many have been used for centuries to treat illnesses. Today, researchers are looking again at these plant compounds to see if they can help fight viruses like SARS-CoV-2. Some of these compounds can block the virus from entering cells, multiplying, or escaping the immune system. For example, plants from the Clerodendrum genus, which have been used to treat breathing problems, have shown activity against SARS-CoV-2 in computer studies. Compounds like taraxerol, friedelin, and stigmasterol from these plants bind well to important viral proteins and meet key criteria for drug development. However, more laboratory and animal studies are needed to confirm how well they work. There is also growing interest in using compounds from spices as antiviral agents, since they often have few side effects. Spices like turmeric and black pepper contain curcumin and piperine, which have anti-inflammatory and antiviral effects. Curcumin has been shown to block the replication of SARS-CoV and its main protease, while piperine binds strongly to key SARS-CoV-2 proteins in computer models) [21]. These compounds may help by both stopping the virus and calming the immune system, which is important because severe COVID-19 often involves an overactive immune response. Other medicinal plants, such as *Terminalia chebula* and Neem, also have compounds that can block important viral proteins, with some like seselin working against several targets at once. Computer modeling now makes it easier to screen plant compounds for their ability to block viruses. For example, catechins from green tea, especially EGCG, have been shown to fight viruses like Zika and influenza. In SARS-CoV-2, green tea compounds such as prodelphinidin B-2 3′-O-gallate and TF3 can block a key viral enzyme [46]. Haritaki, a traditional Ayurvedic remedy, also reduces the activity of this enzyme, especially when combined with green and black tea extracts [36, 37]. These findings suggest that using several plant compounds together could boost their antiviral effects. Other plant compounds like berberine, magnolol, and withaferin A also show promise for both fighting the virus and reducing blood clots. Curcumin stands out for its ability to block platelets, lower inflammation, and stop the virus from multiplying (Table 5.2).

5.9 Plant Lectins, Alkaloids, and Other Plant-Derived Substances with Antiviral Potential Against SARS-CoV-2

In the continuous search for effective therapeutic interventions against COVID-19, a considerable body of evidence has emerged highlighting the potential of plant-derived substances—including plant lectins, alkaloids, terpenoids, and polyphenolic compounds—as promising antiviral agents against SARS-CoV-2. These naturally occurring phytochemicals exhibit broad-spectrum antiviral properties by interfering with various stages of the viral life cycle, particularly viral entry, replication, and protein activity. Among these, plant lectins have drawn particular attention due to their strong affinity for glycoproteins present on viral envelopes. Several lectins exhibit potent anti-CoV activity by binding to mannose-rich glycan residues on the viral spike (S) glycoprotein, thus obstructing viral attachment and membrane fusion. Mannose-specific agglutinins, such as those found in *Urtica dioica* L., have demonstrated the ability to inhibit SARS-CoV-2 replication during the viral entry

Table 5.2 Table enlists extracts obtained from a number of medicinal plants that have the potential to stop the replication of SARS-CoV and function as inhibitors of the ACE2 receptor and Mpro

S. no	Names for plants that are usually used	Botanical name	Utilized plant part	Aquatic extract activity inhibition percentage at 100 UG/ML
1	Aloe vera	*Aloe barbadensis miller*	Leaf	0
2	Amala	*Phyllanthus emblica*	Fruit	10%
3	Anise	*Pimpinella anisum*	Fruit	0
4	Antmul	*Tylophora indica*	Leaf	0
5	Bael	*Aegle mamelos*	Pulp of fruit	0
6	Ber	*Ziziphus mauritiana*	Fruit	0
7	Bhilawa	*Semecarpusana cardium*	Fruit	0
8	Bhojpatra	*Betula utilis*	Leaf	0
9	Black tea	*Camellia sinensis*	Leaf	100%
10	Chirata	*Swertia chirayita*	Leaf and stem	0
11	Chopachini	*Smilax glabra*	Rhizome	0
12	Cinnamon	*Cinnamomum verum*	Bark	0
13	Coconut	*Cocos nucifera*	Husk fiber	0
14	Daaruharidra	*Berberi saristata*	Stem	0
15	Deodar	*Cedrus deodara*	Bark	0
16	Elaichi	*Elettaria cardamomum*	Fruit	0
17	Fennel	*Foeniculum vulgare*	Fruit	0
18	Fenugreek	*Trigonella foenum-graecum*	Fruit	0
19	Garlic	*Allium sativum*	Clove	0
20	Giloy	*Tinospora cordifolia*	Stem	13%
21	Ginger	*Zinziber officinale*	Rhizome	0
22	Gojihva	*Onosma bracteatum*	Leaf and stem	0
23	Green tea	*Camellia sinensis*	Leaf	100%
24	Guggul	*Commiphora wightii*	Resin	8%
25	Haridra	*Curcuma longa*	Rhizome	5%

(continued)

Table 5.2 (continued)

S. no	Names for plants that are usually used	Botanical name	Utilized plant part	Aquatic extract activity inhibition percentage at 100 UG/ML
26	Haritaki	*Terminali achebula*	Fruit	100%
27	Kachur	*Curcuma zedoaria*	Rhizome	0
28	Kalmegh	*Andrographis paniculata*	Leaf	0
29	Kantakari	*Solanum surratense*	Fruit	0
30	Lemon	*Citrus limon*	Fruit	0
31	Mamira	*Coptis teeta*	Leaf	0
32	Maricha	*Piper nigrum*	Fruit	0
33	Mulethi	*Glycyrrhiza glabra*	Root	40%
34	Musta	*Cyperus rotundus*	Rhizome	0
35	Nimba	*Azadirachta indica*	Leaf	15%
36	Nirgundi	*Vitex negundo*	Leaf	0
37	Pippali	*Piper longum*	Fruit	0
38	Puskarmool	*Inulara cemosa*	Root	0
39	Rasna	*Pluchea lanceolate*	Leaf, stem	0
40	Sallaki	*Boswellia serrata*	Resin	12%
41	Shalparni	*Desmodium gangeticum*	Root	0
42	Shatavari	*Asparagus racemosus*	Rhizome	0
43	Shleshmatak	*Cordia dichotoma*	Fruit	0
44	Spirulina	*Arthrospira platensis*	Entire plant	0
45	Sunthi	*Zinziber officinale*	Rhizome	10%
46	Surajmukhi beej	*Helianthus annuus*	Seed	5%
47	Suranjaan	*Colchicum luteum*	Rhizome	0
48	Talispatra	*Abies webbiana*	Leaf	0
49	Tilpushpi	*Digitalis purpurea*	Leaf	0
50	Tulsi	*Ocimum sanctum*	Leaf	20%
51	Vasaka	*Adhatoda vasica*	Leaf	0

(Adapted from [44])

phase. Griffithsin, a lectin isolated from *Griffithsia* species, binds with high specificity to glycan moieties of the SARS-CoV-2 spike protein via its three glycan-binding domains (GBDs), showing minimal cytotoxicity and holding promise as a lead compound for drug development. Additional lectins from *Allium porrum* L., *Nicotiana tabacum* L., and *Galanthus nivalis* L. have also been shown to impede viral spread by targeting structural glycoproteins [48].

Apart from lectins, alkaloids—nitrogen-containing phytochemicals commonly found in medicinal plants—have demonstrated antiviral efficacy by targeting both viral enzymes and host cell pathways essential for viral replication. Compounds such as cepharanthine, tetrandrine, and fangchinoline, derived from *Stephania* species, have been shown to reduce viral-induced cytopathic effects in human lung cells infected with coronaviruses. Cepharanthine, in particular, inhibits the SARS-CoV protease and mitigates oxidative stress-induced cell death, offering a dual mechanism of therapeutic action. Further supporting evidence comes from the inhibition of viral entry by polyphenolic compounds. For example, cortex-derived procyanidins (A2 and B1) and cinnamtannin B1, isolated from *Cinnamomum cassia*, disrupt clathrin-mediated endocytosis—a key pathway used by SARS-CoV-2 for host cell entry. Likewise, tetra-O-galloyl-beta-D-glucose and luteolin block the binding of the virus to host receptors such as ACE2, thereby preventing membrane fusion and viral internalization. Marine-derived phytochemicals also show significant promise. Phlorotannins, such as 7-phloroeckol, phlorofucofuroeckol A, and dieckol, from the brown alga *Ecklonia cava*, exhibit potent inhibitory effects on coronavirus entry by preventing viral binding to host cell surfaces.

Several terpenoids—including diterpenes, triterpenes, and sesquiterpenes—have also been identified with potent antiviral activity. Silvestrol, a flavagline from *Aglaia* species, uniquely inhibits the RNA helicase eIF4A, a host factor essential for MERS-CoV and potentially SARS-CoV-2 replication, by impairing translation of viral proteins. Additionally, the inhibition of host proteases involved in viral entry offers another strategic mechanism. Cathepsin L inhibitors derived from plant sources prevent viral-host membrane fusion, effectively blocking SARS-CoV-2 entry. These findings are further supported by computational docking studies and AI-assisted in silico modeling, which have helped identify drug-likeness and pharmacokinetic properties of many of these phytochemicals, ensuring they comply with guidelines such as Lipinski's Rule of Five and Veber's Rule. Collectively, these findings underscore the therapeutic potential of plant-derived compounds in the management of COVID-19. With further validation through in vitro and in vivo studies, these phytochemicals may serve as lead molecules for the development of safe, effective, and plant-based antiviral drugs.

5.10 SARS-CoV-2 Treatment Options Targeting Helicase and Protein Expression Inhibition

The replication and transcription of SARS-CoV-2 rely heavily on viral non-structural proteins, particularly RNA helicases and replication complexes. Among these, the helicase enzyme (Nsp13) is an evolutionarily conserved protein across coronaviruses and a critical therapeutic target due to its indispensable role in unwinding RNA secondary structures during viral genome replication. Notably, SARS-CoV-2 helicase shares high sequence and structural homology with helicases in other nidoviruses, underscoring its importance as a conserved target in coronavirus biology. A number of therapeutic compounds, both synthetic and plant-derived, have been shown to exhibit inhibitory activity against SARS-CoV and SARS-CoV-2 helicases. Although repurposed drugs such as favipiravir and hydroxychloroquine have demonstrated low-affinity interactions with helicase, their clinical effectiveness remains limited. Consequently, there is growing interest in more target-specific natural inhibitors.

One of the most promising phytochemicals identified to date is silvestrol, a flavagline compound extracted from Aglaia species. Silvestrol exhibits potent inhibition of the host RNA helicase eIF4A, a key factor utilized by coronaviruses to translate viral RNA. In studies involving MERS-CoV, silvestrol demonstrated an EC50 of 1.3 nM, while for HCoV-229E, the EC50 was 3 nM, indicating high potency. Mechanistically, silvestrol blocks the formation of replication–transcription complexes by inhibiting eIF4A activity, effectively preventing viral protein synthesis and replication. These effects were confirmed in ex vivo studies using bronchial epithelial cells. In addition to silvestrol, several polyphenolic compounds—notably myricetin and scutellarein—have been shown to inhibit SARS-CoV helicase activity, specifically targeting Nsp13, a helicase with essential ATPase functionality. These compounds are found in medicinal plants such as *Scutellaria baicalensis* (Scutellaria radix), *Torreya nucifera*, and *Isatis indigotica*, with myricetin also present in dietary sources like *Vaccinium oxycoccos* (cranberries), garlic (*Allium sativum*), and *Calamus scipionum.* Myricetin inhibits the enzymatic activity through interference with ATP hydrolysis, while scutellarein blocks ATPase activity, demonstrating a unique mechanism of helicase inhibition without directly binding the helicase active site.

The flavonoid quercetin, structurally similar to myricetin and scutellarein, has shown comparable inhibitory effects on coronavirus helicases, making it another viable compound for therapeutic development. These naturally occurring flavonoids represent a class of broad-spectrum antivirals with the potential to inhibit not just SARS-CoV-2 but also other human coronaviruses through a conserved enzymatic target. Furthermore, tomentins derived from *Paulownia tomentosa* (*Scrophulariaceae* family) have demonstrated the ability to suppress papain-like protease (PLpro) activity in SARS-CoV. These compounds act via reversible allosteric inhibition, adding a second mechanism—alongside helicase inhibition—by which plant-derived substances can disrupt the viral life cycle. Together, these findings suggest that plant-derived polyphenols and alkaloids that interfere with helicase function or host translation factors (like eIF4A) provide a rich source of potential anti-SARS-CoV-2 agents.

Their multi-targeted modes of action, low toxicity (in most cases), and evolutionary conservation of the helicase target make them highly valuable leads for further drug discovery and development against COVID-19 and related viral diseases.

5.11 Inhibition of the Synthesis and Replication of Ribonucleic Acid Genomes

The inhibition of SARS-CoV-2 RNA-dependent RNA polymerase (RdRp) presents a promising therapeutic strategy for COVID-19. This approach focuses on nucleoside and nucleotide analogues that impede the function of viral RNA polymerases [38], offering broad-spectrum antiviral activity. Several compounds under investigation through randomized clinical trials have demonstrated efficacy against coronaviruses. Among these, remdesivir (RDV) acts as a substrate mimic for viral RdRp enzymes and has shown clinical promise. Similarly, tryptanthrin, an alkaloid derived from *Strobilanthes cusia* (Nees) Kuntze, and indigodole B (5aR-ethyltryptanthrin) exhibit anti-HCoV-NL63 activity. These compounds inhibit both RNA synthesis and the PLpro-2 enzyme, crucial during various viral replication stages. Tryptanthrin displays an EC_{50} of 1.52 μM, while indigodole B has an EC_{50} of 2.60 μM, directly inhibiting HCoV-NL63 replication.

Furthermore, aqueous extracts of *Thunbergia* species block both SARS-CoV 3CL protease and RdRp activity. Theaflavins (TFs)—including TF-3-gallate, TF-3′-gallate, and TF-3,3′-di-gallate—are polyphenolic antioxidants derived from green tea leaves. These molecules possess tropolone moieties and are recognized for their polymeric antioxidant properties, playing roles in RdRp inhibition. Compounds such as 13-dien-12-one, 3,12-diacetoxyabieta-6,8,11,13-tetraene, hinokinin, curcumin, and savinin—primarily derived from *Sequoia sempervirens* (D. Don) Endl.—also demonstrate antiviral activity by interfering with viral genome replication. Other inhibitors like lignin, betulinic acid, and desmethoxyreserpine block replication, with desmethoxyreserpine particularly effective at preventing viral entry.

Essential oils from *Laurus nobilis* L. and *Thuja orientalis* (L.) Franco exhibit antiviral properties by inhibiting virus replication. α-Pinene, abundant in *Rosmarinus officinalis* L. (rosemary), and cedrol, a sesquiterpene alcohol in essential oils, are potent antioxidants with additional antiviral benefits. Pyranoxanthones such as blancoxanthone and pyranojacareubin, derived from Calophyllum blancoi Planch. & Triana, effectively inhibit the proliferation of HCoV-229E in host cells. At a concentration of 3 μg/mL in MRC-5 cells, blancoxanthone significantly impedes viral growth. Likewise, saikosaponins—isolated from *Bupleurum, Heteromorpha*, and *Scrophularia scorodonia* L.—act during viral entry and early replication stages to block HCoV-229E activity. Ferulic acid, a hydroxycinnamic acid and lignin precursor, shares properties with caffeic and isoferulic acids. Derivatives such as caffeoyltyramine, feruloyltyramine, and feruloyloctopamine inhibit SARS-CoV

PLpro activity. Similarly, CA derivatives, including khainaosides, 6-O-caffeoyl-arbutin, and vitexfolin, demonstrate high binding affinity to the SARS-CoV-2 spike protein (S-protein), Nsp15, and Mpro enzymes, often surpassing the activity of known drugs like nelfinavir. These findings are supported by docking simulations via Toxtree and SwissADME. A notable triterpenoid, toosendanin ($C_{30}H_{38}O_{11}$), isolated from the bark of Melia azedarach L., exhibits multiple biological activities, including antiviral, antibacterial, and anti-inflammatory properties. It has demonstrated efficacy in blocking SARS-CoV-2 RNA replication by targeting the Nsp-9 enzyme (PDB ID: 6W4B) and inhibiting S-protein binding. Lastly, baicalin, a flavone glycoside, has been identified as a compound with strong binding affinity for both the S-protein and Nsp9, highlighting its potential in natural antiviral therapy design.

5.12 Plant-Derived Therapeutic Candidates Targeting SARS-CoV-2 S-Glycoprotein: A Mechanistic Perspective

SARS-CoV-2, the causative agent of COVID-19, initiates infection through the recognition of host cell receptors by its spike (S)-glycoprotein (Fig. 5.3). This interaction, mediated via the receptor-binding domain (RBD) of the S-protein, facilitates viral entry through endosomal membrane fusion and subsequent release of viral RNA into the host cytoplasm [15]. Given the crucial role of the S-glycoprotein in host tropism, infectivity, and pathogenesis, it serves as a key target for antiviral therapeutics, especially those derived from bioactive plant compounds. AI-based predictions of drug-likeness and synthetic accessibility have guided the exploration of several phytochemicals that interfere with S-glycoprotein activity and related pathways. Alkaloids derived from *Stephania tetrandra* var. glabra Maxim, including bis-benzylisoquinolines such as fangchinoline, tetrandrine, and cepharanthine, have demonstrated potent antiviral properties against coronaviruses. These compounds mitigate virus-induced cytotoxicity by suppressing viral replication and inhibiting the expression of both the S-glycoprotein and nucleocapsid (N)-protein. Moreover, they enhance host immune response by activating the p38MAPK signalling cascade [49].

Sambucus javanica subsp. chinensis Fukuoka (CA), commonly referred to as Chinese elder, has yielded extracts that inhibit the binding of HCoV S-glycoproteins to host cells. Compounds such as 3,4,5-trihydroxybenzoic acid and chlorogenic acid, isolated from this species, exhibit similar inhibitory mechanisms. Additionally, black elderberry (*Sambucus nigra* L.), traditionally used for treating cold and flu symptoms, has been extensively studied for its pharmacokinetics, including bioavailability, adsorption, and plasma therapeutic concentrations [34]. Triterpenoid saponins, particularly saikosaponins extracted from *Bupleurum* spp., *Heteromorpha* spp., and *Scorodonia* spp., display broad antiviral effects. Notably, saikosaponin B2 exhibits the strongest anti-CoV activity with an EC_{50} of 1.7 μM. It impedes viral attachment to host receptors and inhibits subsequent S-glycoprotein-mediated cell entry [9]. These findings align with AI-driven molecular docking predictions, which highlight high binding affinity and favourable pharmacophore features for saikosaponins targeting viral entry points.

The antibiotic actinomycin D, biosynthesized by *Streptomyces parvulus*, was identified to inhibit both the attachment and penetration phases of the CoV lifecycle. This supports its potential inclusion in AI-driven drug repurposing pipelines for host-virus interface disruption. Further supporting evidence comes from the antiviral activity of tylophorine alkaloids, namely tylophorine and 7-methoxycryptopleurine, derived from *Tylophora indica*. These compounds demonstrate inhibitory activity against transmissible gastroenteritis CoVs, downregulating S- and N-protein production and obstructing JAK/STAT signalling via NF-κB suppression, a mechanism also captured by AI-based pathway modelling systems [17]. Additional plant-derived compounds with high predictive drug-likeness scores include dihydrotanshinone, which binds to the SARS-CoV-2 S-glycoprotein and effectively blocks viral entry, and tetra-O-galloyl-D-glucose (TGG) from *Rhus chinensis* Mill., which selectively binds the S2 subunit. Luteolin, another bioactive flavonoid, binds to the S2 domain to prevent membrane fusion and viral entry, corroborated by both in silico and in vitro assays [25].

5.13 Modulation of ACE2 and S-Glycoprotein Interaction: Insights from Phytochemical Inhibitors and Therapeutic Implications

SARS-CoV-2 primarily targets the respiratory epithelium—including alveolar type I and II cells—where angiotensin-converting enzyme 2 (ACE2) exists in both soluble and membrane-bound forms [35]. As a membrane-anchored carboxypeptidase, ACE2 plays a central role in the renin–angiotensin system (RAS), and its interaction with the receptor-binding domain (RBD) of the viral S-glycoprotein is essential for viral entry. This ACE2-mediated viral entry, driven by the structural conformation of the S-glycoprotein, disrupts host immune surveillance and enables infection, making this interface a key therapeutic target. While MERS-CoV does not bind ACE2, lineage B coronaviruses like SARS-CoV and HCoV-NL63 do, demonstrating the shared use of ACE2 by several β-coronaviruses. Intriguingly, a secondary acetylcholinesterase-2 (AChE2)–like receptor has been implicated in facilitating viral entry, potentially through overlapping molecular mimicry with ACE2-binding motifs on the viral spike protein. Targeting both ACE2 and TMPRSS2 (transmembrane serine protease 2)—the latter responsible for S-protein cleavage—may thus offer dual blockade strategies against viral infection (Fig. 5.4).

Multiple plant-derived compounds have shown promise in disrupting ACE2 or its interaction with the S-glycoprotein, although most bind outside the ACE2–RBD interface. Notably, the flavonoid hesperidin, abundant in *Citrus* species, binds to ACE2 and modulates its availability for viral recognition [10]. Additional studies show that compounds like phyllaemblicin G7 from *Phyllanthus emblica* L., xanthones from *Citrus aurantium* L., and neohesperidin also associate with AChE2/ACE2 proteins but do not interfere directly at the ACE2–S-RBD binding site [50]. Molecular docking has shown that chrysin, a flavone from *Oroxylum indicum*, can bind ACE2 at allosteric sites, supporting a potential role in conformational interference. Likewise, emodin, an anthraquinone, directly blocks ACE2–S-protein binding and inhibits viral attachment.

In addition to phytochemicals, small molecule drugs such as troglitazone, losartan, cefmenoxime, ergotamine, and silybin—originally developed for diabetes, hypertension, and hepatic disorders—exhibit binding affinity to AChE2, suggesting repurposing potential [9]. TMPRSS2, which cleaves the S-protein to facilitate membrane fusion, is also inhibited by natural and synthetic compounds. These include kouitchenside I, as well as conventional antibiotics like clindamycin, cefoperazone, hetacillin, and pivampicillin, all of which demonstrate activity against TMPRSS2-mediated viral entry pathways [1]. A well-studied class of compounds with antiviral activity against SARS-CoV includes glycyrrhizin and its derivatives from *Glycyrrhiza* spp. (*G. glabra, G. uralensis, G. inflata*). 18-β-glycyrrhetinic acid, a gut microbiota–derived metabolite of glycyrrhizin, inhibits DNA polymerases and the complement system, suppresses TNF expression, and directly interacts with the viral S-glycoprotein. Structural modifications, such as glycosylation with 2-acetamido-D-glucopyranosylamine, significantly enhance its antiviral potency—reportedly by a factor of ten [12]. Another derivative, licochalcone A, further supports the role of licorice-derived phytochemicals in targeting viral proteins through multiple mechanisms. Together, these findings underscore the potential of phytochemicals in targeting host–virus interactions, particularly the ACE2–S-protein interface and downstream entry mediators like TMPRSS2. AI-guided ligand screening and synthetic accessibility modelling enhance the selection of candidates like flavonoids, anthraquinones, and triterpenoids for further preclinical development.

5.14 Targeting Viral Proteases: Natural Inhibitors of SARS-CoV-2 3CLpro and PLpro

The SARS-CoV-2 genome encodes a large polyprotein containing protease domains essential for the viral life cycle. Two key cysteine proteases—papain-like protease (PL-pro)and3C-like protease (3CL-pro, also known as the main protease or M-pro)—play central roles in processing the polyprotein into functional non-structural proteins required for viral replication [16, 23]. Cleavage at 11 conserved substrate-recognition sites by these proteases enables viral RNA synthesis and assembly, making them critical therapeutic targets.

A. **3CL-pro-: A Key Target for Inhibitor Design**

3CL-pro (~ 34 kDa) is highly conserved across coronaviruses, showing 99.02% sequence identity between SARS-CoV-2 and bat-derived strains and remarkable similarity to the proteases of SARS-CoV, HCoV, MERS-CoV, and BCoV. It orchestrates the initiation of viral RNA replication, and thus, targeting this enzyme is a strategic approach in antiviral drug development. Computational screening and molecular docking have facilitated the identification of several natural product inhibitors. Plant-derived compounds such asviniferin, afzelin, myricitrin, biorobin, hesperidin, lactucopicrin 15-oxalate, taiwanhomoflavone A, and nympholide A have shown promising binding affinities to M-pro. Likewise, caffeic acidphenethyl ester (CAPE) and withaferin A—a derivative of withanolides from *Withania somnifera* (Ashwagandha)—exhibit strong inhibitory potential. Flavonoids such as apigenin, curcumin, kaempferol, catechin, and oleuropeinhave been computationally modeled as effective M-pro inhibitors. Furthermore, jensenone, a cyclic ether monoterpenoid found in *eucalyptus* essential oil, shows potential inhibitory activity. Molecular dynamics simulations have also identified oleanolic acid (from onions), α-tocotrienols (from coconuts), asiatic acid (from guavas), and culinosides (*Eucalyptus globulus*) as potent inhibitors, with oleanolic acid demonstrating high binding affinity to M-pro.

B. **PL-pro-: A Dual-Function Viral Enzyme**

PL-pro is not only involved in polyprotein cleavage at the nsp1–nsp4 junctions but also disrupts host immunity through deubiquitination of host proteins, thereby enabling viral immune evasion (Fig. 5.3). This makes PLpro an equally important antiviral target. Cinnamic amides extracted from *Tribulus terrestris* fruits inhibit PL-pro function. Several plant-based compounds have shown efficacy against 3CL-pro as well, including extracts from *Taxillus chinensis*, *Dioscorea polystachya*, *Cibotium barometz*, *Cassia tora*, and *Gentiana scabra*. Lignins, tannins, and coumarinshave also exhibited inhibitory effects. Specifically, savinin(a lignin molecule) and 8-hydroxyabicta-9(11), 13-dien-12-one(a diterpenoid) are active against 3CL-pro.

C. **Phenolic and Polyphenolic Inhibitors**

Tanshinones from *Salvia miltiorrhiza* act as non-competitive inhibitors by preventing protease isomerization. Rosmariquinone, in particular, blocks slow-binding cysteine protease isomerization. Phenolic components in black tea—including tannic acid, 3-isotheaflavin-3-gallate, and theaflavin-3,3′-digallate [22]—also demonstrate significant inhibitory activity against 3CL-pro. From *Isatis indigotica*, compounds such as emodin, sinigrin, hesperetin, indigo, and β-sitosterol reduce 3CL-pro activity. Moreover, *Torreya nucifera* produces several potent biflavonoids, including amentoflavone, ginkgetin, bilobetin, sciadopitysin, luteolin, quercetin, and apigenin. Many of these also occur in *Ginkgo biloba*, *Chamaecyparis obtusa*, and *Hypericum perforatum*. Notably, biflavonoids like amentoflavone, ginkgetin, bilobetin,

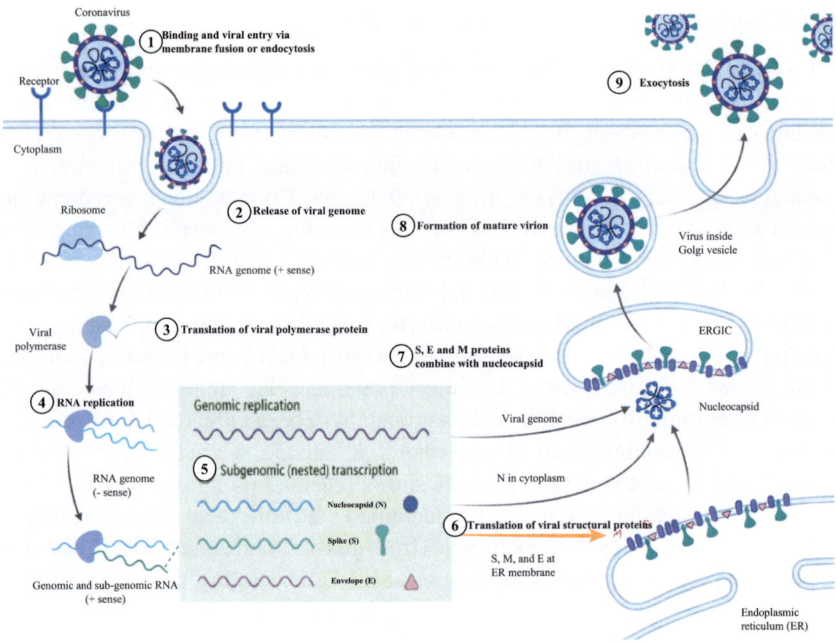

Fig. 5.3 The cycle of life and replication of SARS-CoV 2 (Figure created by Biorender)

and sciadopitysin function as non-competitive inhibitors of 3CL-pro, and compounds such as myricetin and scutellarein have also been shown to decrease 3CL-pro activity.

D. **Inhibitors from Paulownia and Psoralea**

Paulownia tomentosa is a source of tomentin-type compounds such as3′-O-4′-O-methyldiplacone, 3′-O-4′-O-methyldiplacol, diplacone, mimulone, and6-geranyl-4′,5,7-trihydroxy-3′,5′-dimethoxyflavone, which inhibit PL-pro- activity. Similarly, *Psoralea corylifolia* contains polyphenolics likeneobavaisoflavone, bavachinin, isobavachalcone, psoralidin, corylifol A, and4′-O-methylbavachalcone, withpsoralidinshowing significant PL-pro- inhibition.

E. **Triterpenoids and Terpenoids: Potent Natural Antivirals**

In vitro assays using Vero E6 cells demonstrated that terpenoids and lignoids inhibit SARS-CoV replication by over 50%. *Tripterygium regelii* produces quinone-methide triterpenes, such as celastrol (tripterine), tingenone, iguesterin, and pristimerin, which act as competitive inhibitors of 3CL-pro. Ethanolic extracts of *Torreya nucifera* were found to contain 3CL-pro inhibitors, and molecular docking studies identified amentoflavone as particularly effective. Additionally, geranylated flavonoids have been recognized as the most potent PLpro inhibitors identified from natural sources [11].

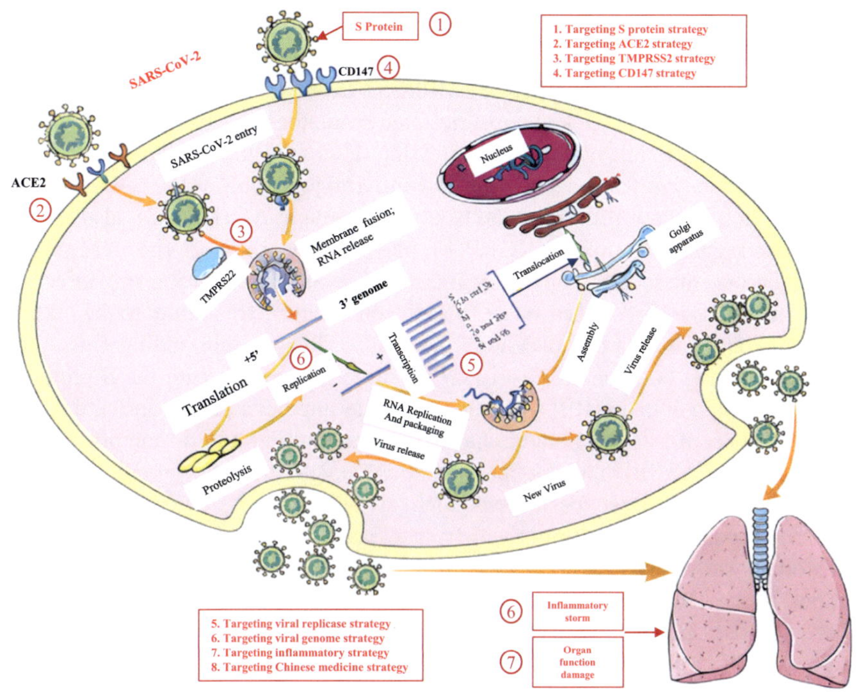

Fig. 5.4 Strategy for intervening in the replication and life cycle of SARS-CoV 2 (Figure created by Biorender)

5.15 Inhibiting Serine Proteases TMPRSS2 and ADAM17: Natural Compounds as Entry Blockers and Inflammation Modulators

SARS-CoV-2 utilizes several host proteases to facilitate entry into target cells, among which transmembrane protease serine 2 (TMPRSS2) and a disintegrin and metalloproteinase 17 (ADAM17) are critical [3]. TMPRSS2 cleaves the viral spike (S) glycoprotein at the S1/S2 site, a step essential for membrane fusion and viral entry. ADAM17, also known as TNF-α converting enzyme (TACE), regulates inflammatory signalling and viral entry by cleaving the ectodomain of ACE2, resulting in a soluble form that alters receptor availability and inflammatory cytokine production. Notably, ADAM17 also processes pro-TNF-α into its active soluble form, contributing to cytokine storms observed in severe COVID-19 cases. Therefore, both TMPRSS2 and ADAM17 are attractive therapeutic targets for preventing viral infection and controlling hyper-inflammation.

Natural products have demonstrated potent inhibitory effects on these proteases. TAPI-2, a broad-spectrum ADAM17 inhibitor, has been shown to block ACE2 shedding and reduce lung inflammation by suppressing TNF-α release. Silymarin, a flavonolignan from *Silybum marianum*, competitively binds to ADAM17's active site, inhibiting its proteolytic function. Cryptotanshinone, a diterpenoid derived from *Salvia miltiorrhiza*, suppresses TMPRSS2 expression by downregulating the androgen receptor (AR)—a key transcriptional regulator of TMPRSS2. This mechanism is particularly relevant in AR-expressing tissues, such as the lungs, and gives cryptotanshinone both anti-androgenic and antiviral potential.

In addition to these proteases, the heat shock protein A5 (HSPA5), also known as glucose-regulated protein 78 (GRP78), plays a critical role in viral recognition and cell surface interaction, especially under stress conditions. During viral infection or cellular stress, HSPA5 translocate from the endoplasmic reticulum to the plasma membrane and forms a complex with the viral S-glycoprotein, aiding entry. The substrate-binding domain (SBD) of HSPA5 directly interacts with the S-protein's receptor-binding domain (RBD) and thus serves as another strategic antiviral target. AI-based molecular docking studies have identified multiple natural compounds that bind to the HSPA5 SBD and disrupt its interaction with the spike protein.

Among these are phytoestrogens such as biochanin A, daidzein, genistein, and formononetin, primarily derived from *Cicer arietinum* L. (chickpea). These compounds show strong affinity for HSPA5 SBD and may block virus-host interactions at the membrane level. Other natural compounds like chlorogenic acid, hydroxytyrosol (from olive leaves), cryptotanshinone, and caffeic acid phenethyl ester (CAPE) exhibit moderate binding affinity, while thymoquinone, palmitic acid, linolenic acid, and cinnamaldehyde have weaker interactions but may act synergistically. Additionally, cis-p-coumaric acid competes for S-protein binding due to similar affinity profiles.

Interestingly, endogenous hormones—estrogen, progesterone, and testosterone—also demonstrate moderate binding to HSPA5 SBD, which may partly explain observed gender-based differences in COVID-19 susceptibility and severity. Among all ligands studied, phytoestrogens and estrogenic compounds were found to be the most effective binders to HSPA5. This supports the hypothesis that dietary or plant-derived estrogens could modulate viral entry by disrupting chaperone-assisted S-protein attachment.

In conclusion, the inhibition of TMPRSS2 and ADAM17 offers a two-pronged approach to preventing SARS-CoV-2 infection and reducing associated inflammation. At the same time, interfering with HSPA5-mediated viral recognition provides an auxiliary mechanism for limiting viral spread, particularly in older or immunocompromised individuals experiencing elevated cellular stress. The identification of natural inhibitors such as cryptotanshinone, silymarin, and various phytoestrogens through AI-driven molecular docking offers promising avenues for therapeutic development and drug repurposing in the context of COVID-19.

5.16 Conclusion

The COVID-19 pandemic has highlighted the need for new ways to fight SARS-CoV-2 and similar viruses. Since these viruses share many features, treatments that target common viral mechanisms could work against several types. Both prevention and treatment are important for controlling outbreaks. Advances in artificial intelligence and computational biology have sped up the search for new drugs, especially those that block the virus from entering or multiplying in cells. AI tools can quickly screen many natural and synthetic compounds to find the most promising ones. Many of these candidates are plant-based compounds with known effects on viral and human proteins. Their variety and proven activity make them good starting points for further improvement with AI. Because SARS-CoV-2 is similar to other coronaviruses, drugs developed for one may work for others. AI can help make these compounds more effective and easier to use. Combining natural products with AI-driven drug discovery offers a promising path toward better and more accessible treatments for COVID-19 and future outbreaks.utbreaks.

Funding Information: This study has not received any Government or Non-government funding.

Conflict of Interest: The authors declared no conflict of interest.

References

1. Abe M, Tahara M, Sakai K, Yamaguchi H, Kanou K, Shirato K, Takeda M. TMPRSS2 is an activating protease for respiratory parainfluenza viruses. J Virol. 2013;87(21):11930–11935.
2. Anand K, Ziebuhr J, Wadhwani P, Mesters JR, Hilgenfeld R. Coronavirus main proteinase (3CLpro) structure: basis for design of anti-SARS drugs. Science. 2003;300(5626):1763–7. https://doi.org/10.1126/science.1085658.
3. Anbardar N, Dixon SL, Munugoti S, Gaddam M, Kashfi K, Kasulis L, Asadipooya K. Thyroid disorders and COVID-19: a comprehensive review of literature. Front. Endocrinol. 2025;16:1535169.
4. Balasubramanian S, Rao NM, Goenka A, Roderick M, Ramanan AV. Coronavirus disease 2019 (COVID-19) in children-what we know so far and what we do not. Indian Pediatrics. 2020;57(5):435–442.
5. Becker RC. COVID-19 update: COVID-19-associated coagulopathy. J Thrombosis Thrombolysis. 2020;50(1):54–67.
6. Bojkova D, Klann K, Koch B, Widera M, Krause D, Ciesek S, Münch C. Proteomics of SARS-CoV-2-infected host cells reveals therapy targets. Nature. 2020;583(7816):469–72.
7. Cavalcanti IDL, Cajuba de Britto Lira Nogueira M. Pharmaceutical nanotechnology: which products are been designed against COVID-19?. J Nanoparticle Res. 2020;22(9):276.
8. Chan JFW, Kok KH, Zhu Z, Chu H, To KKW, Yuan S, Yuen KY. Genomic characterization of the 2019 novel human-pathogenic coronavirus isolated from a patient with atypical pneumonia after visiting Wuhan. Emerg Microbes Infect. 2020;9(1):221–36. https://doi.org/10.1080/22221751.2020.1719902.
9. Chen C, Yin Q, Tian J, Gao X, Qin X, Du G, Zhou Y. Studies on the changes of pharmacokinetics behaviors of phytochemicals and the influence on endogenous metabolites after the combination

of Radix Bupleuri and Radix Paeoniae Alba based on multi-component pharmacokinetics and metabolomics. Front Pharmacol. 2021;12:630970.
10. Cheng FJ, Huynh TK, Yang CS, Hu DW, Shen YC, Tu CY, Ho CY. Hesperidin is a potential inhibitor against SARS-CoV-2 infection. Nutrients 2021;13(8):2800.
11. Cho JK, Curtis-Long MJ, Lee KH, Kim DW, Ryu HW, Yuk HJ, Park KH. Geranylated flavonoids displaying SARS-CoV papain-like protease inhibition from the fruits of Paulownia tomentosa. Bioorganic Med. Chem. 2013;21(11):3051–3057.
12. Cinatl J, Morgenstern B, Bauer G, Chandra P, Rabenau H, Doerr HW. Treatment of SARS with human interferons. The Lancet. 2003;362(9380):293–294.
13. Clerkin KJ, Fried JA, Raikhelkar J, Sayer G, Griffin JM, Masoumi A, Uriel N. COVID-19 and cardiovascular disease. Circulation. 2020;141(20):1648–1655.
14. Cucinotta D, Vanelli M. WHO Declares COVID-19 a pandemic. Acta bio-medica: Atenei Parmensis. 2020;91(1):157–160. https://doi.org/10.23750/abm.v91i1.9397
15. Ejemel M, Li Q, Hou S, Schiller ZA, Tree JA, Wallace A, Wang Y. A cross-reactive human IgA monoclonal antibody blocks SARS-CoV-2 spike-ACE2 interaction. Nature Commun. 2020;11(1):4198.
16. Gul S, Ozcan O, Asar S, Okyar A, Barıs I, Kavakli IH. In silico identification of widely used and well-tolerated drugs as potential SARS-CoV-2 3C-like protease and viral RNA-dependent RNA polymerase inhibitors for direct use in clinical trials. J. Biomolecul. Struct. Dynam. 2021;39(17):6772–6791.
17. He S, Yang J, He M, Yan D, Tang S, Rong L. The risk of future waves of COVID-19: modeling and data analysis. Mathematical Biosciences and Engineering. 2021;18(5).
18. Hoffmann M, Kleine-Weber H, Schroeder S, Krüger N, Herrler T, Erichsen S, Pöhlmann S. SARS-CoV-2 cell entry depends on ACE2 and TMPRSS2 and is blocked by a clinically proven protease inhibitor. Cell. 2020;181(2):271–280. https://doi.org/10.1016/j.cell.2020.02.052
19. Hottz ED, Azevedo-Quintanilha IG, Palhinha L, Teixeira L, Barreto EA, Pão CR, Bozza PT. Platelet activation and platelet-monocyte aggregate formation trigger tissue factor expression in patients with severe COVID-19. Blood The J. Amer. Soc. Hematol. 2020;136(11):1330–1341.
20. Huang C, Wang Y, Li X, Ren L, Zhao J, Hu Y, Cao B. Clinical features of patients infected with 2019 novel coronavirus in Wuhan, China. The Lancet. 2020;395(10223):497–506. https://doi.org/10.1016/S0140-6736(20)30183-5.
21. Jamshidnia M, Sewell RD, Rafieian-Kopaei M. An update on promising agents against COVID-19: secondary metabolites and mechanistic aspects. Current Pharmaceut. Design. 2022;28(29):2415–2425.
22. Khanal LN, Pokharel YR, Sharma K, Kalauni SK. Plant-derived secondary metabolites as potential mediators against COVID-19: a review. Prithvi Acad. J. 2020;1–18.
23. Kumar P, Bhardwaj T, Kumar A, Gehi BR, Kapuganti SK, Garg N, Giri R. Reprofiling of approved drugs against SARS-CoV-2 main protease: an in-silico study. J. Biomolecul. Struct. Dynam. 2022;40(7), 3170–3184.
24. Kunnumakkara AB, Rana V, Parama D, Banik K, Girisa S, Henamayee S, Aggarwal BB. COVID-19, cytokines, inflammation, and spices: how are they related?. Life Sci. 2021;284:119201.
25. Li X, He X, Chen S, Le Y, Bryant MS, Guo L, Mei N. The genotoxicity potential of luteolin is enhanced by CYP1A1 and CYP1A2 in human lymphoblastoid TK6 cells. Toxicol. Lett. 2021;344:58–68.
26. Lippi G, Lavie CJ, Sanchis-Gomar F. Cardiac troponin I in patients with coronavirus disease 2019 (COVID-19): evidence from a meta-analysis. Progress Cardiovascular Diseases. 2020;63(3):390.
27. Livingston E, Bucher K, Rekito A. Coronavirus disease 2019 and influenza 2019–2020. JAMA. 2020;323(12):1122–1122. https://doi.org/10.1001/jama.2020.2633.
28. Llivisaca-Contreras SA, Naranjo-Morán J, Pino-Acosta A, Pieters L, Vanden Berghe W, Manzano P, Cevallos-Cevallos JM. Plants and natural products with activity against various types of coronaviruses: a review with focus on SARS-CoV-2. Molecules. 2021;26(13):4099.

29. Ma X, Yan L, Zhu Q, Shao F. Puerarin attenuates cisplatin-induced rat nephrotoxicity: the involvement of TLR4/NF-κB signaling pathway. PLoS One. 2017;12(2):e0171612.
30. Mohd Nor NH, Othman F, Mohd Tohit ER, Md Noor S. Medicinal herbals with antiplatelet properties benefit in coronary atherothrombotic diseases. Thrombosis. 2016(1):5952910.
31. Paraskevis D, Kostaki EG, Magiorkinis G, Panayiotakopoulos G, Sourvinos G, Tsiodras AS. Full-genome evolutionary analysis of the novel coronavirus (2019-nCoV) rejects the hypothesis of emergence as a result of a recent recombination event. Infect. Genet. Evolut. 2020;79:104212.
32. Preti E, Di Mattei V, Perego G, Ferrari F, Mazzetti M, Taranto P, Calati R. The psychological impact of epidemic and pandemic outbreaks on healthcare workers: rapid review of the evidence. Curr Psychiatry Rep. 2020;22(8):43.
33. Rathore JS, Ghosh C. Severe acute respiratory syndrome coronavirus-2 (SARS-CoV-2), a newly emerged pathogen: an overview. Pathogens Disease. 2020;78(6):ftaa042.
34. Ren Y, Meyer G, Anderson AT, Lauber KM, Gallucci JC, Gao G, Kinghorn AD. Development of potential therapeutic agents from black elderberries (the Fruits of Sambucus nigra L.). Molecules. 2024;29(13):2971.
35. Ripari N, Sartori AA, da Silva Honorio M, Conte FL, Tasca KI, Santiago KB, Sforcin JM. Propolis antiviral and immunomodulatory activity: a review and perspectives for COVID-19 treatment. J. Pharmacy Pharmacol. 2021;73(3):281–299.
36. Roth M, Araya JJ, Timmermann BN, Hagenbuch B. Isolation of modulators of the liver-specific organic anion-transporting polypeptides (OATPs) 1B1 and 1B3 from Rollinia emarginata Schlecht (Annonaceae). J Pharmacol Exp Ther. 2011;339(2):624–32.
37. Roth M, Timmermann BN, Hagenbuch B. Interactions of green tea catechins with organic anion-transporting polypeptides. Drug Metabolism Disposit. 2011a;39(5):920–926.
38. Saul S, Einav S. Old drugs for a new virus: repurposed approaches for combating COVID-19. ACS Infect. Diseases. 2020;6(9):2304–2318.
39. Shenoy V, Grobe JL, Qi Y, Ferreira AJ, Fraga-Silva RA, Collamat G, Katovich MJ. 17β-Estradiol modulates local cardiac renin-angiotensin system to prevent cardiac remodeling in the DOCA-salt model of hypertension in rats. Peptides. 2009;30(12):2309–15.
40. Soni VK, Mehta A, Ratre YK, Tiwari AK, Amit A, Singh RP, Vishvakarma NK. Can curcumin, a traditional spice component, hold the promise against COVID-19?. Europ. J. Pharmacol. 2020;886:173551.
41. Suzuki YJ, Gychka SG. SARS-CoV-2 spike protein elicits cell signaling in human host cells: implications for possible consequences of COVID-19 vaccines. Vaccines. 2021;9(1):36.
42. Tabrez S, Rahman F, Ali R, Alouffi AS, Akand SK, Alshehri BM, Rub A. Cynaroside inhibits Leishmania donovani UDP-galactopyranose mutase and induces reactive oxygen species to exert antileishmanial response. Biosci. Reports. 2021;41(1).
43. Tolulope A. Discovery of OJT009 as a novel inhibitor of severe acute respiratory syndrome coronavirus 2 infection for potential treatment of COVID-19 (Order No. 30573799); 2023. pp. 2861566908. Retrieved from https://www.proquest.com/dissertations-theses/discovery-ojt009-as-novel-inhibitor-severe-acute/docview/2861566908/se-2
44. Upadhyay S, Tripathi PK, Singh M, Raghavendhar S, Bhardwaj M, Patel AK. Evaluation of medicinal herbs as a potential therapeutic option against SARS-CoV-2 targeting its main protease. Phytotherapy Res. 2020;34(12):3411–3419.
45. Wang B, Andraweera P, Elliott S, Mohammed H, Lassi Z, Twigger A, Marshall HS. Asymptomatic SARS-CoV-2 infection by age: a global systematic review and meta-analysis. The Pediatric Infectious Disease J. 2023;42(3):232–239.
46. Xu T, Zhang N, Nichols HL, Shi D, Wen X. Modification of nanostructured materials for biomedical applications. Mater Sci Eng: C. 2007;27(3):579–594.
47. Yang X, Yu Y, Xu J, Shu H, Xia JA, Liu H, Shang Y. Clinical course and outcomes of critically ill patients with SARS-CoV-2 pneumonia in Wuhan, China: a single-centered, retrospective, observational study. Lancet Respir Med. 2020;8(5):475–81. https://doi.org/10.1016/S2213-2600(20)30079-5.
48. Yonesi M, Rezazadeh A. Plants as a prospective source of natural anti-viral compounds and oral vaccines against COVID-19 coronavirus. 2020.

49. Zhang C, Chen H, Sun L, Zhao P, Qi C, Yang Y, Jung YS. Bis-benzylisoquinoline alkaloids inhibit porcine epidemic diarrhea virus by disrupting virus entry. Pathogens. 2023;12(6):845.
50. Zou X, Chen K, Zou J, Han P, Hao J, Han Z. Single-cell RNA-seq data analysis on the receptor ACE2 expression reveals the potential risk of different human organs being vulnerable to 2019-nCoV infection. Front Med. 2020;14(2):185–192. https://doi.org/10.1007/s11684-020-0754-0

Chapter 6
Challenges in AI-Generated Drug Candidates

Anissa Nofita Sari

Abstract Generative AI has made it faster to find promising molecules in early drug development, but turning these into real medicines remains a challenge. This chapter explores key issues, including the limits of ADMET prediction, generalization, and reproducibility. It presents case studies on kinase inhibitors, antimicrobial peptides, and COVID-19 repurposing to highlight where current methods succeed and where they fall short. The chapter also addresses ethical and regulatory topics like transparency and data privacy. Finally, it offers practical strategies—such as explainable AI, federated learning, multi-omics integration, closed-loop automation, and community standards—to help bridge the gap between computer predictions and real-world outcomes, supporting broader adoption of AI in drug development and regulation.

Keywords AI drug discovery · Generative models · De novo drug design · ADMET

Generative AI is rapidly changing drug discovery by allowing researchers to explore chemical space more thoroughly and design new medicines. While traditional methods like QSAR modeling and high-throughput screening have contributed, they are limited by the size of chemical libraries and rely on known structures [3–5]. Generative AI can propose new molecules, improve multiple drug properties at once, and speed up the search for drug candidates. This has led to hopes that AI can make drug discovery faster and more affordable. Still, even as algorithms improve at predicting binding and chemical properties, real biological systems are more complex than these models can capture. Many promising computer-generated candidates fail in lab or animal tests due to poor absorption, unexpected toxicity, or unstable metabolism. Large pharmaceutical companies report that most failures are

A. N. Sari (✉)
Vaccine and Drug Research Center, Health Research Organization, National Research and Innovation Agency (BRIN), Bogor, Indonesia
e-mail: anis032@brin.go.id

A. A. Parikesit and A. N. M. Ansori (eds.), *Recent Computational Techniques in De Novo Drug Design*, Bio-IT and AI, https://doi.org/10.1007/978-3-032-20428-8_6

linked to poor physical properties, safety concerns, or weak pharmacokinetics [23–25, 27–29]. Reviews of AI-based patents also reveal that few candidates have strong experimental support, especially for in vivo or ADMET results. In cancer research, most drug candidates do not advance because they lack sufficient bioavailability, cause side effects, or offer limited clinical benefit [55]. These issues highlight the gap between computer predictions and real-world outcomes, showing the current limits of AI in drug discovery.

Recent studies point out challenges like dataset bias and weak generalization in generative AI models. Many models are trained on limited data, which leads to overfitting and poor results on new targets or diseases. The field also faces a reproducibility crisis, with inconsistent results and little transparency about data, model design, or parameters. These problems slow progress and lower confidence among regulators and industry professionals.

Ethical and regulatory issues are central to AI-driven drug discovery. The lack of transparency in deep learning makes it harder to ensure accountability, interpret results, and build clinical trust [1]. Regulators are working on guidelines for evaluating AI-generated molecules, but this creates uncertainty for companies. Other key concerns include data privacy, algorithmic bias, and the risk of misuse. Tackling these challenges is necessary for transparent and clinically useful AI in drug discovery. This chapter reviews the main scientific and practical issues through case studies, offering a balanced look at generative AI's potential and the advances needed to connect computational predictions with medical practice.

6.1 Limitations of ADMET Prediction

Accurately predicting ADMET (Absorption, Distribution, Metabolism, Excretion, and Toxicity) properties is essential for evaluating the safety and efficacy of potential medications. While AI systems are making progress in forecasting these pharmacokinetic and toxicological characteristics, their precision is still limited [54]. Many sub-domains remain particularly challenging to model despite advances in AI architectures, larger datasets, and computational power [14–16].

1. Absorption: AI models can estimate how well a drug is absorbed in the gut by using features like logP, molecular weight, and topological polar surface area. These help give early predictions about oral absorption. However, absorption is not just about passive movement through membranes. Transporter proteins, such as P-glycoprotein and organic anion transporting polypeptides, play a big role in drug absorption and removal, but AI models still do not capture their effects well. First-pass metabolism in the liver and gut also greatly reduces how much drug reaches the bloodstream, and most models do not include these factors. Other things like gut bacteria, pH levels in the intestine, and whether food is present also affect absorption, but are not well represented in AI systems. Because of

these gaps, many computer-predicted candidates with good absorption scores do not reach high enough levels in the body during real tests.

2. Drug distribution depends on factors like plasma protein binding, how drugs move into tissues, and crossing barriers in the body. AI can predict plasma protein binding, but has trouble with changes caused by diseases, such as low albumin or inflammation, which affect how much free drug is available. Predicting if a drug can cross the blood–brain barrier is also hard because it depends on both the drug's properties and changing removal mechanisms. Drug distribution in tissues is uneven, and conditions like fibrosis, poor blood flow, or changes in blood vessels make predictions even harder, but these are rarely included in models. Even advanced AI-based PBPK simulations are limited by the quality and amount of data. Because of these issues, predictions of how much drug reaches tissues often do not match what is seen in real patients, making it hard to set the right dose.
3. Predicting metabolism is an important but difficult part of ADMET modeling. CYP enzymes, which handle most drug metabolism, vary a lot because of genetics, other drugs, and interactions. Deep learning can help predict which drugs are processed by these enzymes, but the accuracy is uneven, especially for rare enzymes. AI often fails to spot reactive metabolites, which can cause toxicity. While newer models are better at predicting these pathways, they are not widely used in clinics yet. Sometimes, chemicals that seem safe in computer tests can form toxic byproducts in the body, showing that generative AI may suggest new structures without fully checking their metabolic safety.
4. Excretion, which involves removing drugs through the kidneys and liver, is another area where AI struggles. Most predictions are based on simple ideas about filtration and secretion, but real drug clearance depends on things like transporters, age, diseases, and other medications, which AI often ignores. For example, people with kidney disease may build up drugs differently, and AI may not catch this. Changes in liver clearance due to inflammation or drug interactions are also missed. Not considering these factors can lead to wrong predictions about how quickly drugs are cleared and cause dosing mistakes in clinical settings.
5. Toxicity is still the main barrier in drug development. Machine learning can predict short-term toxicity fairly well, but it is much harder to predict long-term problems like genetic damage, cancer risk, or effects on reproduction. The data used to train these models is often limited and inconsistent, which makes it hard for AI to generalize. Even the best deep learning models struggle to predict effects that depend on dose or build up over time. Some liver or heart toxicities only show up after long exposure, but most models use short-term data. Multi-organ toxicity and immune-related side effects are also hard to predict because there are not enough combined datasets. This gap between AI predictions and real biology explains why unexpected toxicities often cause drug candidates to fail in clinical trials.

6.2 Generalization Issues

Generative AI models often have trouble applying what they learn to new situations. A molecule that seems promising in computer tests might fail in real experiments because of biased or limited training data, or because the model has overfit to what it already knows. If a model is trained on narrow data, it may not pick up rules that work for new cases. As a result, compounds that look good in predictions might have side effects, metabolic issues, or fail safety tests that were not part of the training data. The gap between computer success and failure in the lab or in animals shows that these models often do not generalize well. They tend to perform better on well-studied targets but struggle with less-explored areas like transcription factors, protein–protein interactions, or new allosteric sites. This increases bias, as the model learns features that help with common targets but misses patterns important for rare ones. Some strategies, like oversampling underrepresented classes or domain adaptation, try to reduce this bias, but they often do not solve the core problem of limited data. Dataset bias also comes from past drug development trends, where compounds that were easier to make or had common structures are overrepresented in databases. Generative AI can repeat these biases, making it harder to suggest truly new chemical types. Approaches like active learning and encouraging differences from training data have been proposed to help, such as using generative workflows that penalize similarity to the training set to improve generalization and novelty [6].

6.2.1 Overfitting

Even with lots of training data, models can end up just memorizing details from the dataset instead of learning general chemical rules. This means they might do well on test sets that look like their training data, but perform poorly on new types of molecules or tasks. Overfitting is especially a problem when the data is limited or very similar. Deep learning methods use tricks like dropout, regularization, and data augmentation to help, but it's still hard to create good augmentation methods for molecular data. In practice, overfitting can make a model predict strong binding for a molecule just because it looks like something in the training set, not because it truly fits the target. This can waste time and resources. Some studies suggest using contrastive learning or separating out different features to reduce overfitting, but making models that generalize well to many types of molecules is still a big challenge.

6.2.2 Applicability Domain

Many AI-generated compounds do not follow standard drug-likeness rules like Lipinski's Rule of Five, Veber's guidelines, or rules for how easy they are to make. This lowers their chances of becoming real drugs. For example, if a molecule is too big or flexible, it might not dissolve well, be stable, or get absorbed properly. Generative algorithms often focus on binding or novelty and may suggest molecules that are hard or impossible to make or use. Ideally, models should alert users when a candidate is outside their reliable prediction range, but most do not, so users may not realize these risks. Some approaches, like using a retrosynthetic score, help guide the generation of molecules that are easier to synthesize.

The survey on the optimization of generative artificial intelligence-based models indicates that metrics such as QED and SA are used to guide models toward the drug-like space; however, many generated compounds continue to fail drug-likeness criteria, highlighting the disparity between computational goals and practical limitations [22].

Langevin et al. [26], in their study "On the difficulty of validating molecular generative models realistically," emphasize that generative AI models frequently value novelty and binding scores over chemical feasibility and drug-likeness criteria. Consequently, numerous compounds produced exceed the application domain, contravening established drug-likeness criteria and raising concerns about their translational significance.

6.2.3 Cross-Species Differences

A continual constraint of generative AI models is their failure to account for species-specific pharmacokinetic and pharmacodynamic variations [20]. Compounds anticipated to be stable or efficacious in silico may exhibit unpredictable behavior in vivo due to discrepancies in metabolic enzymes, such as cytochrome P450 isoforms, which vary considerably among rodents, primates, and humans. A chemical may exhibit metabolic stability in human cell tests yet experience quick clearance in mice or accumulation in non-human primates, hence diminishing its translational significance [30].

Traditional extrapolation methods, such as allometric scaling and simplified compartmental scaling, can prove inadequate when metabolic or transporter-mediated processes differ among species. Allometric scaling ($Y = a \cdot BW^b$) is commonly used to estimate characteristics such as clearance or volume of distribution; however, its precision is limited, particularly for small compounds undergoing substantial hepatic metabolism [19]. BPK models generally surpass naïve scaling when they integrate species-specific anatomical, physiological, and mechanistic drug characteristics, for instance, a recent cross-species PBPK investigation of a therapeutic protein demonstrated that direct allometric scaling inadequately predicted human pharmacokinetics without species-specific modifications [7].

Recent studies emphasize the significance of integrating cross-species modeling into AI-driven pharmaceutical design. PBPK simulations of niclosamide prodrugs yielded accurate predictions solely by accounting for species-specific metabolic conversion and tissue absorption [53]. Translational studies also highlight the absence of universal generalizations across species, underscoring the need for explicit cross-species evaluations before advancing AI-generated candidates into preclinical or clinical phases [45].

6.3 Reproducibility Crisis

The reproducibility crisis in AI-driven drug design stems from the lack of open data, opaque methodologies, and inconsistent benchmarks. The following is a list of four significant contributors [9–11].

6.3.1 Transparency

The absence of transparency in AI research for drug discovery is a significant impediment to reproducibility. This precludes independent replication, as numerous studies are published without access to the underlying code, data, or detailed methodology. The pharmaceutical industry's emphasis on intellectual property protection and confidentiality exacerbates this issue, rendering critical datasets and trained models inaccessible to the scientific community [40]. It is challenging to verify claims or conduct an impartial comparison of methods in the absence of transparency. Missing details, such as preprocessing pipelines, hyperparameters, or training protocols, can have a substantial impact on outcomes, even when fragmentary data or pseudo-code are released. The regulatory adoption of AI-driven drug discovery is exacerbated by the lack of reproducibility, which erodes trust [12, 43].

6.3.2 Benchmarking

In the field of chemistry, benchmark datasets like MOSES and GuacaMol have become the standard instruments for evaluating generative models. They facilitate more consistent comparisons among studies by providing standardized assignments and metrics. Nevertheless, these metrics are inherently limited, as they focus primarily on simplified molecular endpoints such as novelty, validity, or physicochemical scores [38].

In the context of actual drug discovery, multi-objective optimization is necessary, which includes drug-likeness, toxicity, selectivity, metabolic stability, and synthetic accessibility [36, 37]. These complexities are not adequately captured by current benchmarks, raising concerns that models that perform well on MOSES or GuacaMol may not perform effectively in experimental validation. Therefore, it is imperative to broaden the scope of benchmark design to encompass multi-parameter clinical relevance [44].

6.3.3 Publication Bias

The reproducibility crisis is further exacerbated by publication bias. Journals and conferences frequently emphasize "success stories" in AI applications, disregarding negative or null results. This results in a distorted perception of AI's efficacy, as the literature is disproportionately populated with positive findings [52]. Consequently, researchers may unintentionally seek to expand upon methodologies that are unsuccessful in broader testing scenarios. The community is also unable to learn from past errors due to the absence of published negative results, leading to wasted resources and duplicated failures. This systemic bias could be mitigated by promoting the publication of negative or equivocal studies [2].

6.3.4 Variability in Implementations

Reproducibility is compromised by diversity in implementation, even when code and data are accessible. Minor modifications in hyperparameters, random seeds, or training protocols might result in significantly divergent outcomes. Hardware accelerators (GPUs versus TPUs) and variations in software library versions may lead to performance or output discrepancies.

This sensitivity makes it difficult to reproduce published results and compare different methods fairly. In AI-driven drug design, this kind of variability can cause people to trust methods that may not work as well as they seem. To address this, the field needs clear reporting standards, solid statistical checks, and testing across different platforms to ensure results are reliable and not just due to technical details. For example, research on DDR1-selective inhibitors shows that while AI and structure-guided methods helped develop new compounds, only a few advanced, partly because of challenges like selectivity and drug-likeness. These issues highlight the need for careful validation and consideration of practical factors in drug development [39].

Case Study 1: **Antimicrobial Peptides**

An exemplary case is "Antimicrobial Peptides Design Utilizing Deep Learning and Rational Modifications: Efficacy Against Bacteria, Candida albicans, and Cancer

Cells." In that study, deep learning models generated candidate peptides; of these, 26 were predicted in silico to have antibacterial activity and toxicity. Subsequently, 12 were synthesized, yielding variable findings (some exhibited high potency but varied between species and assays). The phrase "Discovery of naturally inspired antimicrobial peptides using deep learning" refers to the identification of peptides from extensive genomic datasets, followed by optimization. It also indicates that numerous candidates exhibit only moderate activity or require enhancement for efficacy, as well as considerations for toxicity and stability [32].

Case Study 2: COVID-19 Repurposing (and Weak Efficacy/Contradictory Trials)

Numerous studies indicate that various AI-repurposed medicines for COVID-19 did not demonstrate significant clinical efficacy. The study titled "Emerging role of artificial intelligence in therapeutics for COVID-19: a systematic review" indicates that despite several compounds being predicted and tested in vitro, only a limited number advance to animal studies or clinical trials, frequently yielding inconsistent or weak outcomes [21].

The article "The Failure of Drug Repurposing for COVID-19 as an Effect of Excessive Hypothesis Testing and Weak Mechanistic Evidence" outlines numerous repurposing experiments that initially showed promise but ultimately failed under thorough testing, due to insufficient evidence, small sample sizes, or inadequate mechanistic support [31].

6.4 Ethical and Regulatory Perspectives

AI-driven drug design raises important ethical and regulatory questions. Agencies and stakeholders must decide how to evaluate chemicals created by black-box models, determine responsibility if problems arise, ensure fairness and safety, and prevent misuse. The main concerns are ethical use, regulatory science, and data privacy.

6.4.1 Ethical Use

AI in medication design can be very helpful, but it can also be dangerous, especially if biases or dual-use hazards aren't handled correctly. For example, if training datasets include too many people from certain groups (e.g., race, gender, or age), the AI may prefer chemicals or therapy profiles that work better for those groups but are less safe or effective for groups that are less well represented. People are increasingly talking about dual-use concerns. The same technologies that help make new antivirals or antimicrobials could, in theory, be used to make poisons or other dangerous substances. There is also the potential that new chemotypes designed to be very potent can have unforeseen side effects or be hazardous if ethics, safety, and risk assessment are not integral to the design process. These concerns are increasingly being taken into account by ethical frameworks. The FDA's draft guidance "Considerations for the Use of Artificial Intelligence To Support Regulatory Decision-Making for Drug and Biological Products" stresses fairness, avoiding prejudice, making the context of use clear, and evaluating risks [33, 35]. Also, FDA center offices have issued discussion papers that stress the importance of avoiding algorithmic discrimination when datasets are biased and ensuring that automated systems don't unfairly favor one group of patients over another [34].

6.4.2 Regulatory Science

Regulators are starting to catch up, but things are still changing. In January 2025, the U.S. FDA announced draft guidance outlining a risk-based framework for AI models used in medication or biological product submissions. This recommendation states that developers should identify the AI model's context of use (COU), assess the model's risk, document procedures for assessing its credibility, manage modifications throughout the model's lifecycle, and provide evidence of safety, efficacy, and quality. Internationally, similar problems are being dealt with [46–51]. For instance, the MELLODDY group in Europe showed that federated learning may allow drug companies to collaborate while preserving privacy and still produce models that are safe and valid under regulatory requirements. As of mid-2025, a comparison of regulatory procedures also shows that AI/ML model evaluation across different product stages (preclinical, clinical, and post-market) is increasingly focusing on traceability, reproducibility, and bias reduction [8].

6.4.3 Data Privacy

Using molecular data from patients or businesses to train AI algorithms can potentially violate privacy regulations and compromise business secrets. Federated learning is becoming a promising approach because it enables multiple parties to train a model without exchanging raw data. Horst et al. [13] show that federated learning enables collaboration across regulatory boundaries while preserving data privacy and sovereignty. This is especially important in drug research, where datasets are sensitive [13]. A PubMed-indexed study on collaborative drug discovery corroborates that federated frameworks enhance prediction performance in drug-target and drug-drug interaction tasks while maintaining the locality of raw data [17]. Researchers are also exploring other methods, such as safe multi-party computation, differential privacy, and homomorphic encryption. A new assessment in the Journal of Biomedical Informatics discusses the trade-offs between these privacy-preserving methods.

It says that greater privacy usually means higher processing overhead or lower accuracy. These methodologies demonstrate that achieving a balance between privacy and utility is essential for regulatory acceptance of AI in medication design.

6.5 Future Directions

6.5.1 Explainable AI

Black-box models are hard for chemists and doctors to understand because they don't explain how they came to their conclusions. This makes them less useful in drug discovery. To make molecular predictions easier to understand, many have suggested using explainable AI (XAI) methods such as attention visualization, SHAP values, and feature whitening. Huang et al. [18] presented a self-interpretable graph neural network (GNN) framework that emphasizes chemically significant substructures that influence model decisions, thereby enhancing user trust and regulatory approval.

6.5.2 Federated Learning

Pharmaceutical collaborations can use federated learning to train models on datasets that are spread out without having to share private raw data. The MELLODDY consortium, one of the largest cross-industry partnerships in drug research, used federated learning across 10 pharmaceutical companies to predict bioactivity while preserving privacy. These methods show that federated AI can make training data more diverse, improve generalization, and make it possible for several parties to design drugs in a way that follows the rules.

6.5.3 Integration with Omics

Incorporating multi-omics data (genomics, transcriptomics, proteomics, metabolomics) into AI models can enhance disease-specific context and augment predictive accuracy in intricate treatment domains, like oncology and neurodegenerative disorders. A recent review stressed that AI using omics data can tailor drug discovery pipelines by checking molecular predictions against system-level biology to make them more reliable. This integration is expected to reduce translational discrepancies between in silico predictions and clinical outcomes.

6.5.4 Closed-Loop Automation

AI-guided chemical formation, robotic synthesis, and high-throughput biological screening all work together to create closed-loop pipelines that speed up the drug development process. These iterative methods enable quick design-make-test-analyze (DMTA) loops, reducing the time to optimize leads significantly. A recent approach showed that combining generative AI with automated labs may produce new bioactive chemicals much faster than traditional methods. Closed-loop automation will probably be a key part of drug research and development in the future.

6.5.5 Community Standards

In order to resolve the issues of reproducibility and transparency in the design of AI drugs, it is imperative that community standards be established. Cheminformatics and multi-omics datasets are increasingly being aligned with the FAIR (Findable, Accessible, Interoperable, and Reusable) data principles, ensuring that AI models can be consistently benchmarked and validated. Who introduced the FAIR framework, emphasized its critical role in facilitating data reusability across disciplines. Since then, recent studies have expanded its application to drug discovery AI (Nature Scientific Data). There will be no reliable progress without the establishment of FAIR-compliant benchmarks and open datasets [41, 42].

6.6 Conclusion

Generative AI has transformed drug discovery by enabling the design of new molecules and faster identification of promising leads. However, bringing these advances into real clinical use is still limited by challenges like ADMET prediction accuracy, dataset bias, generalization, and reproducibility. Ethical and regulatory concerns highlight the need for AI systems that are transparent, understandable, and protect privacy. Progress will require collaboration across disciplines, thorough validation, and meeting regulatory standards. Adopting explainable and federated AI, integrating multi-omics data, and following FAIR data principles can help make AI more effective in drug development. Ultimately, success will depend on balancing innovation, ethical responsibility, and clinical needs.

Acknowledgements The author thanks colleagues at the Vaccine and Drug Research Center, BRIN, for discussions and collaborations in computational drug discovery.

Competing Interests
The author declares no competing interests.

References

1. Ai D, Cai H, Wei J, Zhao D, Chen Y, Wang L. DEEPCYPs: a deep learning platform for enhanced cytochrome P450 activity prediction. Front Pharmacol. 2023;14:1099093. https://doi.org/10.3389/fphar.2023.1099093.
2. Bespalov A, Steckler T, Skolnick P. Be positive about negatives-recommendations for the publication of negative (or null) results. Eur Neuropsychopharmacol. 2019;29(12):1312–20. https://doi.org/10.1016/j.euroneuro.2019.10.007.
3. Cai Z, Zafferani M, Akande OM, Hargrove AE. Quantitative structure-activity relationship (QSAR) study predicts small-molecule binding to RNA structure. J Med Chem. 2022;65(10):7262–77. https://doi.org/10.1021/acs.jmedchem.2c00254.
4. Deng J, Yang Z, Wang H, Ojima I, Samaras D, Wang F. A systematic study of key elements underlying molecular property prediction. Nat Commun. 2023;14(1):6395. https://doi.org/10.1038/s41467-023-41948-6.
5. Feng B, LaPerle JL, Chang G, Varma MV. Renal clearance in drug discovery and development: molecular descriptors, drug transporters and disease state. Expert Opin Drug Metab Toxicol. 2010;6(8):939–52. https://doi.org/10.1517/17425255.2010.482930.
6. Filella-Merce I, Molina A, Diaz L, Orzechowski M, Berchiche YA, Zhu YM, Guallar V. Optimizing drug design by merging generative AI with a physics-based active learning framework. Commun Chem. 2025;8(1):238. https://doi.org/10.1038/s42004-025-01635-7.
7. Franz M, Jairam RK, Kuepfer L, Hanke N. PBPK-based translation from preclinical species to humans for the full-size IgG therapeutic efalizumab. Front Pharmacol. 2024;15:1418870. https://doi.org/10.3389/fphar.2024.1418870.
8. Fu L, Jia G, Liu Z, Pang X, Cui Y. The applications and advances of artificial intelligence in drug regulation: a global perspective. Acta Pharm Sin B. 2025;15(1):1–14. https://doi.org/10.1016/j.apsb.2024.11.006.
9. Gangwal A, Ansari A, Ahmad I, Azad AK, Kumarasamy V, Subramaniyan V, Wong LS. Generative artificial intelligence in drug discovery: basic framework, recent advances, challenges, and opportunities. Front Pharmacol. 2024;15:1331062. https://doi.org/10.3389/fphar.2024.1331062.
10. Gangwal A, Lavecchia A. Unleashing the power of generative AI in drug discovery. Drug Discov Today. 2024;29(6): 103992. https://doi.org/10.1016/j.drudis.2024.103992.
11. Gold ER, Cook-Deegan R. AI drug development's data problem. Science. 2025;388(6743):131. https://doi.org/10.1126/science.adx0339.
12. Han H. Challenges of reproducible AI in biomedical data science. BMC Med Genomics. 2025;18(Suppl 1):8. https://doi.org/10.1186/s12920-024-02072-6.
13. Horst A, Loustalot P, Yoganathan S, Li T, Xu J, Tong W, Renaudin M. Federated learning: a privacy-preserving approach to data-centric regulatory cooperation. Front Drug Saf Regul. 2025;5:1579922. https://doi.org/10.3389/fdsfr.2025.1579922.
14. Hoyt CT, Zdrazil B, Guha R, Jeliazkova N, Martinez-Mayorga K, Nittinger E. Improving reproducibility and reusability in the journal of cheminformatics. J Cheminform. 2023;15(1):62. https://doi.org/10.1186/s13321-023-00730-y.
15. Hu XM, Hou YY, Teng XR, Liu Y, Li Y, Li W, Ai CZ. Prediction of cytochrome P450-mediated bioactivation using machine learning models and in vitro validation. Arch Toxicol. 2024;98(5):1457–67. https://doi.org/10.1007/s00204-024-03701-w.
16. Hu Y, Ren Q, Liu X, Gao L, Xiao L, Yu W. In silico prediction of human organ toxicity via artificial intelligence methods. Chem Res Toxicol. 2023;36(7):1044–54. https://doi.org/10.1021/acs.chemrestox.2c00411.
17. Huang D, Ye X, Sakurai T. Multi-party collaborative drug discovery via federated learning. Comput Biol Med. 2024;171: 108181. https://doi.org/10.1016/j.compbiomed.2024.108181.
18. Huang D, Ye X, Zhang Y, Sakurai T. Collaborative analysis for drug discovery by federated learning on non-IID data. Methods 2023;219:1–7. https://doi.org/10.1016/J.YMETH.2023.09.001

19. Huh Y, Smith DE, Feng MR. Interspecies scaling and prediction of human clearance: comparison of small- and macro-molecule drugs. Xenobiotica. 2011;41(11):972–87. https://doi.org/10.3109/00498254.2011.598582.
20. Idasiak-Piechocka I, Lewandowski D, Swigut W, Kalinowski J, Mikosza K, Suchowiejski P, Miedziaszczyk M. Effect of hypoalbuminemia on drug pharmacokinetics. Front Pharmacol. 2025;16:1546465. https://doi.org/10.3389/fphar.2025.1546465.
21. Kaushal K, Sarma P, Rana SV, Medhi B, Naithani M. Emerging role of artificial intelligence in therapeutics for COVID-19: a systematic review. J Biomol Struct Dyn. 2022;40(10):4750–65. https://doi.org/10.1080/07391102.2020.1855250.
22. Khater T, Alkhatib SA, AlShehhi A, Pitsalidis C, Pappa AM, Ngo ST, Truong VK. Generative artificial intelligence based models optimization towards molecule design enhancement. J Cheminform. 2025;17(1):116. https://doi.org/10.1186/s13321-025-01059-4.
23. Kong X, Lin K, Wu G, Tao X, Zhai X, Lv L, Yang S. Machine learning techniques applied to the study of drug transporters. Molecules. 2023;28(16). https://doi.org/10.3390/molecules28165936
24. Krstevska A, Duris J, Ibric S, Cvijic S. In-depth analysis of physiologically based pharmacokinetic (PBPK) modeling utilization in different application fields using text mining tools. Pharmaceutics. 2022;15(1). https://doi.org/10.3390/pharmaceutics15010107
25. Kuo YM, Barrett JS. Consideration of the root causes in candidate attrition during oncology drug development. Clin Pharmacol Drug Dev. 2024;13(9):952–60. https://doi.org/10.1002/cpdd.1464.
26. Langevin M, Grebner C, Güssregen S, Sauer S, Li Y, Matter H, Bianciotto M. Impact of applicability domains to generative artificial intelligence. ACS Omega 2023;8(25):23148–23167. https://doi.org/10.1021/acsomega.3c00883
27. Lee H, Kim J, Kim JW, Lee Y. Recent advances in AI-based toxicity prediction for drug discovery. Front Chem. 2025;13:1632046. https://doi.org/10.3389/fchem.2025.1632046.
28. Liao C, Sitzmann M, Pugliese A, Nicklaus MC. Software and resources for computational medicinal chemistry. Future Med Chem. 2011;3(8):1057–85. https://doi.org/10.4155/fmc.11.63.
29. Liu YT, Zhang LL, Jiang ZY, Tian XS, Li PL, Wu PH, Sun C. Applications of artificial intelligence in biotech drug discovery and product development. MedComm 2025;6(8):e70317. https://doi.org/10.1002/mco2.70317
30. Martignoni M, Groothuis GM, de Kanter R. Species differences between mouse, rat, dog, monkey and human CYP-mediated drug metabolism, inhibition and induction. Expert Opin Drug Metab Toxicol. 2006;2(6):875–94. https://doi.org/10.1517/17425255.2.6.875.
31. Maziarz M, Stencel A. The failure of drug repurposing for COVID-19 as an effect of excessive hypothesis testing and weak mechanistic evidence. Hist Philos Life Sci. 2022;44(4):47. https://doi.org/10.1007/s40656-022-00532-9.
32. Mesa A, Orrego A, Branch-Bedoya JW, Mera-Banguero C, Orduz S. Antimicrobial peptides design using deep learning and rational modifications: activity in bacteria, candida Albicans, and cancer cells. Curr Microbiol. 2025;82(9):379. https://doi.org/10.1007/s00284-025-04346-3.
33. Niazi SK. Artificial intelligence in small-molecule drug discovery: a critical review of methods, applications, and real-world outcomes. Pharmaceuticals (Basel). 2025;18(9). https://doi.org/10.3390/ph18091271
34. Nwebonyi N, McKay F. Exploring bias risks in artificial intelligence and targeted medicines manufacturing. BMC Med Ethics. 2024;25(1):113. https://doi.org/10.1186/s12910-024-01112-1.
35. Ocana A, Pandiella A, Privat C, Bravo I, Luengo-Oroz M, Amir E, Gyorffy B. Integrating artificial intelligence in drug discovery and early drug development: a transformative approach. Biomark Res. 2025;13(1):45. https://doi.org/10.1186/s40364-025-00758-2.
36. Ozcelik R, Brinkmann H, Criscuolo E, Grisoni F. Generative deep learning for de novo drug design horizontal line a chemical space odyssey. J Chem Inf Model. 2025;65(14):7352–72. https://doi.org/10.1021/acs.jcim.5c00641.

37. Parrot M, Tajmouati H, da Silva VBR, Atwood BR, Fourcade R, Gaston-Mathe Y, Perron Q. Integrating synthetic accessibility with AI-based generative drug design. J Cheminform. 2023;15(1):83. https://doi.org/10.1186/s13321-023-00742-8.
38. Polykovskiy D, Zhebrak A, Sanchez-Lengeling B, Golovanov S, Tatanov O, Belyaev S, Zhavoronkov A. Molecular sets (MOSES): a benchmarking platform for molecular generation models. Front Pharmacol. 2020;11: 565644. https://doi.org/10.3389/fphar.2020.565644.
39. Sanawar R, Sahayasheela VJ, Sarath P, Dan VM. Discoidin domain receptor 1 inhibitors: advances and future directions for novel therapeutics with aid of DNA encoded library screens and artificial intelligence. Mini Rev Med Chem. 2023;23(15):1507–13. https://doi.org/10.2174/1389557523666230125114921.
40. Schaduangrat N, Lampa S, Simeon S, Gleeson MP, Spjuth O, Nantasenamat C. Towards reproducible computational drug discovery. J Cheminform. 2020;12(1):9. https://doi.org/10.1186/s13321-020-0408-x.
41. Serrano DR, Luciano FC, Anaya BJ, Ongoren B, Kara A, Molina G, Lalatsa A. Artificial intelligence (AI) applications in drug discovery and drug delivery: revolutionizing personalized medicine. Pharmaceutics. 2024;16(10). https://doi.org/10.3390/pharmaceutics16101328
42. Spielvogel CP, Schindler N, Schroder C, Stellnberger SL, Wadsak W, Mitterhauser M, Vraka C. Enhancing blood-brain barrier penetration prediction by machine learning-based integration of novel and existing, in silico and experimental molecular parameters from a standardized database. J Chem Inf Model. 2025;65(6):2773–84. https://doi.org/10.1021/acs.jcim.4c02212.
43. Stegmann JU, Littlebury R, Trengove M, Goetz L, Bate A, Branson KM. Trustworthy AI for safe medicines. Nat Rev Drug Discov. 2023;22(10):855–6. https://doi.org/10.1038/s41573-023-00769-4.
44. Thomas M, O'Boyle NM, Bender A, De Graaf C. MolScore: a scoring, evaluation and benchmarking framework for generative models in de novo drug design. J Cheminform. 2024;16(1):64. https://doi.org/10.1186/s13321-024-00861-w.
45. Toutain PL, Ferran A, Bousquet-Melou A. Species differences in pharmacokinetics and pharmacodynamics. Handb Exp Pharmacol. 2010(199);19–48. https://doi.org/10.1007/978-3-642-10324-7_2
46. Tran TTV, Tayara H, Chong KT. Artificial intelligence in drug metabolism and excretion prediction: recent advances, challenges, and future perspectives. Pharmaceutics 2023;15(4). https://doi.org/10.3390/pharmaceutics15041260
47. Tripathi N, Goshisht MK, Sahu SK, Arora C. Applications of artificial intelligence to drug design and discovery in the big data era: a comprehensive review. Mol Divers. 2021;25(3):1643–64. https://doi.org/10.1007/s11030-021-10237-z.
48. Wang Y, Magar R, Liang C, Barati Farimani A. Improving molecular contrastive learning via faulty negative mitigation and decomposed fragment contrast. J Chem Inf Model. 2022;62(11):2713–25. https://doi.org/10.1021/acs.jcim.2c00495.
49. Waring MJ, Arrowsmith J, Leach AR, Leeson PD, Mandrell S, Owen RM, Weir A. An analysis of the attrition of drug candidates from four major pharmaceutical companies. Nat Rev Drug Discov. 2015;14(7):475–86. https://doi.org/10.1038/nrd4609.
50. Warraich HJ, Tazbaz T, Califf RM. FDA perspective on the regulation of artificial intelligence in health care and biomedicine. JAMA. 2025;333(3):241–7. https://doi.org/10.1001/jama.2024.21451.
51. Watanabe R, Esaki T, Ohashi R, Kuroda M, Kawashima H, Komura H, Mizuguchi K. Development of an in silico prediction model for P-glycoprotein efflux potential in brain capillary endothelial cells toward the prediction of brain penetration. J Med Chem. 2021;64(5):2725–38. https://doi.org/10.1021/acs.jmedchem.0c02011.
52. White N, Parsons R, Collins G, Barnett A. Evidence of questionable research practices in clinical prediction models. BMC Med. 2023;21(1):339. https://doi.org/10.1186/s12916-023-03048-6.
53. Yang M, Wang AQ, Padilha EC, Shah P, Hagen NR, Ryu C, Xu X. Use of physiological based pharmacokinetic modeling for cross-species prediction of pharmacokinetic and tissue distribution profiles of a novel niclosamide prodrug. Front Pharmacol. 2023;14:1099425. https://doi.org/10.3389/fphar.2023.1099425.

54. Zhang R, Wen H, Lin Z, Li B, Zhou X. Artificial intelligence-driven drug toxicity prediction: advances, challenges, and future directions. Toxics 2025;13(7). https://doi.org/10.3390/toxics13070525
55. Zhang X, Han Y, Huang W, Jin M, Gao Z. The influence of the gut microbiota on the bioavailability of oral drugs. Acta Pharm Sin B. 2021;11(7):1789–812. https://doi.org/10.1016/j.apsb.2020.09.013.

Chapter 7
AI-Driven Network Pharmacology in Natural Product-Based Drug Design: A Systems Bioinformatics Perspective

Wisnu Ananta Kusuma, Heru Cahya Rustamaji, Rudi Heryanto, and Irmanida Batubara

Abstract Systems bioinformatics integrates computing, statistics, and systems biology to model complex molecular interactions using network-based methods. This field advances precision medicine by combining multi-omics data to discover disease mechanisms and support. A key application is network pharmacology, which uses a multicomponent-multitarget perspective to investigate compound-target-pathway interactions, which is particularly useful for natural product-based drug discovery. Recent advances in artificial intelligence (AI) have improved this integration by enabling network target identification through relationship mining, target association, and navigation strategies, moving away from descriptive analyses toward predictive and mechanistic insights. This paradigm represents a shift away from the conventional single-compound-single-target model, which is frequently ineffective for multifactorial diseases, and toward synergistic, systems-level strategies. To fully realize this potential, we propose a structured framework that includes in silico prediction, pharmacological evaluation, extract standardization, preclinical and clinical testing, and metabolomic analysis. Building on this, we develop a workflow for precision medicine using natural products that includes disease network analysis, target protein selection, genotype–phenotype association, machine learning-based

W. A. Kusuma (✉)
Bioinformatics Study Program, Faculty of Mathematics and Natural Sciences, IPB University, Bogor, Indonesia
e-mail: ananta@apps.ipb.ac.id

H. C. Rustamaji
Department of Informatics, Faculty of Industrial Technology, UPN Veteran Yogyakarta, Yogyakarta, Indonesia
e-mail: herucr@upnyk.ac.id

R. Heryanto · I. Batubara
Department of Chemistry, Faculty of Mathematics and Natural Sciences, IPB University, Bogor, Indonesia
e-mail: rudi_heryanto@apps.ipb.ac.id

I. Batubara
e-mail: ime@apps.ipb.ac.id

A. A. Parikesit and A. N. M. Ansori (eds.), *Recent Computational Techniques in De Novo Drug Design*, Bio-IT and AI, https://doi.org/10.1007/978-3-032-20428-8_7

enrichment analysis, and virtual screening. This workflow is used to examine the metabolite profile of *Murraya paniculata* and its mechanism of action on adipose tissue. Finally, we introduce the integrative platform for research in the Indonesian medicine ecosystem (I-PRIME), a digital platform that bridges Indonesia's biodiversity with biomedical informatics to accelerate translation into evidence-based clinical therapies. This section focuses on I-PRIME as a technical and conceptual framework for developing natural product-based precision medicine in Indonesia.

Keywords Artificial intelligence · Natural product · Network pharmacology · Precision medicine · System bioinformatics

7.1 Introduction

Systems bioinformatics is an interdisciplinary field that integrates computing, mathematics, statistics, and systems biology to elucidate the intricate interactions occurring within biological systems [1]. This method focuses on analyzing dynamic networks of genes, proteins, and metabolites. By integrating multi-omics data, systems bioinformatics allows for the modeling and prediction of biological processes, laying the groundwork for precision medicine by translating molecular heterogeneity into actionable clinical insights [2].

A key application of systems bioinformatics is network pharmacology, which investigates the relationships between bioactive compounds, molecular targets, and biological pathways to capture systemic therapeutic effects. In contrast to the conventional single-component-single-target paradigm, network pharmacology uses a multicomponent-multitarget perspective, making it particularly useful for the discovery of therapeutics derived from natural products. By systematically mapping compound-target-pathway interactions, this approach promotes the rational development of safer and more effective drugs while advancing precision medicine principles [1].

Precision medicine refers to the tailoring of therapy based on an individual's molecular, genetic, and biological profile, with the aim of improving treatment effectiveness while minimizing adverse effects [3]. This method leverages multi-omics data, including genomics, transcriptomics, proteomics, metabolomics, and phenomics, integrated within multi-level and multi-scale omics models to provide a comprehensive understanding of a patient's molecular state [4, 5]. Such modeling involves systematically analyzing and mapping omics data into molecular networks. By leveraging computational methods, including artificial intelligence (AI), omics data can be transformed into molecular networks that reveal disease mechanisms, identify potential therapeutic targets, and guide personalized treatment strategies [4–6].

Methodologically, the use of AI to identify potential network targets and connections among diseases, drugs, and molecular mechanisms can be classified into three main categories [7]. The first is network relationship mining, which uncovers associations from existing knowledge and omics data, frequently using natural language processing and network embedding methods. The second is network target association, which applies predictive modeling to link candidate compounds with disease-related targets. The third is network target navigation, which identifies functional modules that connect bioactive compounds to the disease network. These strategies show how AI expands network pharmacology beyond simple descriptive associations to predictive and mechanistic models.

At its core, AI-powered systems bioinformatics provides a consistent framework for combining large-scale omics data with traditional knowledge of natural product-based medicines. This integration allows us to uncover complex molecular patterns, predict the interactions of compounds with their targets, and classify disease types, thereby promoting the development of effective and personalized natural product-based therapies [8]. Building on this foundation, the remaining sections will discuss recent methods and practical applications of AI-driven network pharmacology in natural product-based drug discovery.

While systems-level methods reveal the complexities of disease mechanisms, conventional pharmacology has largely relied on a single compound-single target paradigm. Although this reductionist model has proven effective in therapies, it frequently fails to address multifactorial diseases such as cancer, diabetes, Alzheimer's, and autoimmune disorders, where multiple pathways and receptors are simultaneously involved. These shortcomings have prompted a paradigm shift toward multicomponent-multitarget strategies, as demonstrated by traditional medicines that use diverse bioactive compounds acting synergistically to restore systemic balance [9, 10]. These challenges underscore the need for a systems-oriented, multicomponent-multitarget framework, which will be elaborated in the following section.

7.2 Framework for a Multicomponent-Multitarget Approach in the Development of Medicines from Natural Sources

Building on this paradigm shift, the multicomponent-multitarget method provides a more comprehensive framework for drug discovery, particularly from natural sources. Natural products inherently embody systems-level therapeutic principles because their diverse bioactive compounds act synergistically across multiple molecular pathways, not only as primary therapeutic agents but also as immunomodulators, anti-inflammatory agents, and antioxidants [11].

Afendi et al. [12] proposed the hypothesis that *Jamu*, a traditional Indonesian herbal medicine, is composed of four key components: the main component, which is directly responsible for therapeutic efficacy against disease, and three supporting components that act as analgesics, antimicrobials, and anti-inflammatories. Owing

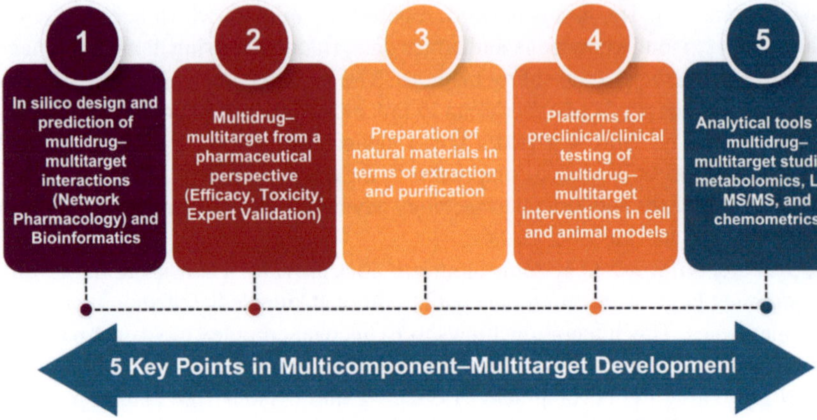

Fig. 7.1 Framework for a multicomponent-multitarget approach to natural-product-based drugs. The framework integrates in silico prediction, pharmaceutical evaluation, extract standardization, preclinical and clinical testing, and metabolomic/chemometric analysis to advance multicomponent-multitarget drug discovery from natural products

to this holistic nature, natural ingredients can comprehensively modulate molecular networks, making them a promising foundation for the development of safer, more effective, and system-based precision therapies [13].

To capitalize on this potential, a structured framework has been proposed to advance the multicomponent-multitarget paradigm. The framework includes the following steps: (1) in silico design and prediction of multicomponent-multitarget interaction, (2) pharmaceutical evaluation of efficacy, toxicity, and expert validation, (3) preparation and standardization of natural product extracts, (4) development of preclinical and clinical testing platforms using cell-based and animal models, and (5) metabolomic and chemometric analysis of multicomponent-multitarget interactions (Fig. 7.1).

Among these strategies, the in silico approach provides a solid foundation by allowing for rapid and comprehensive predictions of how bioactive compounds interact with disease-related molecular networks. This method, which incorporates systems bioinformatics and network pharmacology, allows mapping of the relationships between compounds, targets, and biological pathways, as well as identifying potential side effects early in the drug discovery process.

7.3 Workflow for the Development of Natural Product-Based Drug Using in Silico Methods

To improve the framework for natural product-based drug development, particularly the design and prediction of multicomponent-multitarget drugs using in silico methods, we propose a workflow for precision medicine from natural products, as shown in Fig. 7.2.

This workflow starts with analyzing disease relationships using network methods such as clustering [14] and community detection [15]. Comorbidities have a significant impact on both diagnosis and clinical outcomes, therefore, understanding these relationships is critical [16].

Subsequently, target proteins are screened using several databases, including OMIM—An Online Mendelian Inheritance in Man (https://www.omim.org/), TCGA-The Cancer Genomic Atlas, and GeneCards-The human gene database (https://www.genecards.org). The data were then analyzed using protein–protein interaction networks obtained from STRING, a protein–protein interaction networks database (https://string-db.org/). Protein target screening can be performed using topology-based methods, clustering, community detection, or skyline queries to identify proteins that play an important role in disease regulation [17].

The next step is to identify genomic markers, particularly single nucleotide polymorphisms (SNPs), and analyze their associations with phenotypes. This step ensures the compatibility of the target protein and the patient's genetic profile, as genetic variations can influence disease development and drug response [18]. The identified

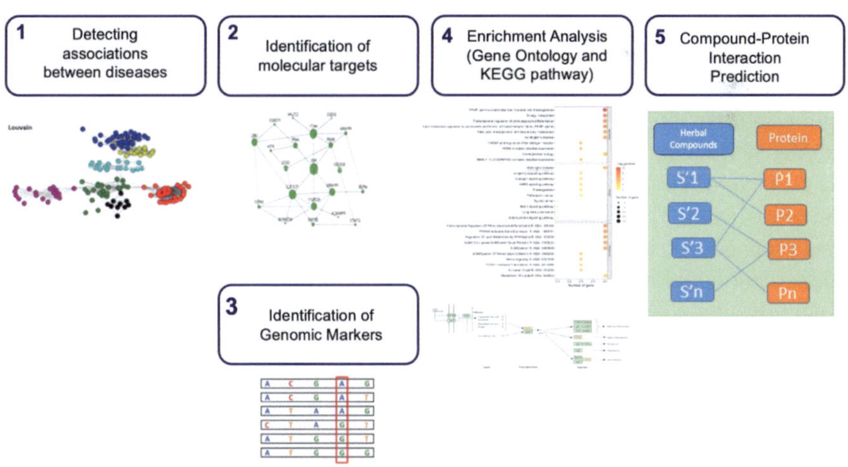

Fig. 7.2 Workflow for developing natural product-based drugs. The framework integrates disease network analysis, target protein selection, genomic profiling, enrichment analysis, and drug-target interaction prediction with in silico validation (docking and dynamics) to accelerate the discovery of multicomponent-multitarget therapeutics from natural products

targets are further examined using enrichment analysis to determine their biological functions [19] and related metabolic pathways [17].

Following this, a drug target interaction (DTI) prediction model is developed using machine learning or deep learning algorithms, with features such as molecular fingerprints and protein descriptors [20–22]. Alternatively, graph traversal of the bipartite plant-compound-protein network can be performed using the branch and bound method [23]. The final stage involves validating interactions using molecular docking and molecular dynamics simulations, which evaluate the affinity, stability, and conformation of compound-protein complexes under physiological conditions.

We used this workflow in several case studies, using different datasets and problem settings at each stage. Finally, we integrated this workflow into a case study of *Murraya paniculata* (kemuning), a natural product candidate with potential anti-obesity activity [17]. The details for each stage of the workflow are provided below.

7.3.1 Disease Network Analysis (Comorbidities)

Communities play an important role in medicine, particularly in understanding and treating complex diseases such as cancer and its associated comorbidities. Comorbidity refers to the presence of long-term health conditions that coexist with a primary disease and have a significant impact on patient prognosis [24]. Identifying the relationships between diseases is therefore essential for selecting accurate protein targets. Target selection should not be limited to protein–protein interaction networks of the primary disease but also include the interaction networks of its comorbidities.

We investigated the suitability of different community detection algorithms for identifying cancer comorbidities [15]. Given the variability of community detection results, the modularity generated by these algorithms was analyzed to determine significant comorbidities. Data on lung cancer comorbidities were collected using PubMed text mining, disease ontology searches, and disease similarity calculations. We then evaluated 20 algorithms using five modularity metrics and 16 fitness function assessments. The results highlighted five algorithms with the highest modularity: label propagation, spinglass, Chinese whispers, Louvain, and RB Pots. These algorithms identified significant comorbidities, including blood vessel disorders, immune system dysfunctions, bone-related conditions, pancreatic and metabolic disorders, atrial septal defect, atrial fibrillation, respiratory system diseases, interstitial lung disease, and diabetes mellitus. The fitness function justifies the results of the community algorithm, and the ones that have a significant effect are average internal degree, size, and edges inside.

This study introduces a new community detection algorithm called disassembly greedy modularity (DGM) [25], which was developed to overcome the limitations of the conventional greedy modularity (GM) algorithm, which frequently becomes trapped in local optima. In contrast to GM, DGM integrates exploration and exploitation strategies, broadening the search space and delivering more reliable community detection results.

DGM improves modularity by introducing exploration through node release and community splitting. These strategies use weak connectivity, non-triad nodes, and conductance to distinguish between stable and unstable communities. This study directly compares GM and DGM, with GM frequently converging to local optima. DGM balances exploration (node splitting, community release) and exploitation, consistently improving modularity.

The evaluation results show that DGM consistently outperforms GM. When compared with 25 other community detection algorithms, DGM ranked among the top ten performers. Its improvement is attributed to the node release and community splitting exploration strategy, which proved particularly effective on the used benchmark datasets such as Karate, Dolphins, Les Misérables, and Polbooks.

Overall, DGM demonstrates high stability and competitive performance when compared with established algorithms such as Spinglass [26], Leiden [27], Louvain [28], and RB Potts [29]. These results indicate that DGM has significant potential as a versatile community detection method applicable across diverse domains involving network data. In particular, DGM shows promise for use in disease network analysis and target protein selection.

7.3.2 *Target Protein Selection*

Target protein selection is the most crucial stage in the development of natural product-based drugs. The systems bioinformatics method analyzes protein–protein interaction networks using topology analysis, clustering, and community detection techniques. In this study, we used the community detection technique with DGM. We also developed a new centrality measure, community consideration centrality (CCC) [30], which considers the influence of nodes in local communities as well as global networks. In the context of protein networks, this method is important because biological interactions occur within specific subnetworks, therefore, identifying potential target proteins must consider their local role within the community. CCC consists of community consideration degree centrality, community consideration betweenness centrality, and community consideration closeness centrality. These three measures are equipped with parameter α (0.0–1.0) that regulates the balance between the contribution of the community (local network) and the global network to the centrality value of the node based on Eq. (7.1) [30].

$$\mathrm{CCC}_{\alpha}(\mathrm{i}) \leftarrow (\mathrm{d}(\mathrm{i} \in \mathrm{C}) + \alpha(\mathrm{d}(\mathrm{i} \in \mathrm{G}) - \mathrm{d}(\mathrm{i} \in \mathrm{C})) \quad (7.1)$$

where:

CCC_{α} *Community Consideration Centrality.*
$\mathrm{d}(\mathrm{i} \in \mathrm{C})$ Centrality value that only considers community networks.
$\mathrm{d}(\mathrm{i} \in \mathrm{G})$ Centrality value that only considers global networks.

If $\alpha = 0.0$, it indicates that the centrality calculation only considers the local community structure. If $\alpha = 1.0$, the centrality calculation only considers the entire network.

The evaluation of the Zachary karate Club, Dolphins, and lung cancer protein datasets shows that the combination of the DGM algorithm and CCC is more effective in identifying important nodes, including cancer marker proteins, compared with traditional centrality measures [30]. We applied this method to identify cancer hallmarks, where the DGM-CCC combination achieved optimal performance at α values of 0.1–0.3, with the best results at $\alpha = 0.1$, and 70% of the top 50 genes were recognized as cancer markers. These findings highlight that incorporating community structure in centrality calculations significantly improves the accuracy of detecting biologically relevant genes, with consistent performance trends across various CCC variants. Following the identification of target proteins, it is essential to assess their potential genetic associations. Therefore, the next step focuses on SNP analysis to identify variant-specific interactions.

7.3.3 Genotype–Phenotype Association

Precision medicine requires modeling the relationship between genomic markers and phenotypes, particularly for complex diseases such as type 2 diabetes mellitus (T2DM). This study used the gradient boosting algorithm [31] to identify SNPs associated with the T2DM phenotype using data from the Mouse Phenome database. The model achieved a mean squared error (MSE) of 0.061 and mean absolute error (MAE) of 0.171, and identified 30 potential SNPs related to insulin tolerance, 22 of which were verified from the Mouse Genome Informatics database [32].

This study summarizes a number of SNPs associated with T2DM as reported by [32]. Several SNPs in the Akt2 gene (thymoma viral proto-oncogene) and Igf1r gene (insulin-like growth factor I receptor) are associated with insulin resistance, which is a key feature of T2DM.

SNPs in the Cdkal1 gene (CDK5 regulatory subunit-associated protein 1-like 1) play a crucial role in protein translation accuracy; its dysfunction results in impaired insulin synthesis. The Iapp gene (islet amyloid polypeptide) has variants associated with impaired glucose tolerance, whereas Irs1 (insulin receptor substrate 1) is involved in direct insulin signaling regulation.

In addition, SNPs in the Fto gene (fat mass and obesity associated) show an association with obesity, cardiovascular disorders, and insulin resistance. Meanwhile, variants in the Insr gene (insulin receptor) are also directly associated with insulin resistance. Overall, these SNPs reflect the involvement of insulin regulatory pathways, protein translation, and energy metabolism in T2DM pathogenesis and have the potential to serve as molecular targets for more precise diagnosis and therapy. Once genetic associations are established, the workflow progresses to enrichment analysis, which contextualizes these targets within biological functions and pathways.

7.3.4 Machine Learning-Based Enrichment Analysis

This study combines ontology-based gene enrichment analysis and transcriptomic data analysis to uncover the biological mechanisms underlying hyperinflammation. The analysis was conducted on the gene functions, cellular components, and biological processes of the related proteins represented in three bipartite graphs, as well as protein–protein interactions using unsupervised learning methods, such as fuzzy k-partite clustering [33] and data from the STRING database. Transcriptomic data were obtained from Gene Expression Omnibus (GEO) and analyzed using Orange Data Mining [34].

The experimental results showed the involvement of proteins in various molecular functions, cellular components, and biological processes related to the cytokine storm in COVID-19, as well as the identification of important interferon-related genes such as Interferon Alpha/Beta Receptor subunit 1 (IFNAR1), Interferon Alpha Inducible Protein 6 (IFI6), Interferon Induced Protein With Tetratricopeptide Repeats (1IFIT1), and Interferon Induced Protein With Tetratricopeptide Repeats 3 (IFIT3). Enrichment analysis showed that 13 of the 18 Gene Ontology (GO) terms associated with cytokine storms were present in the clusters generated by fuzzy K-partite clustering. These findings indicate that interferon signaling plays an important role in COVID-19 hyperinflammation and severity [34].

According to GO analysis, this gene cluster is associated with cytokine and growth factor activity, with a predominance of extracellular and membrane components. These genes support cytokine signaling, immune response, inflammation, and gene expression regulation through the Signal Transducer and Activator of Transcription (STAT) and Mitogen-Activated Protein Kinase (MAPK) pathways. No significant involvement in non-specific cell proliferation was detected, confirming the focus on immune regulation and cytokine-mediated gene expression. With the enriched pathway information in place, the next step is virtual screening to identify bioactive compounds capable of modulating these targets.

7.3.5 Machine Learning Based Virtual Screening

The lack of compound-target interaction data poses a challenge in screening natural compounds, resulting in the development of a DTI prediction model. Erlina et al. [20] proposed a layered virtual screening method that integrates machine learning, pharmacophore modeling, and docking to identify candidate anti-COVID-19 compounds from Indonesian medicinal plants. This method addresses the limitations of conventional methods, which are typically conducted separately, and is expected to yield bioactive compounds with high potential as SARS-CoV-2 antivirals.

Three machine learning methods, Support Vector Machine (SVM), Multi-Layer Perceptron (MLP), and random forest, were used to predict compound interactions with SARS-CoV-2 3C-like protease (3CLPro), each with a unique decision-making method that complements the others. The results showed 98% prediction accuracy with a high F-measure. Ligand-based pharmacophore modeling also performed well, with a hit rate of 27.17%, an area under the curve calculated from all screening data, and an enrichment factor in the top 1% (EF1%) of 13.4.

Screening tests yielded 14 potential compounds from HerbalDB, which were then screened by molecular docking against 3CLpro using lopinavir as a control. Six compounds, including hesperidin and kaempferol, exhibited affinity comparable to that of lopinavir. These compounds are found in *Moringa oleifera* and *Psidium guajava*, potentially serving as herbal therapies for prevention or palliative care of COVID-19.

This study identified a number of Indonesian herbal compounds that have high affinity for the target protein 3CLPro, the primary enzyme in SARS-CoV-2 virus replication. The compound hesperidin, found in *Psidium guajava* and *Citrus aurantium*, showed a binding energy of − 8.72 kcal/mol, approaching the effectiveness of lopinavir (− 9.41 kcal/mol), which was used as a positive control for antiviral drugs.

Other compounds that also exhibited high affinity include kaempferol 3,4'-di-O-methylether (ermanin) from *Zingiber aromaticum* (− 8.51 kcal/mol), peonidine 3-(4'-arabinosylglucoside) from *Ipomoea fistulosa* (− 8.52 kcal/mol), and quercetin 3-(2G-rhamnosylrutinoside) from *Clitoria t*ernatea (− 8.56 kcal/mol). In addition, compounds from *Moringa oleifera*, such as myricetin-3-glucoside (− 8.26 kcal/mol) and rhamnetin 3-mannosyl-(1-2)-alloside, which are also found in *Cassia alata*, showed potential as 3CLPro inhibitors, with a binding energy of − 8.48 kcal/mol. These results indicate that several compounds derived from Indonesian herbal plants have promising antiviral potential against SARS-CoV-2 and could be further developed as bioinformatics-based natural drug candidates.

7.3.6 Integrated Workflow: A Case Study of **Murraya Paniculata**

This study uses a workflow for the development of precision medicine from natural ingredients, with a focus on the metabolite profile of the active compound *Murraya paniculata* and its mechanism of action on adipose tissue. The method integrates network pharmacology, molecular docking, molecular dynamics simulation, and in vitro testing. Although genotype–phenotype association analysis was not included, this study has a significant strength of in vitro validation, which reinforces the reliability of in silico results and adds applied value for developing natural product-based anti-obesity therapies.

Pharmacological network analysis using the skyline query technique and the Block-Nested Loop (BNL) algorithm identified three potential targets in obesity treatment, namely E1A Binding Protein P300 (EP300), Peroxisome Proliferator-Activated Receptor Gamma (PPARG), and Peroxisome Proliferator-Activated Receptor Gamma Coactivator 1-Alpha, also known as **PGC-1α** (PPARGC1A), all of which are associated with thermogenesis, energy metabolism, fat cell differentiation, and glucose homeostasis. The metabolite profile of *M. paniculata* leaves revealed 16 active compounds, 10 of which were selected based on drug-likeness and Absorption, Distribution, Metabolism, and Excretion (ADME) parameters for docking against PPARG and EP300. All compounds exhibited strong binding affinities (≤ -5.0 kcal/mol). Molecular dynamics simulations showed that the PPARG-trans-3-indoleacrylic acid complex exhibited the highest stability (RMSD 0.99 Å), outperforming the original Beecham Research Laboratories (BRL) ligand (2.02 Å). In vitro testing on 3T3-L1 adipocytes confirmed that the ethanolic extract of *M. paniculata* leaves significantly reduced the expression of PPARG and EP300 genes at concentrations as low as 62.5 and 15.625 μg/ml, respectively. These results indicate the strong potential of *M. paniculata* as an anti-obesity agent, acting through the inhibition of adipocyte maturation and regulating intracellular lipid metabolism.

The application of this workflow to other natural products has also been demonstrated through collaborative studies, including investigations on anticancer snake peptides [22, 35], phenolic acids [36], Kratom [37], fermented pineapple juice [38], and the bioactivity of *Sesbania grandiflora* [39]. Together, these efforts create a research ecosystem that integrates bioinformatics methods with biological diversity, providing a foundation for the sustainable development of nature-based medicines.

7.4 I-PRIME (Integrative Platform for Research in Indonesian Medicine Ecosystem)

Since 2014, the integrative platform for research in the Indonesian medicine ecosystem (I-PRIME) platform has been developed and continuously refined. Its origins trace back to two previously developed bioinformatics applications: (1). Indonesia Jamu Herbs (IJAH) analytics (https://ijah.cs.ipb.ac.id) is a web application that predicts interactions between bioactive compounds and target proteins using the bipartite local model with neighbor-based interaction-profile inferring (BLM-NII) algorithm. The prediction results are visualized to show the systemic connections between plants, compounds, proteins, and diseases; (2). Integrated Single Nucleotide Polymorphism Pipeline (ISNIP) (https://isnip.cs.ipb.ac.id) is a pipeline for identifying SNPs by comparing Next Generation Sequencing (NGS)-based sequencing results against reference genomes. This SNP information is crucial in genome-wide association studies, a statistical, and now also machine learning-based, study that links genetic variation to complex phenotypes such as drug response or disease clinical parameters.

Moving forward, I-PRIME is designed to be an increasingly comprehensive platform (Fig. 7.3), with the following key features: (1) An integrated repository covering phenotype data (HPO), genotype data (SNPs from dbSNP), protein data

(OMIM, UniProt, GeneCards, TCGA), protein interactions (STRING), Differentially Expressed Gene (DEG) data, active compounds (DrugBank, BindingDB, KnapSack), biological pathways (Gene Ontology, KEGG, DAVID), and Indonesian plant and natural material data; (2) Bioinformatics analytical functions include disease association identification, protein target screening from protein–protein interaction networks (I-PrimeTarget, under development), SNP-phenotype association using ISNIP, ontology-based gene and metabolic pathway enrichment analysis, and active compound screening against biological targets using IJAH analytics; (3) Interactive visualization supporting systematic data interpretation, including disease networks, therapeutic targets, and bioactivity of natural compounds; and (4) Collaborative facilities for discussion and knowledge sharing, supporting interpretation of results, diagnosis, and the design of more precise therapies.

I-PRIME is more than just a technical platform; it represents a new paradigm for collaborative, cross-disciplinary, and systemic approaches to therapeutic discovery. This platform bridges the study of Indonesia's biodiversity to modern biomedical information systems, accelerating the translation of research into evidence-based clinical therapies. The ultimate goal is to build an integrated, digital, and locally rooted precision medicine ecosystem in Indonesia.

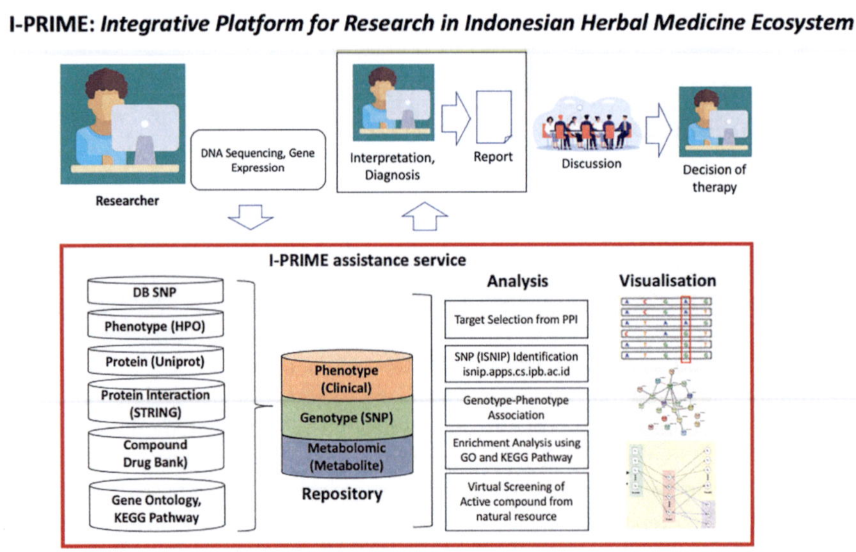

Fig. 7.3 I-PRIME: integrative platform for integrative platform for research in indonesian medicine ecosystem. I-PRIME is a comprehensive bioinformatics platform integrating diverse biomedical and natural product data (phenotype, genotype, protein, pathways, and compounds) with analytical tools for target screening, SNP-phenotype association, enrichment analysis, and compound-target prediction. It provides interactive visualization and collaborative facilities to support systemic, cross-disciplinary therapeutic discovery, bridging Indonesia's biodiversity with precision medicine development

7.5 Conclusion and Future Work

The integration of systems bioinformatics, precision medicine, and natural resource wealth is a holistic and contextual scientific strategy that has the potential to address health-related challenges. This method combines technological advances with a holistic understanding of biological systems, as well as local wisdom that views disease as a manifestation of systemic imbalance. We are building a solid foundation for the development of safe, effective, and relevant therapies that use biodiversity by combining systems bioinformatics, multi-omics analysis, network pharmacology, algorithms, and machine learning.

However, various challenges, including limitations in bioactive, multi-omics, and clinical data; a lack of breakthroughs in multicomponent-multitarget drug testing protocols; and a lack of industry participation in the development of natural-based medicines, still remain. Thus, the I-PRIME platform will be further developed and refined to serve as a collaborative platform for the development of natural product-based drugs. In addition, researchers in various multidisciplinary fields should increase discussions regarding the possibility of designing testing protocols for multicomponent-multitarget drugs by considering the contribution of systems bioinformatics and AI as part of these protocols.

Acknowledgements We express our sincere gratitude to Chairunnisa Nur Amanda for her invaluable assistance in preparing the figures and illustrations, which significantly improved the clarity of this study. We also acknowledge the use of AI-based tools, specifically ChatGPT (OpenAI) for language editing and refinement, and DeepL Translator for preliminary translation support. These tools were used to improve readability, but the authors are solely responsible for the scientific content and conclusions presented herein.

References

1. Oulas A, Minadakis G, Zachariou M, et al. Systems bioinformatics: increasing precision of computational diagnostics and therapeutics through network-based approaches. Brief Bioinform. 2017;20:806–24. https://doi.org/10.1093/bib/bbx151.
2. Jiang W, Ye W, Tan X, Bao YJ. Network-based multi-omics integrative analysis methods in drug discovery: a systematic review. BioData Min. 2025;18:27. https://doi.org/10.1186/s13040-025-00442-z.
3. Ashley EA. The precision medicine initiative: a new national effort. JAMA—J Am Med Assoc. 2015;313:2119–20. https://doi.org/10.1001/jama.2015.3595.
4. Hasin Y, Seldin M, Lusis A. Multi-omics approaches to disease. Genome Biol. 2017;18:83. https://doi.org/10.1186/s13059-017-1215-1.
5. Koumakis L. Deep learning models in genomics; are we there yet? Comput Struct Biotechnol J. 2020;18:1466–73. https://doi.org/10.1016/j.csbj.2020.06.017.
6. Huang S, Chaudhary K, Garmire LX. More is better: recent progress in multi-omics data integration methods. Front Genet. 2017;8:84. https://doi.org/10.3389/fgene.2017.00084.
7. Zhang P, Zhang D, Zhou W, et al. Network pharmacology: towards the artificial intelligence-based precision traditional Chinese medicine. Brief Bioinform. 2024;25:bbad518. https://doi.org/10.1093/bib/bbad518

8. Parikesit AA, Nurdiansyah R, Agustriawan D. Penerapan Pendekatan machine learning pada pengembangan basis data herbal Sebagai Sumber Informasi Kandidat Obat Kanker. J. Teknologi Ind. Pertanian. 2019;29:175–182. https://doi.org/10.24961/j.tek.ind.pert.2019.29.2.175
9. Duval MX. The inadequacy of the reductionist approach in discovering new therapeutic agents against complex diseases. Exp Biol Med. 2018;243:1004–13. https://doi.org/10.1177/1535370218794365.
10. Makhoba XH, Viegas C, Mosa RA, et al. Potential impact of the multi-target drug approach in the treatment of some complex diseases. Drug Des Devel Ther. 2020;14:3235–49. https://doi.org/10.2147/DDDT.S257494.
11. Beigoli S, Boskabady MH. The molecular basis of the immunomodulatory effects of natural products: a comprehensive review. Phytomedicine. 2024;135: 156028. https://doi.org/10.1016/j.phymed.2024.156028.
12. Afendi FM, Heryanto R, Darusman LK, et al. Jamu informatics: a new perspective in Jamu research. CICSJ Bulletin. 2016;34:47.
13. Bhinder B, Elemento O. Towards a better cancer precision medicine: systems biology meets immunotherapy. Curr Opin Syst Biol. 2017;2:67–73. https://doi.org/10.1016/j.coisb.2017.01.006.
14. Ferris J, Fiedeldey LK, Kim B, et al. Systematic review and meta-analysis of disease clustering in multimorbidity: a study protocol. BMJ Open. 2023;13: e076496. https://doi.org/10.1136/bmjopen-2023-076496.
15. Rustamaji HC, Suharini YS, Permana AA, et al. A network analysis to identify lung cancer comorbid diseases. Appl Netw Sci. 2022;7:30. https://doi.org/10.1007/s41109-022-00466-y.
16. Biswas A, Mehta HJ, Folch EE. Chronic obstructive pulmonary disease and lung cancer: interrelationships. Curr Opin Pulm Med. 2018;24:152–60. https://doi.org/10.1097/MCP.0000000000000451.
17. Fatriani R, Pratiwi FAK, Annisa A, et al. Unveiling the anti-obesity potential of Kemuning (Murraya paniculata): a network pharmacology approach. PLoS ONE. 2024;19: e0305544. https://doi.org/10.1371/journal.pone.0305544.
18. Visscher PM, Wray NR, Zhang Q, et al. 10 years of GWAS discovery: biology, function, and translation. Am J Hum Genet. 2017;101:5–22. https://doi.org/10.1016/j.ajhg.2017.06.005.
19. Ashburner M, Ball CA, Blake JA, et al. Gene ontology: Tool for the unification of biology. Nat Genet. 2000;25:25–9. https://doi.org/10.1038/75556.
20. Erlina L, Paramita RI, Kusuma WA, et al. Virtual screening of Indonesian herbal compounds as COVID-19 supportive therapy: machine learning and pharmacophore modeling approaches. BMC Complement Med Ther. 2022;22:207. https://doi.org/10.1186/s12906-022-03686-y.
21. Fadli A, Kusuma WA, Annisa, et al. Screening of potential Indonesia herbal compounds based on multi-label classification for 2019 coronavirus disease. Big Data Cognitive Comput. 2021;5:75. https://doi.org/10.3390/bdcc5040075
22. Kusuma WA, Fadli A, Fatriani R, et al. Prediction of the interaction between Calloselasma rhodostoma venom-derived peptides and cancer-associated hub proteins: a computational study. Heliyon. 2023;9. https://doi.org/10.1016/j.heliyon.2023.e21149
23. Kusuma WA, Habibi ZI, Amir MF, et al. Bipartite graph search optimization for type II diabetes mellitus Jamu formulation using branch and bound algorithm. Front Pharmacol. 2022;13: 978741. https://doi.org/10.3389/fphar.2022.978741.
24. Fowler H, Belot A, Ellis L, et al. Comorbidity prevalence among cancer patients: a population-based cohort study of four cancers. BMC Cancer. 2020;20:2. https://doi.org/10.1186/s12885-019-6472-9.
25. Rustamaji HC, Kusuma WA, Nurdiati S, Batubara I. Community detection with greedy modularity disassembly strategy. Sci Rep. 2024;14:4694. https://doi.org/10.1038/s41598-024-55190-7.
26. Reichardt J, Bornholdt S. Statistical mechanics of community detection. Phys Rev E Stat Nonlin Soft Matter Phys. 2006;74:16110. https://doi.org/10.1103/PhysRevE.74.016110.

27. Traag VA, Waltman L, van Eck NJ. From Louvain to Leiden: guaranteeing well-connected communities. Sci Rep. 2019;9:1–12. https://doi.org/10.1038/s41598-019-41695-z.
28. Blondel V, Guillaume JL, Lambiotte R. Fast unfolding of communities in large networks: 15 years later. J Stat Mech: Theory Exp. 2024;2024:P10008. https://doi.org/10.1088/1742-5468/ad6139.
29. Leicht EA, Newman MEJ. Community structure in directed networks. Phys Rev Lett. 2008;100:118703. https://doi.org/10.1103/PhysRevLett.100.118703.
30. Rustamaji HC, Kusuma WA, Nurdiati S, Batubara I. Community-consideration centrality, a case study of lung cancer proteins. IEEE Access. 2023;11:134340–54. https://doi.org/10.1109/ACCESS.2023.3337375.
31. Friedman JH. Greedy function approximation: a gradient boosting machine. Ann Stat. 2001;29:1189–232. https://doi.org/10.1214/aos/1013203451.
32. Fadli A, Hasibuan LS, Kusuma WA, Heryanto R. Single nucleotide polymorphism and type 2 diabetes mellitus phenotypes association using gradient boosting. In: 2020 International conference on advanced computer science and information systems, ICACSIS 2020. IEEE, 2020. pp. 115–20
33. Hartsperger ML, Blöchl F, Stümpflen V, Theis FJ. Structuring heterogeneous biological information using fuzzy clustering of k-partite graphs. BMC Bioinf. 2010;11:522. https://doi.org/10.1186/1471-2105-11-522.
34. Ramadhani HF, Annisa A, Tedjo A, et al. Combination of enrichment using gene ontology and transcriptomic analysis revealed contribution of interferon signaling to severity of COVID-19. Interdiscip Perspect Infect Dis. 2022;2022:3515001. https://doi.org/10.1155/2022/3515001.
35. Sofyantoro F, Yudha DS, Lischer K, et al. Bibliometric analysis of literature in snake venom-related research worldwide (1933–2022). Animals. 2022;12:2058. https://doi.org/10.3390/ani12162058.
36. Ekowati J, Tejo BA, Maulana S, et al. Potential utilization of phenolic acid compounds as anti-inflammatory agents through TNF-α convertase inhibition mechanisms: a network pharmacology, docking, and molecular dynamics approach. ACS Omega. 2023;8:46851–68. https://doi.org/10.1021/acsomega.3c06450.
37. Wimbarti S, Tallei TE, Kairupan BHR, et al. Probing the molecular mechanisms of Kratom's antipsychotic effects through a multi-modal computational approach. Curr Pharm Des. 2025;31:1370–87. https://doi.org/10.2174/0113816128335217241031033104.
38. Tallei TE, Fatimawali AAA, et al. Molecular insights into the anti-inflammatory activity of fermented pineapple juice using multimodal computational studies. Arch Pharm (Weinheim). 2024;357:2300422. https://doi.org/10.1002/ardp.202300422.
39. Noviany N, Osman H, Mohamad S, et al. Two new naturally dimers constituent from Indonesian Sesbania grandiflora plant and their bioactivity. Pure Appl Chem. 2024;96:379–87. https://doi.org/10.1515/pac-2023-1112.

Chapter 8
Applications of Generative AI in Small Molecule Drug Discovery

Didik Huswo Utomo and Naqiyah Afifah Mulachelah

Abstract Generative artificial intelligence (AI) is changing the way scientists discover new small-molecule drugs. It helps design new molecules, speeds up testing, and predicts how molecules will act by analyzing large data sets. This chapter gives an overview of generative AI, compares it to older AI methods, and explains how deep learning, transformers, and reinforcement learning are used to find, improve, and reuse drug candidates. It also explains why small molecules matter, what makes a good drug candidate, and how formats like SMILES, molecular graphs, and 3D structures are used in AI research. The chapter discusses challenges such as data quality, bias, understanding results, whether molecules can be made in practice, and how AI fits into current drug development. It also covers ethical, legal, and resource concerns. Finally, the chapter introduces new approaches, such as improved ways to describe molecules, graph neural networks, lab automation, and federated learning, all aimed at making AI-driven drug discovery more reliable, private, and repeatable.

Keywords Generative artificial intelligence · De novo drug design · Small-molecule representations · Deep generative models · Federated learning in drug discovery

8.1 Introduction

Generative artificial intelligence, or in short GenAI, is an AI subset that produces new data resembling a given dataset, unlike regular AI, which only recognizes patterns or generates predictions [1, 2]. Recent research demonstrates that AI is being utilized to transform every process of drug discovery. Starting from identifying the drug's target to designing ligands, planning chemical synthesis, selecting trial participants to repurpose existing drugs, analyzing cell images, and predicting the physicochemical and biological properties of a molecule [3].

D. H. Utomo (✉) · N. A. Mulachelah
Bioinformatics Research Center, INBIO Indonesia, Malang, Indonesia
e-mail: didik.huswo@gmail.com

A. A. Parikesit and A. N. M. Ansori (eds.), *Recent Computational Techniques in De Novo Drug Design*, Bio-IT and AI, https://doi.org/10.1007/978-3-032-20428-8_8

Significant developments in Artificial Intelligence and deep generative models have demonstrated their beneficial value in medicinal research, particularly in drug discovery and advancement [4]. Among these, the computational approach offers a promising alternative for preclinical research in screening molecules for drug discovery purposes. The users can choose the most suitable artificial intelligence (AI), such as deep generative models, to address the existing challenges, along with a genuine protocol and parameters. This approach bridges multidisciplinary fields that integrate chemistry, biology, pharmacology, computational science, and disease therapy [4].

A considerable promise emerges in both fields of healthcare education and clinical practice. This is caused by a useful Generative AI tool like ChatGPT. In other areas, such as pharmacy, GenAI promotes efficiency by supporting tasks such as prescription evaluation, drug interaction testing, and observing for adverse drug reactions. These approaches, altogether, offer a contribution to patient healthcare. Nevertheless, the implementation problems caused the exploration of GenAI in pharmacy to be limited. GenAI has the possibility to expand its utility in medical studies, not only by aiding in clinical work, but also by generating a big dataset for training model predictions [5].

8.1.1 Generative AI Versus Regular AI

Generative AI systems are not restricted to creative tasks such as creating artwork that resembles the work of human illustrators or writing in the style of particular authors. They can also be used as sophisticated tools to help people by intelligently answering questions, such as in IT support roles or for routine tasks like giving basic medical advice or cooking advice. Industry studies estimate that generative AI could automate up to 300 million knowledge-based jobs and increase the global GDP by 7%. In contrast to conventional AI, generative AI can create brand-new data that is identical to its training data [6]. By understanding the underlying patterns in data, these models are able to produce unique, artificial content, promoting innovation across diverse areas of industry [2].

On the other hand, there are many challenges in the traditional drug discovery and development process. The main obstacles include rising expenses and lengthened timelines, which frequently result in failure. Moreover, the lengthy and difficult process of finding suitable drug targets, the expensive and time-consuming high-throughput screening of possible compounds, and the iterative refinement procedure to increase a drug's efficacy and safety are some of the other challenges often found. There are additional challenges associated with conducting clinical trials, including recruiting participants, obtaining reliable data, and conducting thorough analysis. The effectiveness and success rate of introducing novel treatments to the market are hampered by these issues taken together [5].

8.1.2 Introduction to Small Molecules

Small molecule drugs are synthetic medicinal chemicals created to imitate, improve, or inhibit the body's natural chemical processes. They are able to be customized for particular therapeutic goals due to their comparatively straightforward and adaptable structures. In most cases, these medications are stable and don't need particular storage conditions. The medication is frequently administered orally and is simple for patients to use because of its typically predictable behavior in the body. They can be used to treat a wide range of illnesses because of their effective bloodstream and cell membrane crossing capabilities. Small molecule drugs are very adaptable because they can be administered in a variety of ways, including tablets, inhalers, suppositories, and injections [7].

8.1.3 Recent Milestones in AI Drug Discovery

Finding new medications throughout history primarily relied on laborious screening procedures and trial-and-error methods. Although these methods have produced useful results, they frequently lack accuracy, are ineffective, and increase the likelihood of failure. Recently, thanks to current developments in artificial intelligence, large-scale chemical and biological datasets can now be processed by machine learning algorithms. These tools can enhance drug distribution and absorption characteristics, forecast molecular interactions, and spot chances to repurpose already-approved medications. They are all the main necessity factors for drug discovery. Pharmaceutical workflows, from automated screening to formulation, have been greatly accelerated by robotics (in addition to AI). This effort significantly increases speed and consistency. A new era in drug development is being shaped by the convergence of AI, robotics, and big data. This is an era where intelligent technologies may not only help, but also completely change the way medications are found and customized for specific patients and intricate disease processes [1].

The biotechnology company Insilico Medicine made a historic announcement in January 2023. It revealed that INS018_055, its main AI-generated drug candidate, had shown promising outcomes in Phase I clinical trials. The compound exhibited promising pharmacokinetic properties. In a seven-day period, it was found to be safe, well tolerated, and no significant accumulation in the body over. After being designated as an Orphan Drug by the FDA in February 2023, Insilico moved forward with Phase II trials. These started in June 2023, when the first patient in China and the US received a dose. In order to assess the safety, tolerability, pharmacokinetics, and early efficacy of a 12-week oral regimen of INS018_055 in patients with idiopathic pulmonary fibrosis (IPF), a randomized, double-blind, placebo-controlled trial with four groups is currently underway. An estimation of 60 patients will be enrolled in the trial at about 40 locations. As the first clinical testing of a drug that was fully discovered and designed by generative AI, the start of these Phase II trials represents a significant milestone in pharmaceutical innovation, giving patients with IPF new hope and possibly opening the door for future AI-driven therapies (https://insilico.com/blog/first_phase2). (Insilico Medicine. (2023, Jul 1). First Generative AI Drug Begins Phase II Trials).

8.1.4 Challenges of AI in Drug Discovery

The scarcity of good, well-annotated datasets for model training is a significant barrier to AI-driven drug discovery. This field frequently uses data from a variety of sources, which leads to complexity and inconsistency. The data could be molecular string, clinical trials, biological analysis, and many more. It is challenging and time-consuming to integrate these diverse datasets into a single, consistent format suitable for AI training. Furthermore, biases in training data may jeopardize the models' precision and applicability. For example, the model may generate skewed results that perform poorly in more general real-world applications if the dataset disproportionately represents a particular population group or disease type. Careful data curation, efficient pre-processing techniques, and strategies to mitigate bias while ensuring that the data accurately represents the larger population and its context are necessary to overcome these challenges [5].

8.2 Basics of Small Molecules in Pharmaceutical Science

8.2.1 Small Molecules

Chemical engineering is used to create small molecules that interact with particular biological targets in a specific manner. Researchers can modify these molecules to produce specific biological responses while reducing unwanted reactions by altering their atomic structure. Small molecules have a distinct advantage over other therapeutic approaches because of their versatility. This property enables researchers to investigate a wide "chemical space." Years of meticulous planning and synthesis are required to develop a small molecule medication that functions as intended and has minimal adverse effects.

One of the most popular targets in drug development is proteins. Proteins are big, intricate biomolecules made up of amino acids. The shape and biological function of a protein are determined by its particular amino acid sequence. These proteins can be impacted by small molecules in a variety of ways. Some of the general methods are explained below.

- Enzyme inhibitors: proteins that have a function to quicken chemical reactions inside the body system called enzymes. Disease pathways can be disrupted by small molecules that inhibit specific enzymes. For instance, statins are a type of enzyme inhibitor. This inhibitor drug can lower the risk of cardiovascular disease by blocking an enzyme in the liver that produces cholesterol. The decreasing cholesterol accumulation in the body means a lower risk of heart disease and stroke.

- Receptor antagonist/agonists: These medications work by interacting with proteins on the cell surface known as receptors. The interaction could be antagonist or agonist. While antagonists block them, which prevents activation, agonists mimic natural molecules to activate the receptor. The example of a drug that works in this type of way is Albuterol. Asthma agonists such as albuterol open airways by activating pulmonary receptors.
- Ion channel modulators: another crucial protein in drug mechanism is the Ion channels. These proteins are essential for functions like heart rhythm and nerve signaling because they regulate the movement of ions across cell membranes. These channels can be influenced by small molecules, which can modulate their opening or closing to treat epilepsy.

Similar to a lock and key, the majority of the mentioned actions rely on the active site, a particular region of the protein that is structurally suited to bind only specific molecules. Researchers can create a molecular "key" that fits this "lock" precisely by knowing its structure. This effort minimizes interactions with unwanted targets and maximizes the intended effect. This approach creates a highly customized drug design, which, in response, minimizes any undesirable side effects. Small molecule drug design now incorporates even more advanced techniques due to advancements in technology and a better understanding of disease mechanisms. To increase the number of diseases that can be treated, many efforts have been made. These attempts include focusing on protein–protein interactions, creating bifunctional compounds that break down disease-related proteins, and creating molecules that stabilize advantageous protein structures.

8.2.2 Roles in Drug Discovery

Due to their ability to enter cells and precisely interact with biological targets such as proteins and enzymes, small molecules are crucial in the drug discovery process. These substances can be absorbed effectively when administered orally. This is possibly due to the relatively low molecular weight. In order to find small molecules that can bind to disease-associated targets and alter their biological function, researchers frequently search in the large libraries of small molecules in the first phases of drug development. After a potential candidate is identified, its structure is enhanced to optimize its chemical stability, reduce side effects, and boost therapeutic efficacy. Numerous effective drugs are used to treat a range of ailments, such as cancer, infectious diseases, and chronic illnesses. All have been made possible by small molecules approaches.

8.2.3 Common Chemical Properties (Lipinski Rule of Five)

Before creating a new bioactive compound, researchers use mathematical models to speed up the drug development process and cut expenses. These models predict the relationships between a molecule's structure, physicochemical properties, and biological activity both qualitatively and quantitatively. According to IUPAC, lipophilicity is one of the most widely used molecular descriptors in predicting a compound's bioactivity. Lipophilicity is a molecule's affinity for lipid environments. The most well-known bioavailability rule, Lipinski's Rule of Five, serves as a guide for theoretical assessments of drug-likeness alongside lipophilicity. A bioactive compound's profile determines its course once it enters the body. The critical aspects for in account are the molecule's absorption, distribution, metabolism, excretion, and toxicity. Assessing and improving a compound's function and efficacy requires an understanding of its pharmacokinetics.

The body's effective permeability in the jejunum acts as a pharmacokinetic predictor of intestinal absorption. This happens because the majority of medications are not administered intravenously. Higher lipophilic molecules have a tendency to flow more readily through the phospholipid bilayer of enterocytes. This action means that the molecule's characteristic has a significant impact on its absorption.

A significant amount of work is devoted to assessing the "drug-likeness" of molecules in early-stage research to facilitate progress in drug discovery. Although there are other approaches, Chris Lipinski and his associates at Pfizer developed the Lipinski Rule of Five (ROF), which is currently the most popular. ROF, also known as Lipinski rules, offers a set of criteria for determining if a substance with possible pharmacological activity is likely to work well as an oral medication in individuals. Generally speaking, a molecule must not contravene this rule in order to be deemed appropriate for oral administration. The biologically active molecule must implement five conditions to be potentially used as a drug for oral administration. Poor absorption or permeation is most likely if: molecular mass more than 500 Daltons, total of Hydrogen bond acceptors more than 10, total of Hydrogen bond donors more than 10, A LogP value greater than 5 (MlogP above 4.15).

According to the ROF framework, an ideal orally active compound should score between "0" and "4". The number indicates that there should be no more than one infraction of the four criteria. Nevertheless, Lipinski emphasized that substances that do not strictly adhere to these guidelines should not be disqualified from further development immediately. In reality, a lot of effective medications don't strictly follow the ROF. The Rule of Five has a number of drawbacks, despite its extensive application. First of all, it gives each criterion equal weight, which might not accurately represent its true significance. Second, there is no flexibility in the criteria; they are strictly defined. Furthermore, ROF ignores important metabolic factors and does not take into account natural products or molecules that are derived from biological sources [8].

8.2.4 Importance of 3D Structure, Chemical Space, and Target Binding

Understanding the three-dimensional (3D) structure of the drug compound and its biological target is crucial for small molecule drug discovery. According to the "lock and key" principle, a drug molecule's 3D conformation dictates how well it fits into the target active site of a protein. A precise structural match enhances therapeutic potency while reducing side effects. This is possible due to the action of promoting stronger, more selective binding. There are countless opportunities for innovation due to the enormous variety of potential chemical configurations, also known as chemical space. Researchers can create or find molecules with the best physico-chemical and biological characteristics by navigating this chemical space. Scientists can find previously undiscovered compounds with promising therapeutic potential by methodically investigating it. In this process, target binding is an essential factor. The function acts as a verificatory whether a potential molecule can interact both physically and functionally with a disease-related protein. Structure-based drug design is currently a key component of contemporary pharmaceutical research. This valuable approach greatly enhances a drug's potency, selectivity, and overall performance by improving these molecular interactions.

8.3 Overview of Generative AI: Concepts and Algorithms

8.3.1 Introduction to Deep Learning in Chemistry

Deep learning has become a game-changing method in drug discovery and chemistry in recent years. Deep learning is a branch of artificial intelligence (AI) that automatically extracts patterns and representations from massive, intricate datasets by using multi-layered neural networks. Deep learning models can comprehend molecular structures, forecast chemical properties, and simulate drug-biological target interactions with previously unheard-of accuracy in the context of chemistry. Conventional drug discovery, which involves screening large compound libraries through trial and error, is frequently resource- and time-intensive. By facilitating virtual screening, de novo molecule generation, and the prediction of pharmacokinetic and toxicity profiles, deep learning accelerates this process. These models are able to learn subtle patterns that may be difficult for humans to decipher because they can be trained on a variety of data sources, including molecular graphs, chemical fingerprints, SMILES strings, and even 3D conformations. Researchers can significantly reduce expenses, expedite timelines, and increase the success rate of identifying promising drug candidates by incorporating deep learning into the drug discovery process.

8.3.2 Generative Models

8.3.2.1 Variational Autoencoders (VAEs)

One of the most popular frameworks for unsupervised learning, especially when employing complex data distributions, is Variational Autoencoders (VAEs). These models have already proven to be successful in producing a variety of data types, including human faces, handwritten numbers, and image segmentations. Gómez-Bombarelli and associates were among the first to use the VAE architecture in the context of molecular generation [9]. An autoencoder (AE) consists of two neural networks: the encoder, which compresses the input into a lower-dimensional latent vector, and the decoder, which uses the latent vector to reconstruct the input. The latent space produced by traditional AEs is optimized to replicate the input as precisely as possible. However, VAEs introduce a regularization step by substituting probability distributions for precise points in the latent space, thereby preventing issues such as overfitting and discontinuities.

VAEs were utilized in a groundbreaking study to generate molecules, establishing the foundation for a new method of de novo drug design. The encoder is trained to map molecules into a compact latent space, usually assumed to have a Gaussian distribution, such as those represented by SMILES strings. From these latent representations, the decoder subsequently recreates the initial molecular structures. Due to this configuration, molecules are encoded as distributions within the latent space rather than as fixed points. A method is needed to ensure the decoded outputs resemble the probability distribution of the training data. Therefore, the encoder and decoder are jointly optimized during training.

Learning disentangled representations within VAEs now is an increasingly popular topic in chemistry, computational science, and many related fields. This is achieved by assigning each latent vector component to capture a unique and independent feature of the data. If this were to be successfully applied to molecular drug design, by altering the corresponding latent variable, researchers could modify one molecular property while still maintaining others [4].

A common drawback in Generative AI for drug discovery is that the samples used for molecules are usually in a distinct format, such as SMILES strings. The conversion of this discrete format to a continuous one, which neural networks can employ, is often challenging. VAEs make an excellent model for encoding molecular structures into a navigable continuous domain because continuous latent spaces provide more flexibility. Exploring uncharted areas of chemical space requires such continuous transitions. An important benefit in drug discovery workflows is that VAEs can be customized to predict molecular properties, enabling the optimization of particular features during the generation process [10].

Creating new compounds from scratch, or de novo, is the ultimate goal of molecular generation. It is executed by learning a joint probability distribution over all variables, including inputs and outputs. The process of creating data in the real world is replicated by a generative model. Such a model can create completely new types of data, from text and images to music, but it can only do so when the model has been well trained. Novel molecular chemicals can be created in drug discovery by teaching generative models, such as VAEs, to build molecules that weren't in the original dataset. The novel output molecules are totally difference and never existed in the training dataset. This method may enable researchers to investigate the enormous, mainly unexplored universe of hypothetical molecules that could be created, extending beyond the comparatively small subset of known compounds.

The encoder and decoder built the VAE, as well as performing its significant function. The encoder is also known as a recognition model, while the decoder is known as a generative model. The model learns stochastic mappings between observed real-world data and this abstract latent domain by connecting these elements with a vector that lies in the latent space. The characteristic is frequently a continuous and lower-dimensional space. A subset of VAEs, known as "deep VAEs," utilizes the concepts of deep neural networks and features several hidden layers for both encoding and decoding. Navigating between the continuous nature of latent variables and neural networks, as well as the discrete representations of molecules (such as SMILES or molecular graphs), is a major technical challenge in molecular VAEs.

Indeed, developments like molecular grammars and graph-based molecular generation have made this translation possible. Due to SMILES strings having context-sensitive grammar, they pose particular challenges. In contrast to protein sequences represented by one-letter amino acid codes (such as FASTA format), which are more resilient to sequence breaks, randomly splitting a SMILES string frequently does not produce valid molecular fragments. Simplifying the SMILES grammar to remove its most prone parts is one potential solution to this problem. DeepSMILES, a modified SMILES syntax intended to improve compatibility with generative models and minimize syntactic errors during molecular generation, is an example of such an improvement [11].

8.3.2.2 Generative Adversarial Networks (GANs)

Generative adversarial networks present as a useful deep learning framework to produce highly realistic data compared with other AIs. The framework combines the striking neural networks: generator and discriminator. The generator generates synthetic data, while the discriminator assesses its originality. To distinguish real data from synthetic ones, the generator employs iterative adversarial training. The generator progressively improves its products by competing against a discriminator. This practice enables applications in medical imaging, high-resolution imaging, and data augmentation. Demonstrating its use in medical diagnostics and video surveillance, Super-Resolution GAN has achieved success in improving image resolution [5].

GANs' emergence in the deep generative class is a groundbreaking invention. The other model, such as VAE, offers an explicit probability density function. Meanwhile GANs presents the framework based on adversarial training framework. The discriminator and the generator operate as a competitive framework to produce the best results with real data. They are trained through an adversarial zero-sum process. The most convincing result of the generator is obtained when the discriminator is no longer able to distinguish "real" from synthetic molecules. The discriminator trains a categorized model with the purpose of maximizing the generator's error rate of the resulting molecules. This means resembled the real data [4].

The early invention of neural networks focused on training data (structured or unstructured), and the output was produced based on the training sets. In newer neural networks, like GANs, rather than classifying it, they generate new data. GANs are presented as an improved framework in deep learning, advancing the generation of artificial data in a competitive learning algorithm by utilizing two neurons against each other. As a result, GANs have proven to be a useful tool for the automated processing of information in multiple domains, including computer vision, natural language processing, healthcare, and design generation [10].

Due to GANs' advantages, the applications have found widespread use in the medical and healthcare sectors. Examples of the significant applications include image-to-image translation, image segmentation, style transfer, and classification. Artificial Intelligence represents the crucial technological accomplishment of the twenty-first century. AI's potential to integrate with healthcare systems is one of the main focuses of researchers. In this field, AI is able to identify a pattern that a regular program is not capable of. Generative AI, especially GANs, is a type of machine learning that produces synthesized data after being trained with real data. The use of fake data is advantageous in clinical studies, as it helps maintain patient confidentiality.

GANs make it possible for an anonymous dataset and no trace-back data for any person. Therefore, real patient data can be replaced, and the data can be improved and balanced. GANs allow doctors to specify their diagnosis and therapy, while the patients can be more involved. Generative AI can also assist in encouraging patient involvement by generating clever question-and-answer models, presenting medical guidance, and providing customized treatment plans [10].

Using GANs, it is now possible to generate new compounds with specific properties, streamlining the process of de novo drug design discovery. DruGAN is one of the GAN models that is able to generate specific molecule drug designs based on particular criteria. This model integrates GANs with autoencoders. This novel method allows scientists to investigate large chemical spaces and find possible therapeutic candidates that conventional approaches may fail to identify [1].

8.3.2.3 Transformers

Transformer models are significant in NLP due to their self-attention mechanism. In sequential data, this system effectively performs long-range dependencies. Transformers, as opposed to RNNs, process sequences in parallel, improving performance on tasks like speech recognition, text generation, and machine translation. One of the popular language models in computational biology, called Protein Language Models

(PLMs), is inspired by transformer models, such as Generative Pre-trained Transformer (GPT), Bidirectional Encoder Representations from Transformers (BERT), and Text-To-Text Transfer Transformer (T5).

The basic transformer architecture is employed by PLMs to analyze biology-related sequences. The system views the proteins as "text" made up of amino acid "alphabets." PLMs facilitate tasks like protein structure prediction (e.g., AlphaFold2), function annotation, protein design, and mutation effect prediction by training on large protein databases. The advantages they offer are varied and all beneficial for leveraging the user's necessity in research. Features to process 3D structural data are possible through transfer learning using pretrained models such as Evolutionary Scale Modeling-2 (ESM-2) and Protein Generative Pre-trained Transformer 2 (ProtGPT2). Moreover, unsupervised learning from databases like UniProt and geometric awareness can be performed by models like OmegaFold.

Although challenges like interpretability and computational costs still exist, future directions include energy-based optimization for de novo protein design and multimodal protein models. The ESM-2 model from Meta AI, for instance, demonstrates the transformative potential of transformers in decoding biological "language" by using 15 billion parameters to predict protein structures from sequences. Additionally, NLP can examine EHRs to find trends and connections between patient attributes, the course of the disease, and the results of treatment. This may result in the identification of new biomarkers, the forecasting of patient reactions to particular treatments, and customized treatment regimens [5].

8.3.2.4 Reinforcement Learning in Molecule Generation

In the field of artificial intelligence (AI), deep reinforcement learning (Deep RL) has become one of the most potent and promising toolboxes, especially when it comes to optimizing complex objectives in environments with large solution spaces. The creation of AlphaGo, which defeated expert Go players by navigating an incredibly large and computationally demanding decision space, was a noteworthy innovation that demonstrated the potential of reinforcement learning. The chemical space involved in drug discovery is one of the similarly complex domains where researchers have been motivated to apply reinforcement learning by this achievement narrative.

In terms of scale, the solution space of the game Go is frequently compared to the chemical space, which comprises all potential molecules that could possibly exist. Reinforcement learning offers an intriguing method for investigating and optimizing in this area. This is possible due to its size and complexity. RL enables the methodical investigation of molecular structures in the direction of particular, user-defined goals by conceptualizing molecule generation and optimization as a sequential decision-making process. The environment, the reward function, and the agent are the basic elements of reinforcement learning. All three factors play important roles in this process. In general, a neural network model serves as the "agent" in this framework, deciding what molecular change to make next. The "reward function" assesses the effectiveness of the agent's actions in terms of target properties like druglikeness or synthetic accessibility, while the "environment" consists of domain-specific rules or constraints that dictate what is chemically feasible.

In order to learn the syntactic rules of molecular representations, a common pipeline starts with training a generative model on a sizable dataset of molecules. The format often employs SMILES notation, as seen in transformer-based models or recurrent neural networks. Following this initial training stage, the generative model can be refined using reinforcement learning to generate molecules with specific physicochemical or pharmacokinetic characteristics. Synthetic accessibility (SA), binding affinity, solubility, quantitative estimate of drug-likeness (QED), and even toxicity profiles are a few examples. The Policy Gradient for Forward Synthesis (PGFS) algorithm is a frequently cited example. A neural network serves as the agent in this approach, and its actions are described as chemical transformations. These changes include a molecule's atoms or bonds being added or eliminated. In order to encourage the agent to favor molecules that are more likely to be synthesizable in a laboratory, the reward function is made to evaluate how synthetically accessible the final molecule is.

Reinforcement learning is a branch of artificial intelligence where an agent learns how to act in a given environment by acting and getting feedback in the form of rewards or penalties. Over time, the agent can improve its policy thanks to this feedback loop. RL has gained popularity as a de novo molecular design technique in drug discovery. The goal of de novo drug design is to create completely new molecules from the ground up that haven't been found before and have the desired chemical or biological characteristics. This contrasts with conventional drug discovery pipelines, which frequently concentrate on altering well-known molecules or repurposing already-approved medications [10].

A study applied reinforcement learning to refine molecular graphs toward target qualities in the context of de novo drug generation using a graph-based deep generative model. A more expressive and natural representation of molecular structures is made possible by graph-based models, which depict molecules as graphs of atoms and bonds rather than linear SMILES strings. These generative models are then refined using reinforcement learning to create new compounds with ideal characteristics, like low toxicity or a high binding affinity to a specific protein [10]. Another noteworthy strategy is the adaptation of large language models (LLMs) such as GPT for molecule generation, which takes advantage of their strong sequence prediction capabilities. Reinforcement learning, specifically Reinforcement Learning from Human Feedback (RLHF), can be used to refine these models in order to match the generative process with the preferences of human experts or desired chemical criteria. By allowing human-in-the-loop training, RLHF bridges the gap between expert intuition and computational efficiency by allowing domain experts to direct the model toward producing more appropriate or realistic molecular suggestions.

Another crucial tool in 3D drug discovery is molecular docking. The optimization of this tool can be performed by reinforcement learning. A basic computational method for predicting how small molecules (ligands) will interact with biological targets (receptors), like proteins, is called molecular docking. Finding possible drug candidates requires accurate docking. In this situation, the search for ligand conformations that best match the target's binding site can be optimized through reinforcement learning.

A fragment-based reinforcement learning framework utilizing an actor-critic model was presented by previous research [10]. Long Short-Term Memory (LSTM) neural networks are used in this model for both the critic and actor parts. The model selects replacements based on anticipated improvements in molecular properties after performing fragment replacement from an initial set of lead molecules by arranging fragments in a balanced binary tree. This method preserves structural viability while enabling the creation of novel molecules with optimized properties.

RL is also being used in drug repurposing, sometimes referred to as drug repositioning, which is the process of finding new therapeutic uses for medications that are already on the market or have received approval. Because it avoids many of the early stages of the conventional drug development pipeline, the primary benefits of drug repurposing are its speed and cost-effectiveness. Repurposing can be represented as a decision-making problem in the RL framework. The "state" could contain pharmacological and biological information like gene expression profiles, protein targets, or disease types. The "action" is the choice or alteration of potential medications, and the "reward" function assesses results like patient response, side effect profile, or therapeutic efficacy. For example, medications that demonstrate effectiveness against a disease with few side effects may be given a high reward.

Drug repurposing has the benefit of lowering drug discovery costs when compared to creating new medications from the ground up. The following definitions apply to the state, action, and reward basic components of reinforcement learning in the context of drug repurposing:

- State: The surroundings, which could contain details about illnesses, biological variables, and molecular data.
- Action: The choices the RL agent makes, like which molecules to interact with.
- Reward: One essential component of RL is the reward. Rewards may differ depending on the drug's clinical evaluation, side effects, or effectiveness in treating a condition.

A technique that makes use of a graph traversal algorithm was suggested by a study [10]. Multi-agent approaches in reinforcement learning (RL) entail several agents interacting with one another in a common setting. To increase its reward, each agent engages with the environment and other agents. Repurposing drugs may benefit from the use of multi-agent RL. By coordinating the actions of multiple agents, multi-agent reinforcement learning (ML) can overcome the difficulties posed by biological datasets and their complex interactions. Investigating the drug-disease space, in which each agent stands for a distinct technique or strategy for simulating drug-disease interactions, is one promising strategy.

8.4 Methods and Tools for Generative AI in Small Molecules

8.4.1 *Types of Data and the Tools*

The types of data in GenAI for processing small molecule data can be diverse. Different data is processed using different tools, which generate a particular result.

Although there are many types of data, in general, it can be classified into three types of data. The data can be tailored based on the purpose of the research. Three types of data consisted of SMILES, molecular graph, and 3D visualization. In addition, bioactivity and ADMET data serve as important annotations or labels for model training, filtering, and validation.

8.4.1.1 Simplified Molecular-Input Line-Entry System (SMILES)

In a simpler analogy, just like DNA with its sequence databases, SMILES represents a molecule with its own type of sequence data. Molecular-Input Line-Entry System Simplified with Self-Referencing Atoms, bonds, and stereochemistry are all contained within embedded strings, which depict a molecule's structure as a string. Canonicalization algorithms produce a single representative canonical SMILES string, even though each molecule may have multiple SMILES strings.

However, due to syntax errors, not all SMILES are valid. This is where SELFIES, an alternative, occur. A SMILE-based tool called SELFIES ensures that each string in the correct syntax matches a legitimate molecule. This tool offers a robust alternative [2].

8.4.1.2 SELFIES

SELFIES (self-referencing embedded strings) is a 100% robust molecular string representation system that was first released in 2020. This implies that a chemically valid molecular structure will always be represented by any sequence created from the SELFIES alphabet. Consider contrasting this idea with a natural language, such as English, to help visualize it. In general, a random arrangement of Latin letters (a–z) won't result in a word that is both recognizable and legitimate in English. This demonstrates how SELFIES retain robustness in the chemical domain while English lacks it. At its core, SELFIES functions as a formal grammar or automaton, guided by certain derivation rules—like a small computer program that uses very little memory to guarantee completely legitimate molecular derivations. Preventing the creation of syntactically and semantically incorrect molecules is the primary goal of grammar, especially in generative modeling scenarios [12].

By guaranteeing that each sequence of SELFIES symbols can be decoded into a molecule that meets valency requirements, SELFIES successfully tackles issues of chemical invalidity. This makes it particularly suitable for molecular optimization strategies utilizing genetic algorithms such as GA + D, STONED, and JANUS, as well as chemical language models that produce molecular strings. Regardless of the optimization strategy, the representation improves string-based molecular generation by directly integrating valency rules into the format.

In addition to improving tasks like image-to-string and string-to-string translation, as well as molecular generation in data-scarce scenarios, studies have shown that SELFIES improves distribution learning for language models. Furthermore, new but structurally related molecules can be created by merely adding, replacing, or deleting SELFIES strings—a crucial mechanism for genetic algorithms. These changes enable molecular interpolation when paired with Tanimoto similarity, as demonstrated by Nigam et al.'s STONED method, which has since been used as a crossover mechanism in algorithms such as JANUS. Identifying decision boundaries in molecular classifiers is facilitated by this type of interpolation, which also helps determine counterfactuals [13].

Group SELFIES is the result of combining the flexibility of string-based representations with the structural intuition and clarity of fragment-based techniques, as well as the robustness of SELFIES. By adding tokens that represent functional groups or molecular substructures, this enhanced representation broadens the scope of SELFIES. By leveraging in-depth cheminformatics research, users can create custom groups or extract significant fragments from chemical datasets. The inclusion of these group tokens is the primary difference between SELFIES and Group SELFIES. Atoms and bonds with particular attachment points define each group and determine how the group can create new bonds. These points are labeled by the attachment indices, which help the encoder and decoder handle them correctly [13].

Neural architectures like Recurrent Neural Networks (RNNs), which include Gated Recurrent Units (GRUs), Long Short-Term Memory networks (LSTMs), or Transformers, are commonly used to process these molecular string representations. To predict molecular properties, these models can be trained through supervised learning techniques. Additionally, many studies use self-supervised learning (SSL) techniques to leverage large amounts of unlabeled data. Rebuilding the original strings, creating different SMILES representations of the same molecule, or anticipating masked elements within the molecular strings are common tasks for SSL frameworks [14]. The widely used tools for SMILES, SELFIES, and group SELFIES processing are RDKit, SELFIES, DeepChem, REINVENT, ChemTS, MolBERT, ChemBERTa, Open Babel, etc. A more detailed explanation can be seen in the next chapter.

8.4.1.3 Molecular Graph

Aside from its string, another primary building block of molecule is its topography. No matter how accurate the result of the string analysis, in SMILES and SELFIES, important topology structures of molecular graphs are ignored. Therefore, we need a 2D visualization to represent the molecule's accuration. This tool can be used to solve our problem, enhance our understanding, and also support validation analysis to validate the main data.

The foundation of molecular graph representation is the notion that atoms and chemical bonds in a molecule can be abstracted into a mathematical structure of nodes and edges to model the molecule's internal structure. In its simplest and most widely used form, every atom in the molecule is regarded as a node, and the chemical bonds that bind these atoms together are shown as edges connecting the nodes. Through this abstraction, intricate molecular structures can be reduced to computationally manageable graph-based representations that are ideal for applications in cheminformatics and artificial intelligence. Although atoms and bonds are used in the standard mapping, graph theory's flexibility allows for the possibility of other mappings based on the particular task or the desired molecular features to be highlighted.

Nodes are frequently represented as spheres or circles in visual graph representations, and the edges are shown as lines joining them. However, the nodes in molecular graphs typically carry extra information. They are frequently identified by chemical symbols that correspond to the type of atom (for example, C for carbon, O for oxygen, and N for nitrogen) or are just shown as vertices where lines intersect, particularly when carbon atoms are assumed implicitly. Despite being formally a two-dimensional (2D) mathematical object, molecular graphs are widely used to depict chemical information that is intrinsically three-dimensional (3D), including atomic coordinates, bond lengths and angles, torsional angles, and even stereochemistry or chirality. Nodes are not given spatial positions in the mathematical definition of a graph, and only their pairwise relationships are defined. Therefore, the construction of a graph is usually accomplished by encoding spatial and stereochemical data as attributes or features associated with the graph's nodes and edges.

Despite this drawback, cheminformatics platforms and contemporary visualization tools have created strong methods for efficiently representing 2D and 3D molecular graphs. These molecular graphs can be rendered in both flat and spatially accurate formats using a variety of popular software packages. For producing 2D representations of molecules that are suitable for publication. One example of the frequently utilized programs is ChemDraw.

Other programs, such as Mercury, Avogadro, VESTA, PyMOL, and VMD, offer a range of features for 3D molecular visualization. These programs provide features that can be used to analyze small molecules as well as larger macromolecules like proteins or nucleic acids. Fortunately, a number of these tools are free or open-source. This advantage enables researchers from a variety of fields to use them. These graphical representations serve as the foundation for many AI-driven drug discovery initiatives. It provides a crucial function to encode molecules in a way that is heavy in chemical meaning and readable for a machine. Other tools that can be utilized are RDKit, Open Babel, DGL-LifeSci, PyG, NetworkX, OGB, PyTorch Geometric, MolGAN, and JT-VAE.

8.4.1.4 3D Conformers

Despite the fact that molecular docking is still a popular method in drug discovery, the limitations of force fields and scoring functions inevitably limit its accuracy. On the other hand, machine learning (ML) techniques present a strong substitute. This is possible especially because of the much faster rate of inference. However, the amount and caliber of available data have a significant impact on machine learning performance. Given enough training data, neural networks can potentially learn extremely complex patterns. An example of this is the connection between a molecule's structure and biological characteristics. Unfortunately, datasets involving successful drug compounds are usually small. This is because finding ligands that bind strongly and specifically to target proteins remains very challenging.

Designing neural network architectures that are optimized to function well with little data becomes crucial to solving this problem. The potential of machine learning in drug discovery has been significantly expanded by recent developments in deep learning, especially with models that work directly on molecular graphs. These models are frequently called message-passing neural networks (MPNNs). They learn to produce continuous, data-driven molecular fingerprints by representing molecules as graphs. Neural networks then analyze these fingerprints to forecast pertinent characteristics like toxicity or binding affinity.

These fingerprints are mainly generated by current deep learning models using static three-dimensional structures or two-dimensional molecular graphs. To predict how well a drug molecule binds to a protein target, for instance, some models are particularly proficient at using 2D-graph-based data. Meanwhile, other models, such as PotentialNet, are made to combine 2D connectivity and 3D spatial data.

However, in actual cases, molecules are not actually fixed 3D shapes or even flat 2D diagrams. Rather, every molecule is a collection of numerous 3D conformations that are always changing in space. Given the inherent three-dimensional nature of the drug-target binding process, it is crucial to take into account the entire spectrum of three-dimensional structures that a molecule can take on. Although such ensembles have occasionally been used to compute fixed descriptors in traditional computational drug discovery tools, MPNN-based deep learning frameworks have not yet fully incorporated this idea.

Mentioning that we're talking about intelligence tools, any problem seems close to a solution. It is possible that 2D graph-based models already indirectly capture some ensemble-related information. These deep learning models may be learning more about molecular flexibility than previously thought. This is because a molecule's molecular graph determines the range of 3D conformers it can adopt. In contrast to models that explicitly incorporate one or more 3D conformers, the degree to which 2D models can implicitly encode and utilize this ensemble information is still up for debate. Useful tools that are widely used for 3D conformer are diverse. All offer different features that cater to the user's demands and needs. These advanced programs are bridging the gap between scientific question and the answer. Some of those widely used tools are RDKit (ETKDG), Open Babel, AutoDock, Chimera, PyMOL, DiffDock, etc.

8.4.2 The Tools and Libraries

Many kinds of tools can be used for the particular goal of your analysis. For data mining and/or training, different libraries can also be utilized. In summary, all are explained in Table 8.1.

8.4.2.1 RDKit

A software provided by GitHub and SourceForge. This tool is beneficial for creating variables that represent molecules and retrieving information about molecules. RDKit is commonly used by user who prefer to perform their data in raw SMILES strings. Its fast C++ backend and versatile Python API make this tool widely used. The various functions of RDKit are also what make it distinguishable, ranging from 1D, 2D, and 3D data can be processed using this software. Furthermore, this software also provides a bug tracker feature. RDKit provides a complete description and guide of its software on https://rdkit.readthedocs.io/.

8.4.2.2 SELFIES

SELFIES is an official Python library to convert SMILES to SELFIES. This tool ensures every string is valid in terms of molecular syntax and semantics. SELFIES is highly adaptable and can be used in various Machine Learning models. Various tasks such as molecular drug discovery, fingerprinting, reaction detection, and similarity analysis can be performed by this tool. The valid representation of the SELFIES chemical structure is possible due to its formal grammar and automatons to define its ruleset for molecular string construction. To access the official website, here is the link https://github.com/aspuru-guzik-group/selfies. The library for SELFIES is called SELFIES documentation. It can be accessed through https://selfies.readthedo cs.io/.

8.4.2.3 Open Babel

This tool is used for conversion between chemical formats, including XYZ, MOL2, PDB, and SMILES to 3D structure, and geometrical optimization. Since OpenBabel can be used in the command line or inside a program (C++ library), this software can be inserted in the IA/ML pipeline. This software can be run on Windows (64-bit or 2-bit), Linux, and macOS. It is available in a binary package or to compile their own OpenBabel. This software is also integrated with Python, which is called PyBEL. The official website for OpenBabel can be accessed here: https://openbabel.org/.

Table 8.1 The tools and libraries for the analysis of small molecules in drug discovery

No	Tool/Library	Function	Representation	Note
1	RDKit	SMILES conversion to structure, conformer optimization, molecule features	1D, 2D, and 3D model	Most popular and widely used in drug discovery
2	SELFIES	SMILES conversion to SELFIES (valid representation)	1D model	Prevent invalid string with chemically valid grammar
3	Open babel	Format converter, 3D generation, energy minimization	1D, 2D, and 3D model	Alternative for RDKit, flexible for various format output/input
4	DeepChem	Framework ML for drug discovery	1D and 2D model	Support SMILES integration and bioactivity prediction
5	DGL-LifeSci	Deep learning with GNN (Graph neural network) for molecule	2D model	Specialized for molecular graph learning based on deep graph library
6	PyTorch Geometric (PyG)	Graph neural network training	2D model	For molecule after conversion to graph
7	NetworkX	Graph visualization and manipulation	2D model	Graph structure exploration
8	SELFIES library	Encoding/decoding SELFIES	1D model	Official implementation by Aspuru-Guzik
9	AutoDock tools	Molecule preparing for docking	3D model	Commonly used for AutoDock Vina for simulation
10	PyMOL	3D molecule/protein structure visualization	3D model	Conformer and docking result visualization
11	ETKDG (via RDKit)	3D conformer generation	3D model	More realistic modern method compared with geometry distance conventional
12	MolBERT/ChemBERTa	Pre-trained transformer models for SMILES	1D model	Property prediction or generative tasks based on SMILES
13	OGB (Open graph benchmark)	Dataset for graph learning	2D model	Converting molecule into graph, suitable for benchmarking

8.4.2.4 DeepChem

DeepChem is a high-quality open-source software. This tool focused on a broad range of applications, from molecular machines to drug discovery, and also bioinformatics, material science, and computational physics. It supports three kinds of packages, including pip, conda, and Docker. Each package requires an individual installation. DeepChem also provides detailed tutorials, which guide the user from the very beginner level to the proficient level, including examples and tutorials. The tutorials can be downloaded in PFD format and performed in Google Colab. More information can be explored through the website https://deepchem.io.

8.4.2.5 DGL-LifeSci

DGL-LifeSci is a Python package for graph neural networks in chemistry and biology. The functionalities of this software include methods for graph construction, featurization, evaluation, model architecture, training scripts, and pre-trained models. The installation can be performed on Linux, macOS X, and Windows 10. It supports Python 3.6+, DGL 0.7.0+, and PyTorch 1.5.0+. Here is the website https://github.com/awslabs/dgl-lifesci.

DGL-LifeSci is particularly beneficial for tasks like drug-target affinity estimation, protein–ligand interaction modeling, and molecular property prediction. Rapid experimentation and benchmarking are made possible by the built-in datasets. These approaches are frequently used in bioinformatics research. Its modular design, which makes it simple for users to modify graph encoders or create new architectures for domain-specific applications, is one of its advantages. It is a good option for both inexperienced researchers and seasoned developers working at the nexus of AI and life sciences due to its ease of use and compatibility with deep learning frameworks.

8.4.2.6 PyTorch Geometric (PyG)

This tool is a library based on PyTorch to generate and train a Graph Neural Network (GNN). In particular, the application provides strong support for geometric deep learning. It enables models to learn from non-Euclidean data like molecular structures and three-dimensional point clouds. It can handle multiple small graphs and a single large-scale graph with ease thanks to its user-friendly mini-batch loaders. This feature makes it versatile for a range of graph-based tasks. Support for multiple GPUs is incorporated to improve performance and scalability during training. Additionally, the application works with DataPipes to optimize data handling workflows and torch compile for model optimization. Numerous benchmark datasets are easily accessible, and users can create custom datasets with ease thanks to user-friendly interfaces. This software is a complete platform for sophisticated graph learning applications. It offers a wide range of data transformation tools designed for both general graphs and specialized data types like 3D meshes and point clouds. It supports Linux, macOS, and Windows. The package is available in pip and conda. The website for this software can be found at https://pytorch-geometric.readthedocs.io/.

8.4.2.7 NetworkX

NetworkX is originally a software for creating, manipulating, and studying the structure, dynamics, and function of complex networks, which in this case is molecular data. This Python package enables the exploration and manipulation of molecular graph structures. The nodes are flexible, ranging from text, images, XML records, etc. The official website can be accessed through https://networkx.org/.

NetworkX is particularly helpful in the context of molecular analysis and drug discovery because it can represent molecules as graphs, with bonds serving as edges and atoms as nodes. This enables scientists to compute crucial graph-based metrics that are connected to molecular characteristics, such as stability or reactivity, including centrality, clustering coefficients, and shortest paths. Additionally, NetworkX facilitates integration with other scientific libraries, including Matplotlib, SciPy, and NumPy, allowing for smooth data processing and visualization. It is a useful tool for cheminformatics research and education due to its adaptability and simplicity.

8.4.2.8 AutoDock Tools

AutoDock is a famous docking software to visualize the 3D structure of molecules, as well as how they interact on a particular basis. This software is intended to estimate the size of the molecules, such as substrates or drug candidates, that bind to a receptor of known 3D structure. AutoDock can be applied for virtual screening (HTS), ligand/protein docking, X-ray crystallography, lead optimization, chemical mechanism studies, structure-based drug design, and combinational library design.

Three types of AutoDock are AutoDock 4.2.6 (Legacy version), AutoDock Vina, and AutoDock GPU. The classic version (AutoDock 4.2.6) and the latest version (AutoDock GPU) both utilize a grid in their calculations. The docking process is performed based on two basic programs: Autodock and Autogrid. Autodock runs the ligand binding to a set of grids that describe the protein target. Autogrid precalculated the grids. The visualization of the atomic affinity of the grids is also available. Meanwhile, AutoDock Vina runs without a grid. The atom types are calculated only when it is necessary, unlike the other versions, which performed instantly. The user does not require pre-calculated grid maps for them. A more detailed explanation can be perceived in Table 8.2. For more information, please visit https://autodock.scripps.edu/.

Table 8.2 AutoDock types comparison

No	Feature	AutoDock 4.2.6	AutoDock Vina	AutoDock-GPU
1	Speed	Slow	Fast	High speed
2	Algorithm base	Genetic Algorithm	Empirical & knowledge-based	Genetic algorithm (GPU)
3	Autogrid	Yes	No	Yes (but faster)
4	GPU support	Yes	No	Yes
5	Visualization (ADT)	Yes	Yes	Yes
6	Beginner friendly	No	Yes	No
7	Virtual screening suitability	Limited	Yes	Yes (high performance)

8.4.2.9 PyMOL

PyMol is a widely used 3D molecular software to visualize, analyze, and explore the structure of proteins, nucleic acids, and other molecules. This software also allows the user to prepare a ligand before it binds to another ligand/protein. This tool is lightweight and fast, but still provides a high-quality resolution. The important features of PyMol are that it shows essential interaction, as well as helpful representation, key labels for any area of interest, produces stunning ray-traced images that add depth, shadowing to any structure, creates consistent figures that are easy to reproduce and modify, and many more. The visualization example of PyMol can be seen in Fig. 8.1.

8.4.2.10 ETKDG (via RDKit)

Xperimental-Torsion Knowledge-based Distance Geometry, also known as ETKDG, is an RDKit feature that provides various functions related to molecular graphs and 3D structures. The main feature they propose is exploring more chemically and physically valid though distance geometry approach using the molecular properties. The properties in consideration are ETDG (experimental torsion angles) and K (basic knowledge). ETDG allows a more reasonable torsion arrangement of the conformer generation, while basic knowledge serves as an anchor to maintain constraint validity. In summary, the ETDG explores the distance geometry experiment, while the K guarantees that the conformers produced follow fundamental chemical laws.

Essentially, the ETKDG workflow entails: (1) Determining the molecule's distance bounds matrix using the ETKDG rules and its connection table; (2) Applying a triangle-bounds smoothing algorithm to the bounds matrix; (3) In order to satisfy the smoothed bounds, a random distance matrix must be generated; (4) To acquire atomic coordinates, embed this distance matrix in three dimensions. A "distance geometry

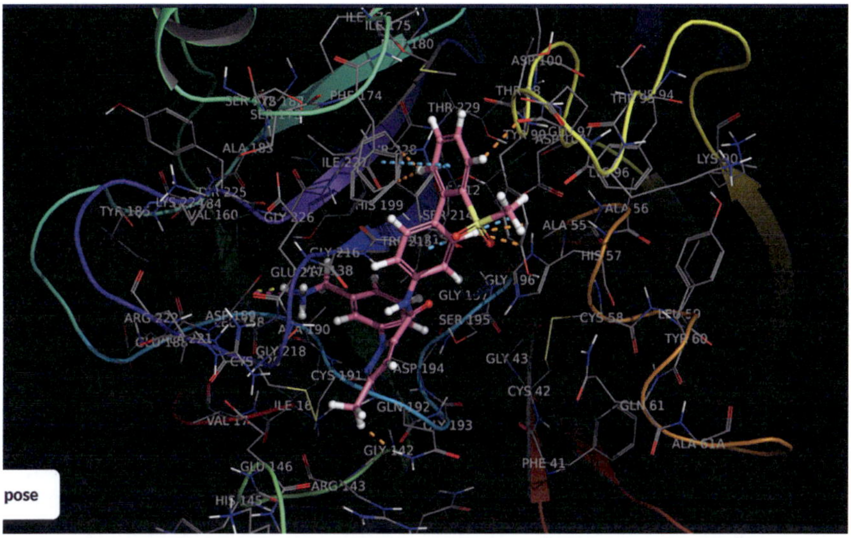

Fig. 8.1 The example of docking using PyMol (retrieved from the PyMol website)

force field" that takes into account the bounds matrix's distance constraints is used to refine the produced coordinates.

8.4.2.11 MolBERT/ChemBERTa

MolBERT and ChemBERTa are two models of machine learning that are frequently used for molecular representations. They use natural language processing (NLP) methods to comprehend chemical structures. The molecular representation mechanisms are performed as reading molecular SMILES strings as "chemical sentences." These models are able to pick up on atom-bond relationships and patterns. MolBERT predicts molecular properties and creates new molecules by using a BERT-based architecture that has been pre-trained on extensive chemical datasets. Developed from another transformer-based model, RoBERTa, ChemBERTa takes a similar approach. For tasks like virtual screening, activity prediction, and drug-likeness evaluation, both tools are helpful because they allow computers to "read" and interpret molecules similarly to how they process human language.

8.4.2.12 OGB (Open Graph Benchmark)

An established tool named Open Graph Benchmark (OGB) was intended to assess and contrast graph-based machine learning models. Many datasets used in drug discovery, such as networks of protein interactions or molecular structures, can be visualized as graph representations. OGB helps researchers test the performance of their graph algorithms consistently by offering baseline models, curated datasets, and transparent evaluation metrics. They provide many features for processing molecular graphs. Node classification, link prediction, and graph property prediction are just a few of the many tasks it covers. Scientists and developers can make sure their graph neural network models are reliable, comparable, and appropriate for practical uses in biology, chemistry, and other fields.

8.4.3 Method

In fundamental principle, drug discovery in small molecules can be performed based on two types of methods: from scratch or drug repurposing. But in general, the pipeline of generative AI-driven drug discovery is detailed in Fig. 8.2.

De novo drug design involves building new molecular structures from scratch. These AI-driven generative models have drastically changed the development of drug design in order to leverage the drug discovery process. The examples of this advanced method include generative adversarial networks (GANs), autoencoders, recurrent neural networks (RNNs), and reinforcement learning (RL)-based systems. These models are increasing chemical libraries and making it possible to create compounds with particular, desired qualities. Therefore, it supports a significant impact in optimizing important molecular properties like toxicity and bioactivity. The limits of protein–ligand docking and molecular modeling are being pushed by state-of-the-art tools such as DiffDock, which uses diffusion models, MoleculeGen for molecule generation, and ESM-2 for protein structure prediction. Simultaneously, AI-powered platforms like Atomwise and BenevolentAI represent how powerful AI can be in transforming the early stages of drug discovery. These companies analyze large biological and chemical datasets, find new drug targets, and create promising candidate molecules using cutting-edge deep learning (DL) and machine learning (ML) techniques. For example, Atomwise uses AI to find lead compounds more quickly by screening large compound libraries against specific protein targets. In order to better understand disease mechanisms and identify novel therapeutic opportunities, BenevolentAI integrates a variety of data sources, such as genetic information, clinical trial results, and scientific literature. These platforms have produced encouraging outcomes, highlighting AI's revolutionary role in speeding up drug discovery procedures. The notable outputs achieved, such as the identification of new drug candidates for difficult illnesses like Ebola and COVID-19 [5].

Drug repurposing, which finds novel therapeutic applications for already-approved medications, provides a time- and money-efficient approach to drug development while lowering the expenses of early experimental phases. AI has a big impact on drug repurposing because it makes it possible to quickly analyze large datasets and find current medications that may have therapeutic applications for novel illnesses. In contrast to MultiDCP (https://github.com/qiaoliuhub/MultiDCP), which uses multi-view DL to integrate various data sources, including molecular

Fig. 8.2 Flowchart of a generative AI-driven drug discovery pipeline, illustrating the stages from molecular dataset curation and pre-processing to model training, molecule generation, docking/validation, and final candidate selection

structures and gene expression profiles, to improve prediction accuracy for drug-disease associations, AI-driven frameworks such as DeepPurpose use DL models with customizable architectures to predict drug-target interactions, enabling high-throughput screening and ranking of repurposed drug candidates efficiently. Leading companies in this field include BenevolentAI and IBM Watson. These platforms identify possible drug candidates for repurposing by analyzing biomedical literature, clinical trial data, and other relevant information using AI algorithms, such as ML and NLP. With the help of artificial intelligence, baricitinib, an oral Janus kinase (JAK)1/JAK2 inhibitor that was first licensed for the treatment of rheumatoid arthritis, was effectively repurposed as a COVID-19 treatment. Repurposing drugs is fraught with legal and patent issues. Because repurposed medications frequently obtain weaker method-of-use patents, which restrict commercial incentives and market exclusivity, patent issues arise. Complex approval processes, such as the requirement for comprehensive clinical trials to validate new uses even in cases where the drug has previously been approved, are examples of regulatory barriers [5].

Repurposing existing drugs for new drug discovery offers a cost-effective approach that speeds up drug development while minimizing experimental expenses. AI significantly affects drug repurposing by enabling the rapid analysis of vast datasets to identify existing drugs with potential therapeutic uses for new diseases. AI-driven frameworks like DeepPurpose utilize DL models with customizable architectures to predict drug-target interactions, enabling high-throughput screening and ranking of repurposed drug candidates efficiently. MultiDCP (https://github.com/qiaoliuhub/MultiDCP) employs multi-view DL to integrate diverse data sources, such as molecular structures and gene expression profiles. This feature enhances prediction accuracy for drug-disease associations. Companies such as IBM Watson and BenevolentAI are at the forefront of this field. These platforms utilize AI algorithms, including ML and NLP, to analyze biomedical literature, clinical trial data, and other pertinent information to identify potential drug candidates for repurposing. An example of the application is Baricitinib. Baricitinib is an oral Janus kinase (JAK)1/JAK2 inhibitor that was first licensed for the treatment of rheumatoid arthritis. With the help of artificial intelligence, this drug was effectively repurposed as a COVID-19 treatment. However, a challenge may still appear in this field. Repurposing drugs is fraught with legal and patent issues. The reason for this is that repurposed medications frequently obtain weaker method-of-use patents, which restrict commercial incentives and market exclusivity. This is where the patent issues arise. Complex approval processes, such as the requirement for comprehensive clinical trials to validate new uses even in cases where the drug has previously been approved, are examples of regulatory barriers [5].

8.5 Challenges, Limitations, and Ethical Considerations

Despite the enthusiasm for AI integration, many challenges and limitations still arise. The use of AI still has many controversies. In medicine, the main factor is always the limitation of data training since patients' information is generally confidential. In drug delivery, many factors remain a challenge. In this chapter, all the challenges, limitations, and ethical considerations regarding generative AI in drug medicine are discussed.

8.5.1 Data Quality and Bias in Datasets

A significant obstacle in drug discovery using AI is the limited dataset for model training [15]. Another notable challenge is data heterogeneity. This occurs when data originates from various sources like chemical structures, biological assays, and clinical trials. It can be difficult and time-consuming to combine and match these various data sources into a single format for AI training. Moreover, the biases in the training data could influence the performance and dependability of the model. For instance, the resulting model may have biases that restrict its accuracy and generalizability in practical applications if the dataset primarily represents a particular disease subset or demographic. Careful data curation, strong data pre-processing methods, and the creation of strategies to reduce bias and guarantee data representativeness are all necessary to meet these challenges [5].

The performance of generative AI adoption in healthcare and medicine depends on trust and validation. The quality and veracity of ChatGPT's responses have varied widely, and most significantly, in an unpredictable manner. This "unpredictability" is the primary obstacle to favorable implementation. This happens because we are unable to predict when generative AI will provide a good response and when it will provide an incorrect or misleading one. In other words, we are unable to determine when to trust generative AI and when not to, particularly when the user lacks the necessary skills to evaluate the quality (accuracy and completeness) of a particular response. For instance, ChatGPT has a history of fabricating content by creating and referencing fictitious scholarly articles. This phenomenon is referred to as "generative AI hallucinations". A technique named Retrieval Augmented Generation (RAG) can counter this hallucination phenomenon. Additionally, depending on how the training models are, GenAI is susceptible to different types of bias and may not always function as well in different languages. Clinical safety and reliability are related issues that are brought up in this trust discussion. Any significant medical application of generative AI will be hampered by these interconnected concerns of trust, safety, and dependability until we have a properly medically trained and validated AI (ChatGPT, for instance, is not specifically medically trained). This model must meet specific criteria in order to gain the definition of "medically trained". For example, a model that has been thoroughly and precisely trained using a corpus of high-quality, evidence-based medical texts that adequately cover a particular medical specialty is valid, which can be referred to as medically trained.

8.5.2 Lack of Interpretability and Black-Box Concerns

The innate complexity and lack of interpretability of AI systems are two of the main obstacles to their widespread adoption, especially in delicate domains like drug development and healthcare. Many artificial intelligence (AI) models, particularly deep neural networks (DNNs), are frequently referred to as "black boxes," with difficult-to-examine internal decision-making processes. Lack of transparency can erode trust, create problems with accountability, and make it more likely that biases will go unnoticed. For instance, to guarantee informed decision-making and protect patient outcomes in clinical settings, healthcare professionals must comprehend how an AI model makes a diagnosis. Similarly, the interpretability of AI-generated predictions is equally important in drug discovery. This area involves complex relationships among compounds, biological targets, and interconnected systems [5].

Although deep learning-based and generative models hold great promise in drug discovery, these systems frequently rely on identifying surface-level statistical patterns in data rather than revealing mechanistic insights. This challenge could result in inaccurate or deceptive conclusions. End users must understand the models' structure, the data they use, and the degree of reliability in order to lessen this. Model design and interpretation can be enhanced by cooperation between biologists, clinicians, and AI researchers. Perturbation-based approaches, in which model parameters or input variables are systematically changed to observe changes in output and uncover causal relationships, may be used to increase interpretability.

Latent space disentanglement is a promising method that allows the targeted generation of molecules with particular desired properties by decomposing the internal representation of the model into discrete and interpretable factors, such as toxicity, bioactivity, or synthesizability. Furthermore, increasing the transparency of models enables their predictions to be explained in a way that is understandable to human users. Semantic-rich representations could help users comprehend how specific structural characteristics affect drug-like qualities by providing insight into the algorithm's logic. To more meaningfully direct the creation of molecules, causal graphs that depict the underlying biological and chemical relationships could be used. Last but not least, hierarchical generative frameworks may make it possible to trace decisions step-by-step, improving communication between humans and AI and facilitating more manageable, objective molecular design. Enhancements in interpretability and transparency are essential for ensuring that biomedical AI models are not only potent but also reliable and useful [4].

8.5.3 Difficulty in the Synthesis of Generated Molecules

Large datasets are a major component of current AI approaches. However, the available data is frequently unbalanced because of privacy concerns, ethical dilemmas, security issues, or limited cases like rare diseases. These concerns lead to sparse clinical data, particularly with regard to toxicity and bioactivity. Few-shot learning strategies can be used to get around this; by incorporating past knowledge, machines can learn efficiently from small samples. The data source is where a workable plan starts. Data augmentation is a popular technique for increasing sample size. This is performed by altering the starting atom or branching order in SMILES strings and taking advantage of their structural non-uniqueness. By carefully adding or deleting edges, one can add variation to graph-based data, such as 3D conformational information or multiscale features like atomic, pharmacophore, and toxicophore properties.

The lack of target-specific training data is a persistent problem in tasks such as de novo molecular generation, particularly for peptides or proteins. By applying knowledge from one domain to improve performance in another related domain, transfer learning provides a solution. In general, smaller target-specific data is used to refine models that have already been trained on larger datasets. The foundation is the pre-trained parameters, which improve learning in situations with limited data. Zero-shot learning can be used in the absence of active molecules. This enables models to forecast characteristics or actions that have never been seen before by depending on learned information rather than actual data. Using large databases such as ChEMBL to learn compound-target interactions and direct molecule design through disease-relevant targets (instead of general molecular patterns) can be useful in rare diseases or understudied targets.

This approach places a strong emphasis on mechanistic comprehension of drug-target interactions. Target-based molecule generation can be reframed as an image captioning problem. The protein's distance map or 3D image serves as the input, and the output is a SMILES string, since AlphaFold has predicted structures for 98.5% of human proteins. Molecular SMILES can be generated even with incomplete or unbalanced data by utilizing pipelines that combine a language model (to generate molecular SMILES) and a visual encoder (for the target structure) [4, 16].

8.5.4 Integration into Existing Drug Development Processes

Although the vast, promising features of AI, many obstacles still need to be overcome when integrating AI tools into current drug development pipelines. Rigid procedures and a heavy focus on tried-and-true techniques are common features of traditional pharmaceutical workflows. It might be necessary to make considerable adjustments to integrate AI into the current workflows, infrastructure, and knowledge. Adoption of these technologies may also be hampered by concerns about intellectual property, data privacy, and the possible effects of AI on jobs in the pharmaceutical sector.

8.5.5 Resource Availability

In physical systems, thermodynamics theory explains that the higher the entropy, the higher the energy requirement (more disorder demands more energy to create or maintain order). In other words, the more complex a system is, the more energy is needed. This analogy can be applied to bioinformatics. Greater molecular complexity demands more energy, as well as resources. Ranging from the capacity of the device, the quantity and quality of the datasets, the time needed to operate, to the cost of the software. When one desires complex data with a complex result, more effort, resources, time, and financial investment are inevitable. This challenge is rather pronounced since not all of the resources are available.

However, on a larger scale, we must not overlook the enormous cost of drug production, which frequently far outweighs computational costs, even though the software can be expensive. Optimizing research and development during the drug discovery process requires the integration of bioinformatics tools and resources. To reach its full potential, this integration necessitates a balanced investment of time, money, and technology infrastructure.

8.5.6 Reproducibility and Open Science Ethics. Ethical Risks: Dual-Use, IP Issues, Dataset Privacy

AI integration in pharmacy practice shows a lot of potential, but it also raises a lot of moral questions. A recent survey of pharmacy professionals in the Middle East and North Africa (MENA) area identified a number of urgent issues. The issues include the lack of clear legal regulations (67.0%), cybersecurity risks (58.9%), potential job displacement (62.9%), and patient data privacy (58.9%). These results demonstrate how important ethical precepts, such as beneficence, justice, informed consent, and transparency, are in directing the application of AI. Curiously, those who were more tech literate also showed higher levels of ethical awareness, suggesting that education and ethical frameworks are crucial for navigating these issues in a responsible manner. Ensuring patient safety and conducting clinical trials present another significant ethical dilemma.

In order to maintain the integrity of trials, participants must be fully informed of the risks involved through comprehensive informed consent procedures. Inadequate performance in this area may compromise safety and violate human rights. Even though organizations like the Food and Drug Administration (FDA) enforce guidelines to safeguard trial participants, issues like the exploitation of weaker populations continue to exist. Transparent governance, interdisciplinary collaboration, and an unwavering focus on patient welfare are crucial to ensuring AI is used in medicine in an ethical manner. Algorithmic bias, opaque decision-making procedures, data security concerns, and unclear accountability are additional moral and legal obstacles.

The urgency for uniform regulations that protect patients without impeding innovation is highlighted by the fact that regulatory frameworks are still evolving as they attempt to keep up with the quick advances in AI. Artificial intelligence's success in patient care and pharmaceutical development, as it develops further, hinges on striking a balance between creativity and moral rectitude to maintain public confidence [5].

8.6 Future Directions in AI-Driven Drug Discovery

8.6.1 *Integration with Robotics and Multimodal AI*

Since the industrial revolution, the introduction of robotics has had a substantial impact on a wide range of industrial processes, including drug development and discovery. Machines are increasingly replacing human labor in processes that have historically involved human labor, which has many advantages. Minimal human interaction contributes to the ultra-high purity needed for pharmaceuticals in a number of upstream and downstream drug production processes, such as discovery, development, formulation, and production. Furthermore, as demonstrated during the COVID-19, Ebola, polio, and cholera outbreaks, robotics facilitates the faster and more effective synthesis, testing, and throughput of compounds. This integration allows for a quicker response to urgent medical needs.

Robotic systems automate the movement of microplates between various equipment and storage areas. This action significantly cuts down on manual handling time and allows for prolonged, unattended screening. Combining robotics with intelligent detection technologies such as imagers, plate readers, and high-content screening systems improves the gathering and examination of large datasets. These systems can quickly process, analyze, and store vast amounts of data when paired with specialized software. This action speeds up the development of possible drug candidates and the identification of active compounds.

Scalable workflows are significantly made possible by the modular design of robotic platforms. This advantage lets the researchers modify capacity in accordance with the demands of particular projects. By predicting how molecular changes may affect compound performance and speeding up the search for viable candidates, the combination of robotics and machine learning eventually optimizes lead compounds. One of machine learning's advantages is its capacity to learn and adjust in response to fresh data, which enhances strategic planning and prediction accuracy as robotic systems gain more knowledge.

Furthermore, real-time data analysis is made possible by ML combined with cloud technology. This advancement gives researchers the ability to instantly monitor experiments and make data-driven decisions, improving flexibility and responsiveness in drug discovery. Companies like Recursion Pharmaceuticals, for instance, demonstrate how well robotics and AI can be integrated into drug discovery. Millions of cell samples are treated, stained, and imaged by robots at Recursion. Automating these procedures is so beneficial. The machine learning algorithms can swiftly and effectively assess morphological changes in cells. Similar to this, Evotec and Exscientia showed how robotics can have a big impact on small-molecule drug development by finding a novel anticancer compound in eight months, compared to four to five years in the past. A company named Arctoris has further transformed automation and enabled quick and accurate experimentation by utilizing robotic systems throughout the entire drug discovery pipeline. From high-throughput screening to compound management and data analysis, all are integrated accordingly.

While Pfizer uses automation for high-throughput screening and data analysis to improve the speed and accuracy of hit selection, Insilico Medicine uses robotics, AI, and deep learning to design and optimize new drug candidates through predictive modeling. By automating crucial tasks and offering reliable data analysis, predictive modeling, and experimental design, the combination of robotics and machine learning has the potential to significantly increase the success rates of pharmaceutical projects.

Although the processes of drug development, delivery, and discovery have benefited greatly from AI and robotics, certain drawbacks still exist in their use. For instance, the cost of the Vinci Surgical System is one of its primary drawbacks. For certain hospitals and healthcare facilities, this system may be unaffordable. Adoption may be significantly hampered by the system's high cost, which may restrict its availability and accessibility. AI can solve problems effectively when trained and reliable databases with a wealth of information are available. Another obstacle to using AI in drug delivery and discovery is the lack of large databases with reliable data. AI may produce skewed conclusions or inaccurate or inconsistent predictions in the absence of reliable data. This implies that training AI-assisted drug discovery models requires consistently accurate, objective, representative, and high-quality datasets. The selection of AI algorithms for a given problem is a crucial factor to take into account when using AI in drug delivery and discovery. Inaccurate results will be obtained if the incorrect AI algorithms are used to solve a given problem. Thus, in order to use AI, researchers must be aware of the right computer algorithms for a given chemical or biological task. Equally crucial are the explanations and interpretations of AI-generated outcomes. The results produced by AI algorithms must be interpretable by scientists.

Like other application domains, artificial intelligence (AI) models, especially those employed in robotics for pharmaceutical tasks, frequently depend on enormous volumes of private or sensitive data in order to operate efficiently. A major ethical and technological challenge is maintaining data privacy during the collection, storage, and processing of this information. Although safe data handling procedures are crucial, they may also restrict the amount and kind of data that can be used to train and enhance models. Additionally, robotics applications often require AI to allow for rapid, real-time modifications in response to changing conditions or unforeseen obstacles. Finding a balance between computational efficiency and the precision of AI-driven decisions is a persistent technological challenge. The action is usually hampered by hardware limitations and the need for robust sensor integration. Real-time adaptation is challenging since it necessitates fast processing speeds and dependable algorithms that can manage dynamic data.

8.6.2 Federated Learning for Privacy-Preserving Research

Drug discovery is a complex and time-consuming process that involves several steps, from the identification of new therapeutic compounds to their development. Molecular modeling techniques and high-throughput screening technologies have greatly increased the availability of large-scale Quantitative Structure–Activity Relationship (QSAR) datasets in recent years. In the context of drug discovery, QSAR streamlines the screening and design stages by acting as a predictive tool for assessing compound characteristics like toxicity, biological activity, and other pharmacokinetic parameters. More effective molecular design and synthesis are made possible by QSAR modeling. This model uses the structural and activity-based information of molecules to quantitatively link molecular characteristics to biological responses.

Drug discovery methods have historically relied on centralized data analysis systems. The past system employed a mechanism in which datasets were compiled from multiple sources on a single server in order to train models. However, there are a number of issues brought about by this centralized structure. The issues are particularly with regard to data security and privacy. Since research organizations and the pharmaceutical industry frequently hold extremely sensitive and proprietary information, direct data sharing between entities becomes a significant concern. Collaborative analysis frameworks have drawn interest as viable substitutes to get around these restrictions. Modern cryptography advances have led to the development of privacy-preserving computation methods that guarantee safe multi-institutional cooperation. Secure multiparty computation (MPC), which enables several parties to collaborate on computing on encrypted data without disclosing the underlying inputs, is one such method. Participants in MPC systems encrypt their data before sending it to a reliable third party, who does the calculation while maintaining confidentiality.

The practical application of MPC in drug discovery settings has been the subject of numerous studies. For example, some researchers have developed MPC-based systems specifically designed to analyze drug-target interaction (DTI) and QSAR data. Others have suggested methods for evaluating drug-drug interactions (DDI) that use secret-sharing techniques. Despite these advancements, participants may still be reluctant to submit data. The increased political and ethical concerns are still a problem. Even when encrypted, some laws expressly prohibit cross-border or inter-organizational data transfers due to the increased enforcement of data protection laws worldwide.

Federated learning (FL) has become a promising solution to these problems. FL is a distributed machine learning framework that runs without disclosing raw data. Institutions can work together to train models using this tool. Using their own data, each participating institution creates a local model. Only the model parameters or updates are sent to a central server for aggregation. Since the original datasets never leave local servers, this paradigm protects the confidentiality of the data. FL is highly adaptable. Depending on how the data is distributed across institutions, the machine can be divided into transfer learning, vertical learning, and horizontal learning.

In applications related to drug discovery, federated learning has demonstrated considerable promise. For instance, some research used horizontal FL on QSAR data and produced remarkable results. Another historic initiative allowed for unprecedented data sharing without jeopardizing proprietary data through a collaborative FL project involving several pharmaceutical companies. Subsequent studies have validated the project's real-world implementation. Federated learning's usefulness in the field has also been further highlighted by other researchers' successful demonstration of its integration into molecular discovery process.

Nevertheless, the majority of the previously mentioned techniques are predicated on the idea that QSAR data is independently distributed (IID) and identical across institutions. In practical situations, this presumption is rarely accurate. Non-IID datasets are frequently the consequence of differences in molecular structures, compound sources, and experimental procedures between institutions. This heterogeneity presents additional difficulties for collaborative drug discovery. Models developed on data from one institution might not generalize well to data from another. Therefore, it is essential to create strong methodologies that can handle non-IID QSAR data in order to advance cooperative drug discovery efforts.

8.7 Conclusion

The pharmaceutical industry has seen significant change as a result of artificial intelligence. From quicker identification of promising drug targets to better clinical trial procedures, all are integrated with AI. By interpreting large and complicated datasets, artificial intelligence helps researchers identify possible compounds, forecast treatment outcomes, and save time and money. In these processes, machine learning, deep learning, reinforcement learning, and even generative AI all have distinct functions. AI in clinical settings can enhance trial patient selection, facilitate real-time monitoring, and even tailor treatment regimens according to an individual's genetic profile. Despite the encouraging developments, there are still a number of obstacles to overcome. These include providing access to high-quality data, increasing the transparency of models, and guaranteeing that AI tools can actually be incorporated into the conventional drug development workflows.

In the future, it will take more than just creating more intelligent algorithms. Cross-sector cooperation, responsible AI governance, and training the next generation of researchers are essential for making real progress. Only with consistent research, ethical supervision, and careful application can AI in drug development reach its full potential. These efforts will ultimately result in quicker discoveries and better patient outcomes. Small molecule drug discovery is changing as a result of generative AI, which makes it possible for scientists to investigate chemical space more effectively and imaginatively than in the past.

Generative models like transformer-based architectures, generative adversarial networks (GANs), and variational autoencoders (VAEs) can learn from existing molecular datasets to create novel compounds with desired properties. This workflow is in contrast to traditional approaches that mainly rely on rule-based algorithms or exhaustive screening. These models aid in scaffold hopping, de novo drug design, and the optimization of pharmacokinetic and pharmacodynamic profiles, as well as speeding up lead identification.

From hit generation to lead optimization, the use of generative AI spans several phases in the drug discovery process and has the potential to greatly cut costs and time. But there are still issues, especially with guaranteeing the chemical validity, synthesizability, and biological significance of compounds produced by AI. Furthermore, to reduce the risks of data biases or unexpected results, incorporating generative AI into experimental workflows necessitates thorough validation, interdisciplinary cooperation, and responsible governance.

Gazing into further direction, hybrid systems that integrate deep learning and domain expertise will be key to the future of generative AI in drug discovery. This allows for more dependable and explicable molecular design. Generative AI has the potential to revolutionize the way we find, develop, and administer small molecule therapies as instruments advance and data ecosystems expand, bringing us one step closer to quicker, more intelligent, and more individualized medications.

References

1. Oyejide AJ, Adekunle YA, Abodunrin OD, Atoyebi EO. Artificial intelligence, computational tools and robotics for drug discovery, development, and delivery. Intell Pharm. 2025;3(3):207–24.
2. Kanakala GC, Devata S, Chatterjee P, Priyakumar UD. Generative artificial intelligence for small molecule drug design. Curr Opin Biotechnol. 2024;89: 103175.
3. Gangwal A, Ansari A, Ahmad I, Azad AK, Kumarasamy V, Subramaniyan V, et al. Generative artificial intelligence in drug discovery: basic framework, recent advances, challenges, and opportunities. Front Pharmacol. 2024;7(15):1331062.
4. Zeng X, Wang F, Luo Y, Kang SG, Tang J, Lightstone FC, et al. Deep generative molecular design reshapes drug discovery. Cell Rep Med. 2022;3(12):100794.
5. Kant S, Deepika, Roy S. Artificial intelligence in drug discovery and development: transforming challenges into opportunities. Discov Pharm Sci. 2025;1(1):7.
6. Feuerriegel S, Hartmann J, Janiesch C, Zschech P. Generative AI. Bus Inf Syst Eng. 2024;66(1):111–26.
7. Southey MWY, Brunavs M. Introduction to small molecule drug discovery and preclinical development. Front Drug Discov. 2023;30(3):1314077.
8. Ivanović V, Rančić M. University of Niš, Faculty of Science and Mathematics, Department of Chemistry, Arsić B, University of Niš, Faculty of Science and Mathematics, Department of Chemistry, Pavlović A, et al. Lipinski's rule of five, famous extensions and famous exceptions. ChemN. 2020;3(1):171–81.
9. Tang B, Ewalt J, Ng HL. Generative AI models for drug discovery. In: Saxena AK, editor. Biophysical and computational tools in drug discovery [Internet]. Cham: Springer International Publishing; 2021 [cited 2025 Jul 31]. pp. 221–43. (Topics in Medicinal Chemistry; vol. 37). Available from: https://doi.org/10.1007/7355_2021_124.
10. Zamanifar A, Faezipour M, (eds). Application of generative AI in healthcare systems [Internet]. Cham: Springer Nature Switzerland; 2025 [cited 2025 Aug 1]. Available from: https://doi.org/10.1007/978-3-031-82963-5.

11. Kell DB, Samanta S, Swainston N. Deep learning and generative methods in cheminformatics and chemical biology: navigating small molecule space intelligently. Biochem J. 2020;477(23):4559–80.
12. Krenn M, Häse F, Nigam A, Friederich P, Aspuru-Guzik A. Self-referencing embedded strings (SELFIES): a 100% robust molecular string representation. Mach Learn: Sci Technol. 2020;1(4): 045024.
13. Cheng AH, Cai A, Miret S, Malkomes G, Phielipp M, Aspuru-Guzik A. Group SELFIES: a robust fragment-based molecular string representation. Digit Discovery. 2023;2(3):748–58.
14. Zang X, Zhao X, Tang B. Hierarchical molecular graph self-supervised learning for property prediction. Commun Chem. 2023;6(1):34.
15. Bilodeau C, Jin W, Jaakkola T, Barzilay R, Jensen KF. Generative models for molecular discovery: recent advances and challenges. WIREs Comput Mol Sci. 2022;12(5): e1608.
16. Huang D, Ye X, Zhang Y, Sakurai T. Collaborative analysis for drug discovery by federated learning on non-IID data. Methods. 2023;219:1–7.

Chapter 9
Generative AI for Antimicrobial and Antiviral Drug Design

Richard Sutejo

Abstract This chapter looks at how generative artificial intelligence and graph neural networks are changing drug discovery for antimicrobial and antiviral resistance. It covers the global challenges from ESKAPE bacteria, multidrug-resistant tuberculosis, and resistant viruses like HIV and SARS-CoV-2, and explains why flexible AI methods are important. The chapter introduces key generative models, such as variational autoencoders, generative adversarial networks, diffusion models, transformers, and different types of graph neural networks. It explains how these models help represent molecules, design new compounds, and predict several properties at once. The chapter also reviews main data sources, including genomic, structural, and chemical databases, and discusses the difficulties in combining these data to create strong training sets. It describes full generative pipelines for finding antibacterial and antiviral agents, from choosing targets and generating molecules to optimizing and testing them in the lab. Case studies, like AI-designed antibiotics such as abaucin and inhibitors of the SARS-CoV-2 main protease, show how using AI with experiments can lead to promising drug candidates. The chapter also points out the need for good data, ongoing resistance monitoring, and global teamwork.

Keywords Generative artificial intelligence · Graph neural networks · Antimicrobial resistance · De novo drug design · Antiviral pipelines

The continuous development of fresh infections alongside reactivated infections continues to disrupt worldwide healthcare systems. Drug-resistant bacteria have become increasingly problematic during the recent years because of overuse of antibiotics and industrialization and environmental pressure. *Enterococcus faecium, Staphylococcus aureus, Klebsiella pneumoniae, Acinetobacter baumannii, Pseudomonas aeruginosa*, and *Enterobacter* species [105, 107, 111] are among the most dangerous ESKAPE pathogens. These bacteria have developed resistance to current antibiotics which makes them a major cause of infections found in hospitals.

R. Sutejo (✉)
Faculty of Biomedicine, i3L University, Jakarta, Indonesia
e-mail: richard.sutejo@i3l.ac.id

A. A. Parikesit and A. N. M. Ansori (eds.), *Recent Computational Techniques in De Novo Drug Design*, Bio-IT and AI, https://doi.org/10.1007/978-3-032-20428-8_9

Multiple viral pathogens are rapidly mutating to develop antiviral resistance which makes worldwide treatment of diseases increasingly difficult. Studies have documented drug-resistant strains of HIV-1 and hepatitis C virus (HCV) which creates difficulties for patient care and demonstrates the requirement for novel treatment approaches [36]. The reported cases demonstrate the critical requirement to develop new antibiotic and antiviral medications as well as treatment methods that combine existing drugs.

The antibiotic and antiviral research field uses generative AI systems to generate new content starting from large datasets through learning processes. The implementation of generative AI in drug development accelerates antibiotic and antiviral research through molecular interaction modeling and compound structure optimization and new drug candidate generation. The design process of novel antibiotics effective against multidrug-resistant bacteria including *Acinetobacter baumannii* has been validated through experimental methods with the aid of generative models [105, 107, 111]. Deep learning frameworks have discovered bifunctional antimicrobial peptides with antibacterial and antiviral properties [36].

Research on antimicrobial resistance requires immediate integration of generative AI tools because they help fight drug-resistant bacteria and viruses [36, 88, 105, 111]. The essential role of generative AI in antimicrobial research emerges because these technologies convert bacterial and viral data into drug design and discovery processes at a speed that matches the emergence of new resistant species.

9.1 Global Landscape of Antimicrobial and Antiviral Resistance

Modern medicine encounters a major difficulty because drug-resistant microbes continue to rise at alarming rates thus rendering first-line antimicrobial treatments ineffective in both hospital and community settings [30, 95, 117]. The recent years have shown that multidrug-resistant (MDR) bacteria cause severe infections that force doctors to choose between second-line treatments and costly last-resort medications [95, 117]. The weakened effectiveness of these treatments puts at risk the therapeutic advancements that affect surgical procedures and cancer treatments and intensive-care interventions [95, 117]. The existing antiviral drug pipeline shortage coupled with drug-resistant viral strains including HIV, hepatitis B and C and influenza makes the situation worse [95, 117]. The current situation requires immediate development of innovative drug design methods alongside alternative therapeutic approaches to monitor and fight both antibiotic and antiviral resistance [39, 95, 117].

The crisis continues because of three main factors which include antibiotic misuse in agricultural and aquacultural sectors together with substandard medical practices in healthcare facilities and inadequate infection control measures [59, 95, 117]. The movement of newly developed resistant strains between countries through international connections and global supply chains [95, 117]. Antimicrobial resistance hotspots no longer stay contained within geographic boundaries because urban areas and conflict zones and industrialized regions have become affected [95, 117]. Antiviral resistance clusters have been found in HIV treatment areas as well as areas with prolonged influenza circulation [39].

The list of priority pathogens is expanding. ESKAPE bacteria—*Enterococcus faecium, Staphylococcus aureus, Klebsiella pneumoniae, Acinetobacter baumannii, Pseudomonas aeruginosa, and Enterobacter spp.*—dominate as nosocomial infections [30, 95]. In the viral realm, resistance in HIV-1, hepatitis B and C viruses, and influenza A subtypes complicates therapeutic decision-making [39, 95]. The emergence of multidrug-resistant *Mycobacterium tuberculosis* further exemplifies the adaptability of these microbes under pharmacological pressure [95, 117].

Effective response depends on robust surveillance but monitoring frameworks continue to show inequalities. Global initiatives such as WHO's GLASS for bacterial pathogens have improved data harmonization, yet many regions still lack routine access to culture-based diagnostics or sequencing-based analysis, particularly in Sub-Saharan Africa, Central Asia, and Oceania [95].

Countermeasures are progressing. Antimicrobial stewardship programs and infection-prevention protocols have shown the ability to curb the spread of resistance, though implementation is uneven [95, 117]. New antimicrobial drugs and therapeutic approaches involving nanomaterials and antimicrobial peptides together with phage therapy and CRISPR-based methods are under development to match the evolutionary adaptability of pathogens [18, 74, 82, 91, 132, 140]. The successful implementation of high-resolution genomic surveillance with big-data analysis and generative AI-driven molecular design requires equitable global access together with rigorous validation and coordinated governance [95, 117, 119].

9.2 Foundations of Generative AI and Graph Neural Networks

The rapid advancement of artificial intelligence has ushered in a new era of computational drug discovery, with generative AI and graph neural networks standing as two foundational pillars that are revolutionizing how we design antimicrobial and antiviral compounds. The understanding of these fundamental technologies provides the foundation for assessing their transformative capability to solve the previously discussed resistance crisis.

9.2.1 *Generative Artificial Intelligence: Theoretical Foundations*

Generative AI is fundamentally changing the approach to drug discovery. Previously, models were limited to predicting patterns in existing data. Now, they can generate entirely new molecular structures for novel molecules [44, 79, 112, 138]. At its core, this technology works by training the AI to understand the fundamental patterns and rules of existing chemical data, which then allows it to generate new, scientifically plausible examples of its own [44, 79, 112, 138]. The ability to generate novel molecular structures with desired therapeutic properties makes this capability highly valuable for drug discovery [44, 79, 112, 138].

The mathematical foundation for this lies in probability theory. Two components are essential: the first, called Kullback-Leipler (KL) divergence, acts like a ruler to measure the difference between the AI's creations and the real-world examples. The second, the Evidence Lower Bound (ELBO), is the practical technique the model uses to minimize that difference and continuously refine its designs [44, 138]. In practice, this is achieved using specific model architectures like Variational Autoencoders (VAEs) and Generative Adversarial Networks (GANs). These frameworks provide the structure for the AI to learn from complex scientific data and efficiently produce designs for new potential medicines [44, 79, 112, 138].

9.2.2 *Core Generative AI Architectures: VAE and GAN*

Applications in molecular design are built upon several core generative architectures, each with unique strengths. One of the most influential is the Variational Autoencoder (VAE), which combines the pattern-recognition power of neural networks with principles from probabilistic modeling. A VAE is composed of two main neural networks. The first, an **encoder**, analyzes input data (like an existing molecule) and compresses it into a simplified, abstract summary called a **latent space**. The second, a **decoder**, then uses that summary as a blueprint to generate a new, similar data point from scratch [14, 138]. VAEs serve as popular tools for creating chemically valid diverse molecular structures for drug discovery research [14, 138].

GANs consist of two components, a generator and a discriminator, that train adversarially. The generator improves its ability to create authentic samples, while the discriminator distinguishes real from synthetic data. GANs effectively generate high-quality synthetic data for drug discovery, including molecular structures [14, 67, 107, 138]. MolFilterGAN represents a recent advancement enhancing evaluation of AI-generated chemical compounds [67].

Following the success of GANs, newer generative modeling approaches have emerged to overcome their limitations in stability and diversity. Among these, diffusion models have gained attention for their robust ability to reconstruct data by reversing a gradual noise addition process through iterative denoising. These models not only excel in realistic image generation but are increasingly applied to molecular synthesis tasks [107, 138].

Autoregressive models further advance the generative paradigm by predicting each component of data sequentially based on preceding outputs. In molecular generation, they represent chemical structures as SMILES sequences and create novel compounds through token-by-token prediction [14, 107, 138].

Extending these innovations, transformer-based large language models (LLMs) integrate natural language processing with molecular design. Recent studies demonstrate their ability to generate chemically valid structures, propose molecular modifications, and optimize compound properties using language-driven multi-agent frameworks [104, 107, 109, 138].

9.2.3 Graph Neural Networks: Fundamental Concepts

Graph Neural Networks (GNNs) represent a distinct neural network architecture which specializes in processing graph-structured information while diverging from standard neural networks that handle fixed vectors and regular image grids [122, 128]. Their capacity to manage irregular graphs with variable nodes and edges makes them particularly suitable for molecular learning, as chemical compounds naturally form molecular graphs with atoms as nodes and bonds as edges [40, 125, 128, 133].

The core mechanism of GNNs relies on message passing, a process that transfers information between connected nodes within a graph. Each propagation step allows a node to aggregate information from its neighbors, update its representation, and then relay messages to adjacent nodes. This iterative operation enables GNNs to capture both local chemical environments and holistic molecular features [100, 122, 128]. Node update functions combine self-representations with neighboring data, while edge update functions integrate bond-specific characteristics [100, 128].

GNNs are now widely employed for molecular property prediction, docking evaluation, compound optimization, and de novo molecule generation with high accuracy [40, 125, 128]. The integration of geometric and motif-based information further enriches molecular representation learning [40, 133]. Additionally, GNN performance improves through self-supervised and contrastive learning strategies, particularly under limited labeled data conditions [40, 125, 128].

9.2.4 GNN Architectures and Variants

The development of diverse GNN architectures has addressed multiple demands in graph-based molecular property prediction. Graph Convolutional Networks (GCNs) apply convolutional operations to molecular graphs, enabling efficient extraction of structural features [87, 125]. Building upon the principles of GCNs, Graph Attention Networks (GATs) introduce attention mechanisms that allow nodes to weigh the importance of neighboring connections, improving the interpretation of complex molecular relationships [125]. Further extending this evolution, Message Passing Neural Networks (MPNNs) unify multiple GNN formulations under a generalized message-passing framework and consistently demonstrate superior performance in molecular property prediction compared to other deep learning methods [108, 125].

The expressive capability of GNNs has been theoretically characterized through connections to the Weisfeiler-Leman graph isomorphism test, which provides bounds on the types of graph structures that different GNN architectures can distinguish [47, 125]. Higher-order GNNs, such as k-dimensional GNNs, have been proposed to capture more complex structural patterns essential for understanding molecular behaviour [47, 125].

Recent progress includes hybrid and multi-view GNN architectures that integrate the strengths of multiple frameworks to enhance molecular prediction accuracy [47, 83]. Additionally, self-supervised and motif-based learning approaches have been developed to enhance the ability of GNNs to capture substructure information and generalize in low-data regimes [125, 136, 141].

9.2.5 Molecular Representation Learning

The integration of generative AI with Graph Neural Networks (GNNs) has introduced advanced frameworks for molecular representation learning. Traditional molecular representations, such as SMILES (Simplified Molecular Input Line Entry System) strings, encode molecules as text sequences suitable for sequence-based generative models. However, SMILES notations are inherently non-unique and fail to capture essential three-dimensional structural information, often limiting their accuracy in modeling molecular properties [90, 135].

Graph-based representations overcome these limitations by directly encoding atomic connectivity and geometric relationships, allowing GNNs to extract features from molecular graphs more effectively [89, 90, 135]. Recent directions increasingly integrate multiple representation modalities, combining sequence-based and graph-based encodings to capture complementary aspects of molecular structure [90, 135]. Moreover, advanced models incorporate conformational space information, accounting for multiple 3D configurations that influence biological activity [38].

Innovative approaches now combine generative self-supervised learning with GNN architectures, enabling richer and more informative molecular representations derived from large-scale unlabeled data [62, 68, 90, 125]. These methods have significantly improved molecular property prediction, drug discovery performance, and novel molecule generation [62, 68, 90, 125]. Additionally, graph transformation layers and hierarchical or motif-based GNNs further enhance the ability to capture multi-level and substructural information within molecules [89, 136].

9.2.6 Integration and Synergies

The convergence of generative AI and Graph Neural Networks (GNNs) has produced hybrid architectures that combine the strengths of both paradigms for molecular property prediction and drug discovery. Among these, graph-based Variational Autoencoders (VAEs) integrate the probabilistic framework of VAEs with the structural awareness of GNNs, enabling the generation of valid molecular structures that retain desired physicochemical properties [131, 138]. Likewise, graph-based Generative Adversarial Networks (GANs) and diffusion models can directly generate molecular graphs, ensuring that synthesized compounds comply with chemical and structural constraints [2, 131, 138].

These integrated approaches have demonstrated superior performance in molecular property prediction, drug–target interaction modeling, and de novo molecular design compared with traditional computational methods [20, 113, 114, 131, 138, 141]. Hybrid models that combine molecular graphs with chemical fingerprints or transformer-based embeddings achieve state-of-the-art results across diverse drug discovery datasets [20, 83, 113, 141]. The capacity to jointly optimize molecular generation and property prediction within a unified framework represents a significant advancement in computational and predictive drug discovery [131, 138].

The foundational concepts developed through these hybrid systems form a strong platform for applying generative AI and GNNs in antimicrobial and antiviral drug design. The combination of robust mathematical foundations and advanced representational capabilities underscores their transformative potential in next-generation computational drug discovery platforms [44, 131, 138].

9.3 Data Foundations: Genomic, Structural, and Chemical Repositories

The success of generative AI and Graph Neural Networks (GNNs) in antimicrobial and antiviral drug design depends on access to high-quality, comprehensive, and well-integrated data repositories. These data foundations provide essential inputs for training and validating AI models, supplying genomic, structural, and chemical information necessary to generate therapeutic compounds that address microbial resistance. Extensive global efforts by researchers have established a vast network of biological and chemical databases. This accumulated scientific knowledge forms a sophisticated information ecosystem that computational models can leverage to accelerate drug discovery and optimization.

9.3.1 Genomic Databases: The Blueprint of Resistance

Genomic databases provide the foundational framework for understanding molecular resistance mechanisms in antimicrobial and antiviral therapies. The National Center for Biotechnology Information (NCBI) GenBank serves as the primary repository for nucleotide sequences, housing millions of bacterial and viral genomes that reveal resistance evolution patterns across pathogen lineages and geographic regions [41, 48, 96]. These resources enable researchers to track global resistance gene dissemination and develop sophisticated analytical tools for resistance monitoring [41, 48, 96].

The Comprehensive Antibiotic Resistance Database (CARD) functions as a specialized antimicrobial resistance research platform. CARD employs the Antibiotic Resistance Ontology (ARO) to organize sequence data, facilitating deeper understanding of resistance mechanisms. This database provides comprehensive information on resistance genes, their protein products, and associated phenotypes, supporting AI model training for resistance pattern prediction [4, 5, 54]. CARD enhancements made include improved curation methods, standardized nomenclature, and automated analysis capabilities [4].

ResFinder maintains extensive antimicrobial resistance gene collections and offers phenotype prediction tools, making it a widely adopted platform for resistance detection in next-generation sequencing data [42, 48, 137]. Generative models trained on these databases learn resistance mechanisms, enabling them to circumvent known escape pathways and support real-time genomic surveillance and predictive modeling [4, 42, 48].

The Global Initiative on Sharing All Influenza Data (GISAID) and specialized NCBI resources maintain databases that enable genomic surveillance of antiviral resistance mutations. These repositories provide critical insights into RNA virus evolution and drug-resistant mutations that compromise antiviral therapy efficacy [48, 96].

9.3.2 Structural Databases: Three-Dimensional Architecture of Drug Targets

The Protein Data Bank (PDB) serves as the primary global repository for three-dimensional structural data of biological macromolecules, encompassing both experimentally determined and computationally predicted structures [12, 85]. This extensive dataset supports structure-based drug design and provides essential training resources for AI models, particularly Graph Neural Networks (GNNs), that utilize pathogen target structures for prediction and optimization [12, 85]. The growing inclusion of experimental and predictive entries continues to expand the PDB's role in AI-driven drug discovery research [12, 85].

The PDB ecosystem has further advanced through enriched annotations and standardized data formats that enhance its analytical utility [12]. Structural classifications provided by Pfam, SCOP2, and CATH add valuable functional and evolutionary context, enabling researchers to identify relevant binding sites and potential therapeutic targets [12]. The European counterpart, PDBe, complements these efforts by offering interactive tools for structural visualization and analysis, while PDBe-KB delivers community-driven annotations that connect structural data with functional insights [12].

Structural repositories play a pivotal role in antimicrobial and antiviral drug development by supplying information on drug–target complexes, resistance-associated mutations, and allosteric binding sites that guide therapeutic optimization [1, 12, 85]. Structural data from wild-type and resistant variant proteins allows AI models to discover resistance mechanisms which leads to the development of compounds targeting resistant strains [1, 12, 85].

Combining structural data with binding affinity measurements has yielded large-scale structure–activity relationship (SAR) datasets crucial for predictive modeling. Databases such as BindingDB and ChEMBL strengthen AI-based drug design by linking structural and functional information, enabling more precise predictions of molecular interactions [12, 85].

9.3.3 Chemical Databases: The "Universe" of Molecular Diversity

AI-driven drug discovery relies heavily on comprehensive chemical databases that provide vast molecular libraries for identifying patterns and generating novel therapeutic candidates. The ZINC database (ZINC-22) represents one of the largest collections of commercially accessible compounds, comprising billions of molecules available for experimental validation. This extensive resource enables generative models to efficiently explore chemical space and identify regions enriched with bioactive compounds [45, 115].

ChEMBL, maintained by the European Molecular Biology Laboratory, contains over 2.1 million compounds linked to detailed bioactivity data. The database uniquely connects chemical structures with quantitative biological assay results, providing essential labeled datasets for supervised learning models. Integrating ChEMBL with structural and genomic data enhances AI capability to analyze structure–activity relationships and molecular target interactions [22, 97, 115].

The chemical repository PubChem operates under NCBI as a major database that contains more than 100 million different chemical structures alongside their biological data. NCBI resources connect through PubChem to create seamless integration between chemical and genomic and bibliographic information for data mining purposes [22, 115].

Specialized chemical repositories dedicated to antimicrobial and antiviral research curate focused collections of active compounds and their activity profiles. These datasets allow AI models to identify structural motifs linked to antimicrobial potency and to pinpoint core chemical scaffolds that serve as promising leads for resistance-evading drug design [25, 45, 115].

9.3.4 Data Integration challenges and Solutions

Integrating diverse chemical, structural, and genomic datasets poses major challenges for AI-driven drug discovery. Chemical databases, such as ZINC-22, offer billions of commercially available compounds, providing essential resources for generative models to explore chemical space and identify bioactive candidates [45, 115]. ChEMBL, managed by the European Molecular Biology Laboratory, enriches this landscape with over 2.1 million compounds and linked bioactivity data, connecting chemical structures to quantitative biological measurements for supervised learning model training [22, 97, 115].

NCBI's PubChem integrates more than 100 million chemical structures with biological and bibliographic records, facilitating seamless cross-referencing and comprehensive data mining. Combining chemical datasets with structural and genomic information enables multi-modal AI models to analyze complex relationships among chemical features, biological activities, and target properties [22, 97, 115].

Specialized chemical resources focused on antimicrobial and antiviral compounds provide curated agent collections and activity profiles, supporting AI models in detecting activity-related molecular motifs and discovering resistant drug candidates [25, 45, 115].

9.3.5 Quality control and Validation

The reliability of AI models and drug discovery applications depends strongly on the quality of their training data. Effective database curation combines automated validation systems with expert manual review to ensure accuracy and consistency [16, 44, 46, 106]. For instance, the CARD database employs automated literature evaluation tools that facilitate active updates and the inclusion of newly identified resistance mechanisms [44].

Beyond initial data curation, maintaining ongoing data quality requires systematic cross-validation and performance benchmarking across databases. Cross-database validation functions as a key quality control technique to detect duplication, errors, and ambiguous entries that may indicate data inconsistencies [16, 44, 106]. Comparative performance evaluations further guide the selection of appropriate data sources for AI training by benchmarking database reliability across distinct research applications [16, 44, 106].

The integration of experimental validation data with computational predictions enables continuous improvement of both database quality and AI model performance through feedback loops. The rapidly changing antimicrobial and antiviral resistance fields require data resources to remain both relevant and accurate thus making this process essential [16, 44, 46, 75, 106]. The increasing size of biological and chemical data demands the creation of advanced data integration systems which are vital for managing complex information [50, 57, 65, 78, 103].

Generative AI applications in antimicrobial and antiviral drug development thus require both comprehensive data repositories and advanced integration systems capable of efficiently extracting and processing multi-source information to accelerate the discovery of future therapeutic agents [46, 50, 57, 75, 78, 103, 106].

9.4 Generative Pipeline for Antibacterial Agents

The integration of modern generative AI with Graph Neural Networks (GNNs) has produced powerful data-driven pipelines that leverage genomic, structural, and chemical data as foundational inputs for antibacterial drug discovery. These generative pipelines employ automated workflows that transform biological, chemical, and structural information into therapeutic candidates, resulting in novel antibacterial compounds with improved efficacy against resistant pathogens [23, 45, 105, 109, 114, 138].

Unlike traditional drug development, which depends on labor-intensive screening of chemical libraries, AI-based discovery systems generate molecules de novo guided by design objectives and biological constraints informed by bacterial resistance profiles [23, 45, 105, 109, 114, 138]. These systems operate through sequential modules encompassing target validation and selection, molecular design and property prediction, multi-objective optimization, and experimental assessment [23, 45, 105, 109, 114, 138].

The feasibility of end-to-end AI-guided antibiotic discovery is now supported by emerging generative frameworks. Models such as SyntheMol and APEXGO can design synthesizable antibiotics and optimize antimicrobial peptides that demonstrate efficacy in both in vitro and in vivo testing [23, 45, 105, 109, 114, 138]. These systems employ structure-based clustering of predicted proteomes to detect essential microbial targets that lack human homologs and apply generative models to balance chemical potential with biological activity [105, 114].

Recent advances confirm the end-to-end feasibility of AI-guided antibiotic discovery, where workflows begin with target identification and progress to compound testing. Generative models such as SyntheMol and APEXGO design synthesizable antibiotics and optimized antimicrobial peptides, demonstrating efficacy in both in-vitro and in-vivo studies [23, 45, 105, 109, 114, 138]. These platforms cluster predicted proteomes based on structure to pinpoint essential, conserved targets lacking human homologs, then use generative models to balance chemical potential with biological effects [105, 114].

The AI-guided AMP-hydrogel-Designer further exemplifies this approach, generating antibacterial biomaterials with potent activity against drug-resistant bacteria and accelerating wound healing in animal models [55]. Such integrated pipelines showcase how generative AI and GNNs tackle antimicrobial resistance by developing new therapies through data-driven design [23, 45, 55, 105, 109, 114, 138].

The first phase of generative pipelines starts with an organized search for vital bacterial targets that support pathogen survival and allow intervention through small molecules. The current genomics pipelines select protein families that appear throughout ESKAPE pathogens and other high-priority bacteria through comparative methods while focusing on targets which are crucial yet non-human-homologous and commonly found in all species [86, 93]. Druggability assessment then draws on structural biology, integrating experimental (PDB) and predicted (AlphaFold) structural data to evaluate binding pocket features, allosteric sites, and structural flexibility [23, 93].

Structure-based virtual screening of essential bacterial proteins further advances hit identification by pinpointing promising inhibitors. Combining structure-based clustering and virtual screening allows AI-guided pipelines to select enzymatic inhibitors efficiently and match new compounds to target-specific requirements [23, 28, 93]. Through generative AI models scientists create new compounds which precisely match the structural requirements and functional properties of validated targets [23, 28, 93].

Target selection now incorporates evolutionary and systems biology to identify critical vulnerabilities less prone to resistance via mutation. Emphasis is placed on highly conserved active sites, allosteric networks, and key protein–protein interfaces essential for virulence and survival [93]. The development of advanced pipelines includes a mechanism to detect potential synergistic effects by identifying multiple targets which can be used for multi-target drug design thus raising the chances of overcoming resistance mechanisms [23, 93].

9.4.1 Molecular Generation and Design Engine

The generative component of contemporary antibacterial pipelines uses advanced AI models to create novel molecules that target bacteria and combat drug resistance. These pipelines combine diffusion models, autoregressive models, graph neural networks, and language models to achieve a balance between chemical validity, synthetic accessibility, and biological activity [23, 93, 105]. Generated compounds undergo rigorous post-processing, filtration, and analog search to yield synthesizable drug candidates [93, 105].

Molecular design further leverages three main generative methods: variational autoencoders, generative adversarial networks, and normalizing flow models. These approaches enhance active learning, improve drug-likeness scoring, enable virtual docking, and assess synthesizability in silico [23]. The SyntheMol model, for example, produces readily synthesizable antibiotics with novel scaffolds showing promising activity against diverse bacterial strains [105].

Recent efforts integrate diffusion models, transformer-based language models, and normalizing flow architectures to explore large chemical spaces for both small molecules and antimicrobial peptides [17, 23, 93, 126]. Diffusion-based pipelines have produced antimicrobial peptides with strong antibacterial and antifungal properties confirmed in experimental studies [126]. AI platforms now enable peptide antibiotic and biomaterial design that demonstrate efficacy against drug-resistant pathogens in preclinical models [55, 126].

Graph neural networks provide unique advantages for antibacterial design by directly modeling molecular structures, preserving chemical constraints, and incorporating three-dimensional features. These models generate molecular graphs that respect valence rules and synthetic feasibility while optimizing for binding affinity, selectivity, and pharmacokinetics [17, 23, 93].

9.4.2 Multi-Objective Property Prediction Networks

Effective antibacterial discovery pipelines require the concurrent optimization of multiple, often conflicting, objectives that determine clinical potential. Clinical development of new compounds prioritizes antimicrobial potency, bacterial versus human target selectivity, pharmacokinetic properties, toxicity profiles, and synthetic accessibility. Advanced pipelines integrate machine learning models, including ensemble and multi-feature systems, to predict compound properties and biological activities using diverse molecular descriptors and graph representations [63, 102, 122].

ADMET prediction models are incorporated to assess drug-like characteristics essential for clinical translation. The mtk-QSBER model demonstrates high accuracy in predicting antibacterial activity and ADMET parameters for training and test compounds [102]. Contemporary pipelines leverage molecular fingerprints, chemical descriptors, and graph-based features to integrate pharmacokinetic predictions into molecular design [63].

Modern property prediction emphasizes the need for interpretability in AI models. Graph neural networks and related machine learning approaches implement attention mechanisms and feature analysis to reveal molecular substructures and functional groups that drive specific properties, thus aligning model outcomes with established SAR principles [63, 125]. The interpretative features of antibacterial design tools help scientists understand structural activity relationships which guides future optimization efforts.

Specialized models are trained on antimicrobial resistance (AMR) datasets covering various bacterial species and resistance mechanisms. Machine learning classifiers, such as random forests and boosting algorithms, accurately predict resistance phenotypes using genomic and transcriptomic information for a range of pathogens and antibiotics [32, 49, 58, 108, 129]. These interpretable features help scientists understand structure–activity relationships, guiding further optimization. Deep learning and transfer learning improve performance for new antibiotics under limited data conditions and help select designs with reduced resistance risk [58, 108, 129]. The models assess the probability that bacterial populations will become resistant to new compounds which helps designers create antibiotics with better resistance profiles.

9.4.3 Iterative Optimization and Active Learning Framework

One of the most critical phases in AI-driven drug discovery is molecular optimization, where generated compounds are refined toward optimal efficacy, safety, and manufacturability. In antibacterial design, this phase forms the bridge between virtual molecule generation and experimentally validated candidates. By integrating reinforcement learning, Bayesian inference, and active learning mechanisms, modern generative frameworks ensure that design iterations progressively improve compound quality and predictive reliability.

Modern generative pipelines employ advanced optimization strategies that iteratively refine molecular designs based on predicted properties and experimental feedback. These systems utilize multi-objective optimization algorithms, incorporating deep reinforcement learning and meta-learning techniques to balance conflicting design objectives and efficiently explore vast chemical or design spaces [56, 101, 123, 142]. The integration of active learning enables these systems to allocate computational effort to the most informative candidates, thus accelerating discovery and minimizing costly experimental evaluations [26, 33, 76, 98].

Bayesian approaches are frequently incorporated into the optimization process, providing uncertainty estimates for predicted properties and enabling the identification of candidates likely to possess favorable characteristics while highlighting areas where additional data collection would be most valuable [19, 33, 76, 116]. This probabilistic framework is particularly valuable in complex design tasks, such as drug discovery, where high-dimensional property spaces and intricate structure–activity relationships can make deterministic optimization approaches susceptible to local minima and overfitting [56, 101].

Recent implementations have demonstrated the value of reinforcement learning approaches, which can learn optimal design strategies through interaction with predictive models and, when available, experimental feedback [43, 53, 60, 71, 123]. These systems adapt their generation strategies based on the success or failure of previously generated candidates, developing increasingly sophisticated design rules that address the specific challenges of the application domain [53, 60, 121].

The IDOLpro platform exemplifies state-of-the-art multi-objective optimization in structure-based drug design. It integrates diffusion models with differentiable scoring functions to guide efficient chemical space exploration [56]. This method generates ligands with binding affinities up to 20% higher than competing approaches and improved synthetic accessibility, achieving over 100× faster performance than exhaustive virtual screening while identifying molecules with superior predicted properties [56].

9.4.4 Experimental validation and Feedback Integration

Experimental validation represents the critical translational stage of generative antibacterial pipelines, confirming that AI-generated compounds exhibit genuine antimicrobial activity under laboratory and biological conditions. This phase bridges computational design with empirical testing, ensuring that predicted candidates demonstrate clinical relevance and safety. Modern workflows employ multi-step validation pipelines, progressing from biochemical and cellular assays to in vivo efficacy studies in animal models [31, 50, 61, 105, 126]. The integration of experimental outcomes back into computational frameworks creates a closed feedback loop, continually refining generative model performance based on real-world data [31, 50, 61, 105, 121].

High-throughput screening and automated synthesis technologies accelerate evaluation of large compound libraries, generating data that strengthens predictive model accuracy and informs design hypotheses [31, 50, 51, 105]. Automated synthesis and rapid experimental validation, as demonstrated by the synthesis and testing of dozens of candidate molecules within weeks, reduce the time lag between computational design and experimental validation [31, 61, 105, 126]. For example, generative models have been integrated with experimental pipelines to design and synthesize novel antimicrobial peptides and small molecules, several of which have shown significant activity against multidrug-resistant pathogens in both in vitro and in-vivo models [31, 51, 61, 105, 126].

Mechanistic studies embedded within validation pipelines examine how newly designed compounds inhibit pathogens. Through molecular modeling, transcriptomic profiling, and proteomic analysis, researchers have elucidated antibacterial mechanisms such as membrane disruption and enzyme inhibition—findings critical for improving future compound design [31, 51, 61, 126]. These insights guide the refinement of molecular scaffolds toward enhanced potency and resistance evasion.

Recent studies confirm that AI-generated compounds can achieve strong antimicrobial efficacy in both in vitro and in vivo infection models, displaying broad-spectrum activity and minimal resistance potential [31, 51, 61, 105, 126]. These achievements illustrate how generative pipelines, through iterative feedback and advanced validation, now produce antibacterial candidates capable of meeting experimental and translational benchmarks essential for clinical development.

9.4.5 *Pipeline Integration and Automation*

As antibacterial discovery pipelines become increasingly data-intensive and modular, efficient integration and automation have become essential to coordinate the complex flow of information between experimental and computational systems. Modern AI-driven frameworks now combine molecular modeling, high-throughput screening, and automated synthesis into unified workflows that accelerate compound discovery while maintaining analytical rigor and reproducibility [11, 51, 124]. Machine learning integration within high-throughput screening platforms allows for rapid identification of promising molecules—especially those active against multidrug-resistant pathogens [51, 124]. This approach enables efficient large-scale screening while preserving predictive precision across chemical diversity.

Cloud-based, high-throughput discovery frameworks support collaboration between multidisciplinary teams and facilitate systematic evaluation of design decisions. Advanced workflow management systems coordinate upstream and downstream analytical processes, ensuring data provenance, reproducibility, and traceability throughout the discovery pipeline [11]. Together, these systems provide a robust computational backbone that enhances discovery efficiency and mitigates data fragmentation.

The combination of intelligent data management with automated synthesis pipelines significantly accelerates the acquisition of high-quality antibacterial data. Machine learning applied to combinatorial chemical libraries assists in screening optimization, helping researchers identify active compounds against resistant strains such as MRSA [124].

The continued evolution of generative and machine learning–driven antibacterial pipelines promises to deliver faster and more reliable therapeutic development. Their deep integration with experimental validation and automated infrastructure marks a paradigm shift toward adaptive, self-improving systems that will broaden drug discovery success over the coming decades [11, 51, 124].

9.5 Generative pipeline for Antiviral Agents

The design of generative pipelines for antiviral agents presents unique challenges that differ fundamentally from antibacterial discovery. Bacteria are autonomous cells with independent metabolic pathways, while viruses function as obligate intracellular parasites entirely dependent on host cell machinery for replication [37]. Effective antiviral therapeutics must selectively target viral processes without disrupting essential host cellular functions [37]. Designing antiviral drugs with generative AI also requires addressing the highly plastic host–pathogen interface, as viruses can rapidly adopt resistance mechanisms through evolutionary adaptation [37, 68, 73].

Recent advances in generative AI—particularly deep learning and generative adversarial networks—have revolutionized antiviral development by enabling the design of novel molecules and peptides with predicted antiviral activities [63, 68, 73]. The approaches help to speed up the process of finding potential lead compounds and enable quick responses to new viral threats when little experimental data are available [68, 73]. Nevertheless, the complexity of viral replication and the need to avoid off-target effects on host cells are major challenges for the application of generative AI in antiviral drug discovery [37, 73].

9.5.1 Architectural Framework for Antiviral Pipeline Design

The architectural foundation of antiviral generative pipelines builds on established antibacterial frameworks but incorporates viral-specific adaptations to address the unique biology of viral replication cycles. Unlike bacteria, which are self-sustaining cellular entities with distinct metabolic pathways, viruses rely entirely on host cellular machinery for replication and assembly [1, 92].

Modern antiviral pipelines target multiple stages of the viral life cycle—viral entry, genome replication, protein synthesis, and protease activity—offering multiple intervention points [1, 92] This multi-target approach enables the creation of combination therapies capable of retaining efficacy even as viral proteins mutate [1].

Recent improvements in AI-driven antiviral discovery integrate structure-based modeling with machine learning optimization, allowing accurate prediction of molecular binding, selectivity, and pharmacodynamics [1, 81, 84, 85]. Such platforms predict protein structures, identify inhibitors, and support drug repurposing through multi-layered data integration [1, 81]. For example, AI-guided systems have identified small fragment inhibitors for the SARS-CoV-2 main protease (Mpro), several of which show in vitro activity comparable to existing antiviral drugs [92]. These approaches use explainable AI and network analysis to provide insights into the mechanisms underlying antiviral activity, facilitating rational design of next-generation compounds [81].

The integration of generative models with structure-based approaches has been particularly helpful for targeting viral proteases, which are essential for viral polyprotein processing and represent successful antiviral drug targets [1, 92]. AI-driven generative pipelines have been able to discover novel inhibitors for SARS-CoV-2 Mpro, with the identified compounds showing strong potency and favorable pharmacokinetic properties [1, 92]. Such successes show that generative AI can speed up the discovery process while ensuring that the generated compounds have drug-like properties that are suitable for clinical development [1, 81, 92].

9.5.2 Viral Target Identification and Validation Module

The identification of viable viral targets within generative design frameworks relies on advanced computational and machine learning systems capable of distinguishing essential viral functions while minimizing interference with host processes. Because viral proteins exhibit high structural and sequence variability, modern pipelines employ dynamic modeling and generative frameworks to predict multiple functionally relevant conformations and sequence variants [27, 69]. The advent of deep generative models—particularly variational autoencoders and generative adversarial networks—has enabled the synthesis of viral protein sequences and prediction of immune-evasive mutations for developing viral vectors and therapeutic candidates resilient to emerging immune responses [15, 69].

AI-based structural and sequence prediction systems, combined with experimental validation, have significantly expanded the range of antiviral targets, particularly for viral proteins that elude traditional characterization methods. Generative models can also simulate viral vector serotypes that preserve natural protein architecture while lacking pre-existing neutralizing epitopes, effectively overcoming a major limitation in gene therapy [69]. Moreover, deep learning frameworks facilitate the rapid design and optimization of antiviral candidates targeting proteins such as the SARS-CoV-2 main protease and spike protein, supporting accelerated discovery of highly selective and potent therapeutic agents [15, 27].

Given the rapid evolution of viral pathogens, target selection strategies must simultaneously account for near-term therapeutic efficacy and long-term resistance durability. Integrating deep mutational scanning with AI-driven escape prediction models allows researchers to assess resistance potential through the simulation and analysis of diverse viral variants [15]. Generative adversarial networks have further enhanced resistance prediction by identifying immune-evasion mutations within viral surface proteins, such as the SARS-CoV-2 spike, thereby improving anticipatory antiviral design [15].

Current target validation workflows leverage machine learning models to analyze extensive viral sequence datasets and generate experimental fitness maps that quantify mutation impacts on viral viability. These models provide early indicators of resistance potential, enabling proactive optimization of therapeutic strategies before clinical failure occurs [15, 27]. The approaches deliver early indicators of resistance development which enables proactive modifications of antiviral strategies.

9.5.3 Molecular Generation Engines for Antiviral Compounds

Key generative modules in antiviral drug discovery pipelines must satisfy strict structural and functional criteria, blocking viral processes while ensuring safety. Modern approaches leverage deep generative architectures—including variational autoencoders, graph neural networks, and transformer-based models—to explore antiviral chemical space effectively, enhancing discovery of drug-like candidates with desired activity profiles [9, 73, 101, 112, 127].

Variational autoencoders (VAEs) excel by learning continuous latent representations, permitting iterative property optimization without compromising molecular validity [9, 99, 112]. Shape-based generative modeling via VAEs facilitates creation of novel scaffolds with desired functional groups, extending chemical space coverage while preserving lead-like properties [99]. The conditioning mechanism of deep generative models allows them to produce targeted compounds based on molecular or protein characteristics [101, 127].

Graph neural network-based generative models are particularly powerful for antiviral design as they incorporate structure-derived constraints and three-dimensional binding properties. These models optimize critical antiviral features such as binding affinity and resilience to viral mutations, producing compounds effective across diverse protein targets [52, 62, 101]. The combination of conformational sampling and graph-based generation methods improves the capability to handle the natural flexibility of viral targets [52, 62].

Recent transformer-based breakthroughs have expanded antiviral design capabilities. The TamGen system demonstrates GPT-like molecular language modeling to generate target-aware compounds, facilitating discovery of potent infectious disease inhibitors [127]. The TransAntivirus model employs IUPAC-based representations and self-supervised pretraining to generate valid, diverse antiviral candidates, successfully designing nucleoside and non-nucleoside analogues for COVID-19 [73]. Such methods utilize extensive chemical literature to innovate scaffolds and chemical modifications that evade resistance while maintaining antiviral efficacy [73, 127].

Together, these deep generative approaches—VAEs, graph neural networks, and transformers—accelerate antiviral compound discovery by efficiently searching chemical space and optimizing multiple drug-like properties simultaneously [9, 73, 101, 112, 127].

9.5.4 Multi-objective Property Prediction for Antiviral Compounds

Accurate prediction of antiviral efficacy requires specialized models addressing unique pharmacological challenges in viral infections. Antiviral agents must penetrate intracellularly to interact with viral components, necessitating precise prediction of cellular uptake and subcellular distribution. Advanced ADMET (Absorption, Distribution, Metabolism, Excretion, and Toxicity) models tailored for antiviral compounds optimize properties guiding clinical success [6, 110].

Antiviral resistance prediction is a critical feature of modern property prediction systems. Machine learning models trained on viral genotypes and evolution data forecast resistance emergence, aiding identification of scaffolds with innate resistance-blocking properties [110]. The AVPIden model uses a two-stage ML approach with multiple descriptors and explainable algorithms to predict antiviral peptide function and prolong efficacy [77]. DeepStacked-AVPs and other deep learning models perform large-scale screenings to identify antiviral activity-related structural features across viral families [3, 118].

Mechanism-of-action prediction is increasingly important for designing resistance-resilient combination therapies. AI models forecast compound interactions with viral proteins, optimizing therapeutic design strategies [72, 110]. Machine learning algorithms simulate multi-drug synergy, supporting development of effective combined regimens [72].

Modern property prediction emphasizes integration of viral-specific biological data. Broad-spectrum antiviral development benefits from models predicting virus-host protein interactions, enabling discovery of compounds disrupting critical viral-host interfaces [64, 110]. The approaches deliver useful knowledge for creating pan-viral therapies which meet the rising requirement for treatments treating multiple viruses.

Machine learning-based property prediction, resistance forecasting, and mechanism analysis accelerate antiviral development through comprehensive compound screening and informed optimization [3, 64, 72, 77, 110, 118].

9.5.5 Optimization Strategies and Active Learning Frameworks

Antiviral compound development requires sophisticated multi-objective optimization to balance viral potency, host specificity, and resistance-blocking properties. Modern optimization frameworks integrate computational and machine learning methods to identify successful compounds by analyzing extensive biological and chemical databases, which guide vital experimental data needs [37, 110]. Active learning approaches and data-driven priority systems direct computational power to generate the most informative molecular structures thus shortening discovery periods and reducing experimental costs [110].

Addressing viral resistance challenges demands optimization techniques that consider evolutionary pressures and escape routes. These frameworks predict viral fitness landscapes and resistance mechanisms to identify compounds targeting evolutionarily constrained sites or exploiting resistance pathways [37, 69, 134]. Contemporary antiviral design employs a triad strategy combining multi-target drug design, conformational constraints, and host factor modulation [37, 134]. Baseline resistance profiling assists clinicians in selecting effective therapies, while combination therapy optimization fosters multi-agent regimens resistant to individual component failures [35].

Advancements in machine learning and computational modeling enable adaptive optimization that incorporates experimental feedback and resistance evolution data [35, 110]. These systems iteratively refine generative design rules by learning from predictive outcomes and empirical testing, developing more sophisticated antiviral compound optimization strategies [35, 110]. Predictive models forecast resistance emergence via viral evolutionary simulations, allowing preemptive refinement of drug design [35, 110].

9.5.6 *Experimental Validation and Resistance Monitoring*

AI-generated antiviral compounds require comprehensive experimental validation frameworks, encompassing biochemical assays, animal studies, and focused assessment of antiviral activity and resistance development. High-throughput platforms integrating AI-driven virtual screening and in vitro assays enable rapid evaluation of antiviral efficacy across diverse viral strains and variants, providing early insights into antiviral spectrum and resistance potential [6, 7, 120]. Real-time resistance surveillance is increasingly recognized for detecting emerging resistance patterns before clinical impact, although explicit reporting in the literature remains limited [92, 120].

Mechanism-of-action understanding is vital for optimizing future therapeutic combinations of AI-generated antivirals [6, 92]. Cutting-edge validation uses multi-parameter evaluations of potency, cytotoxicity, and pharmacokinetics in appropriate cell lines and animal models [6, 13]. The integration of omics strategies and molecular profiling facilitates comprehensive analysis of antiviral mechanisms and host responses, enhancing predictive model accuracy and guiding compound refinement [6].

Validation studies have demonstrated the practical effectiveness of AI-designed antivirals against multiple viral targets. For example, AI molecular profiling identified thioinosinic acid, which showed significant in vitro and in vivo inhibitory effects against SARS-CoV-2, validating the approach [6]. Another AI-discovered compound, Mebendazole, exhibited antiviral activity in cell culture and animal models against SARS-CoV-2 [13]. Such successes underscore the promise of generative AI approaches while emphasizing the essential role of rigorous experimental validation in translating computational predictions into clinical realities [6, 7, 13, 92, 120].

9.5.7 *Integration with Pandemic Preparedness and Surveillance*

Effective antiviral generative pipelines increasingly incorporate real-time viral surveillance data to enable rapid responses to novel viral threats. By coupling genomic surveillance systems with generative design platforms, these pipelines automatically detect new viral variant targets and create compounds counteracting emerging resistance mutations [77, 118, 120]. This capability is critical for rapidly mutating RNA viruses, which quickly develop drug resistance through genetic variation.

Research on broad-spectrum antiviral design remains a priority, producing compounds that disrupt fundamental viral functions and host pathways essential for replication. Integrating viral phylogenetic analysis with structure-based design identifies conserved viral targets across species, facilitating pan-viral therapy development [6].

The fusion of generative pipelines with global health surveillance systems is set to transform viral threat responsiveness. Computational methods integrated with experimental validation and real-time resistance monitoring represent significant progress in antiviral drug development, supporting accelerated production of effective antivirals and improving pandemic preparedness [34].

9.6 Cross-Domain Case Studies (ESKAPE and SARS-CoV-2)

The practical application of generative AI and graph neural network methods in antimicrobial and antiviral drug development is best illustrated through detailed case studies targeting diverse pathogens. These analyses showcase therapeutic discoveries arising from computational frameworks and pipelines while highlighting the distinct challenges and opportunities presented by different pathogen classes. Examining ESKAPE bacterial pathogens alongside SARS-CoV-2 reveals how generative approaches adapt to biological system complexities and inform novel drug design strategies tailored to each domain.

9.6.1 Case Study 1: AI-Driven Discovery of Abaucin Against Acinetobacter baumannii

The discovery of abaucin marks a significant milestone in AI-based antibiotic research, demonstrating that machine learning can successfully identify narrow-spectrum agents targeting high-priority pathogens such as *Acinetobacter baumannii* [10, 29, 67]. Researchers trained a neural network on bacterial growth inhibition data to test 7,500 compounds as *A. baumannii* antibacterial agents [67]. The model swiftly identified suitable candidate molecules which shortened the conventional drug discovery period [66, 67]. Combining supervised learning for initial screening with systematic experimental validation, this methodology effectively addresses antibiotic resistance through integrated AI-driven modules [29, 66].

The generative AI method, SyntheMol, was developed for novel antibiotics design. The workflow is shown in Fig. 9.1. Initially, a training set of approximately 13,000 molecules was curated. Growth inhibition assays were conducted to assess for their bioactivity against *A. baumannii*. These chemical screening results were used to train a property prediction model to estimate antibacterial activity. The molecule generation process utilized a chemical space composed of nearly 30 billion molecules, each synthesizable by applying one of 13 chemical reactions to combine two or three building blocks selected from a set of approximately 132,000 fragments. SyntheMol method explores this extensive chemical space for antibiotic candidates using Monte Carlo Tree Search guided by the property predictor. The process involves iterative selection of building blocks, followed by their combination through a chemical reaction to form molecules scored by the predictor. The scores are then backpropagated along the synthetic route to inform subsequent selections. After 20,000 iterations, the generated molecules were filtered to identify a set of structurally novel and diverse compounds with high scores, which were then synthesized and experimentally tested against a phylogenetically diverse panel of bacterial species in vitro [105].

Minimum inhibitory concentration determinations combined with mechanism of action studies and mouse-based in-vivo efficacy tests were used to validate abaucin's effectiveness against *A. baumannii* infections [66]. The compound abaucin showed strong and targeted effects against *A. baumannii* while causing little harm to other bacterial strains thus reducing microbiome disruptions [10, 67].

The abaucin discovery process shows two essential principles: using pathogen-specific training data to build models that find compounds with limited activity spectrum which is vital for developing future antibiotics [29, 66]. The combination of computational predictions with extensive experimental testing produced actual therapeutic candidates from potential hits [66]. Abaucin disrupts lipoprotein trafficking through LolE mechanisms which researchers can use to optimize future development and resistance strategies [66].

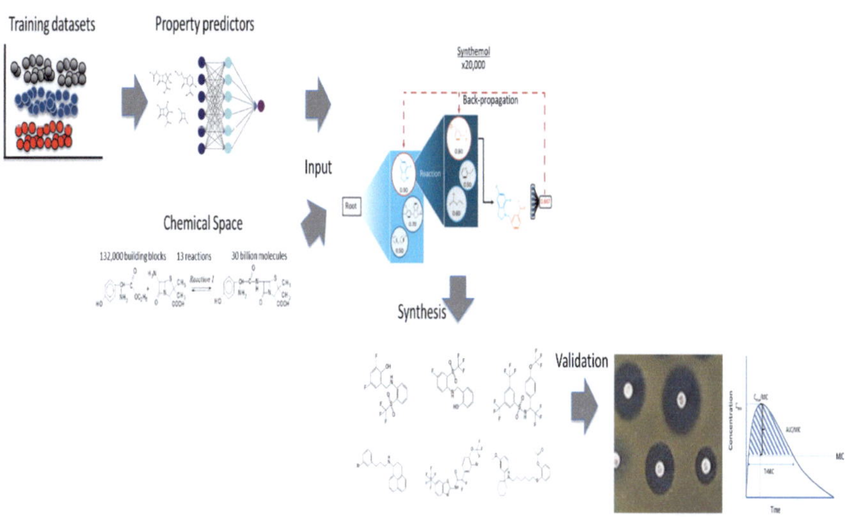

Fig. 9.1 Generative AI based on SyntheMol's method (adapted from [105])

The abaucin case supports AI's transformative role in drug discovery, highlighting its capacity for rapid identification of effective antimicrobials from chemical databases to meet urgent clinical needs [10, 66]. Its pathogen-specific activity enables precision antimicrobials that reduce collateral damage to beneficial microbiota [10, 66].

9.6.2 *Case Study 2: SARS-CoV-2 Main Protease Inhibitor Discovery*

Generative AI and machine learning methods have demonstrated success in discovering potent SARS-CoV-2 main protease (Mpro) inhibitors. Deep learning generative models combined with molecular modeling enable de novo design of small molecules targeting Mpro catalytic activity. Trained on antiviral compound libraries, these models identify compounds with binding affinities comparable to potent inhibitors, offering promising scaffolds for drug development [1, 8, 80, 92].

The workflow of a generative AI model for the discovery of novel antivirals targeting the SARS-CoV-2 main protease (Mpro) is shown in Fig. 9.2. The process begins with transfer learning, where the model is pre-trained on the known chemical space of the Mpro target. This knowledge is fed into a generative model, which operates in a reinforcement learning loop. The model proposes novel chemical structures as generated molecules [92].

These molecules are then evaluated by scoring components, which assess their potential efficacy through criteria such as substructure match and 3D pharmacophore

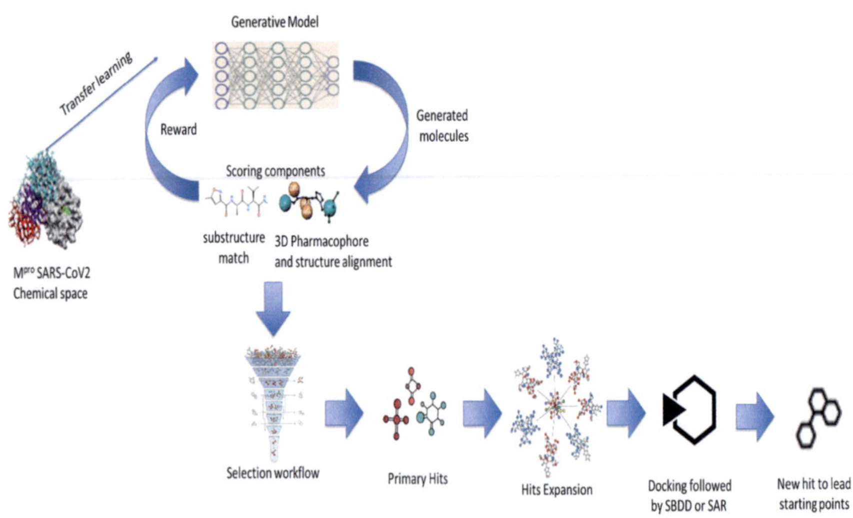

Fig. 9.2 Generative AI workflow for antiviral discovery targetting Mpro SARS-CoV2

and structure alignment. The calculated scores serve as a reward signal that is fed back to the generative model to guide it toward generating more promising structures.

After the generative model produces a refined set of candidate molecules, these compounds enter a downstream selection workflow structured as a funnel to isolate the best candidates. Primary hits are identified, followed by hit expansion through analogue and derivative generation to explore surrounding chemical space. Subsequent computational analyses—including docking, Structure-Based Drug Design (SBDD), and Structure–Activity Relationship (SAR) studies—identify promising hit-to-lead candidates for preclinical advancement [92].

The study revealed five Mpro fragment-like inhibitors with IC_{50} values between 1.5 and 241 μM through AI-driven methods followed by in vitro validation. The most potent compounds showed antiviral effects in cellular models while researchers found reversible and covalent inhibitors using AI methods [92] Deep neural network models applied for small molecule design of the 3CL protease utilized predictive models with generative capabilities to choose candidate molecules based on predicted binding free energy and drug-like properties [8].

The structural mechanism behind Mpro inhibition has been well studied through research that shows effective inhibitors bond covalently with Cys145 and maintain beneficial interactions with additional active site residues [1, 92]. The integration of structural information obtained through crystallographic studies into AI models allows for an iterative process that boosts the performance of the models [1, 80, 92].

The rapid availability of high-quality structural and biochemical data, such as early Mpro crystal structures and biochemical assays, has accelerated AI-driven drug discovery model development for viral targets. This comprehensive target characterization underpins the success and potential rapid deployment of AI-assisted antiviral therapeutics for emergent viruses [1, 80, 92].

9.7 Cross-Domain insights and Methodological Convergence

The review of successful drug design projects for bacterial and viral targets reveals key lessons about generative AI and graph neural network applications in antimicrobial drug development.

9.7.1 Hybrid AI Approaches

Hybrid AI approaches that combine generative models with predictive algorithms and reinforcement learning techniques yield improved outcomes. These integrated methods outperform single-model strategies in molecular generation and optimization, enhancing discovery efficiency and therapeutic potential [24, 70, 112, 139].

A notable example is the Feedback Generative Adversarial Network (FBGAN), which integrates advanced classifiers with generative adversarial networks to optimize antimicrobial peptide (AMP) design. FBGAN uses a classifier to guide the generator toward producing sequences with high predicted antimicrobial activity, effectively increasing the quality and diversity of generated peptides. This framework improves on traditional GANs by reducing selection biases and enhancing the novelty of candidate AMPs, paving the way for innovative therapeutic candidates [139].

9.7.2 Importance of High-Quality Data and Experimental Validation

AI-driven drug discovery critically depends on the quality of training data and comprehensive experimental validation to succeed. Deep learning models, including graph neural networks (GNNs) and recurrent neural networks (RNNs), require extensive, well-curated datasets to predict antimicrobial activity accurately, subsequently leading to experimentally validated candidates[17, 21, 23]. The validation process requires experimental collaboration because it enables researchers to test AI-generated candidates while verifying their effectiveness in real-world conditions [21, 23].

Integration of biological domain knowledge—such as resistance mechanisms, pathogen-specific traits, and pharmacokinetics—into AI model development enhances predictive accuracy. Tailoring models with these specific factors outperforms generic approaches, as domain customization is essential for antimicrobial success [2, 17, 23]. Conditioning GNNs with specified chemical and biological properties facilitates more efficient and targeted molecular generation [2].

Core AI methodologies, including molecular generation, property prediction, and optimization, operate across bacterial and viral pathogen domains. To maximize effectiveness, models require specific adaptations for distinct challenges posed by each target pathogen type [2, 17, 23, 130]. Transferable generative models with fine-tuning capacity enable rapid responses to emerging infectious agents [94, 112].

Generative AI combined with GNN methodologies forms a robust framework for antimicrobial drug discovery, enabling fast responses to novel and drug-resistant pathogens. The development of next-generation therapeutics hinges on integrating experimental validation, domain expertise, and hybrid AI strategies [2, 17, 23, 112, 139].

9.7.3 Conclusion

The future of antimicrobial and antiviral drug development powered by generative AI and graph neural networks promises a transition from supplementary tools to essential components of therapeutic innovation. Advances in technology, methodological rigor in experimental validation, evolving regulatory frameworks, multimodal AI, federated learning, foundation models, and quantum computing collectively enable unprecedented solutions for managing drug resistance.

Achieving this potential requires coordinated efforts across multiple domains, including technological innovation to resolve current limitations, development of validation protocols confirming clinical efficacy, establishment of regulatory pathways ensuring safety, and the creation of global collaborative networks to guarantee equitable access to new therapeutics.

The success of AI-driven drug discovery is exemplified by candidates like abaucin, validating computational approaches through tangible therapeutic outcomes. Continued refinement and overcoming present challenges will enable transformative responses to antimicrobial resistance and empower proactive therapeutic strategies for emerging infectious threats.

Similar successes are evident in antiviral drug discovery, exemplified by AI-driven identification of potent inhibitors targeting the SARS-CoV-2 main protease (Mpro). These AI approaches efficiently leverage high-quality structural and biochemical data to generate clinically relevant therapeutics. Continued advances and resolution of current limitations in AI methodologies promise to revolutionize infectious disease medicine and effectively address the urgent antimicrobial resistance crisis.

The integration of generative AI and graph neural networks establishes a transformative, data-driven, and adaptive framework for globally coordinated infectious disease drug discovery. Artificial intelligence accelerates the development of therapeutics that match the evolutionary dynamics of pathogens by shortening discovery timelines, optimizing compound characteristics, and enhancing resistance prediction.

While significant challenges persist, the synergy among computational power, biological insight, and international collaboration promises a future where AI-driven discovery systems not only combat current antimicrobial resistance crises but also pioneer a precision therapeutic paradigm for infectious diseases.

References

1. Abadi MHHE, Ghasemlou A, Bayani F, Sefidbakht Y, Vosough M, Mozaffari-Jovin S, Uversky V. AI-driven covalent drug design strategies targeting main protease (mpro) against SARS-CoV-2: structural insights and molecular mechanisms. J Biomol Struct Dyn. 2024;1–29. https://doi.org/10.1080/07391102.2024.2308769.
2. Abate C, Decherchi S, Cavalli A. Graph neural networks for conditional de novo drug design. Wiley Interdiscipl Rev Comput Molecul Sci. 2023;13. https://doi.org/10.1002/wcms.1651.

3. Akbar S, Raza A, Zou Q. Deepstacked-AVPs: predicting antiviral peptides using tri-segment evolutionary profile and word embedding based multi-perspective features with deep stacking model. BMC Bioinform. 2024;25. https://doi.org/10.1186/s12859-024-05726-5.
4. Alcock B, Huynh W, Chalil R, Smith K, Raphenya A, Wlodarski M, Edalatmand A, Petkau A, Syed S, Tsang K, Baker S, Dave M, McCarthy M, Mukiri K, Nasir J, Golbon B, Imtiaz H, Jiang X, Kaur K, McArthur A, et al. CARD 2023: expanded curation, support for machine learning, and resistome prediction at the comprehensive antibiotic resistance database. Nucleic Acids Res. 2022;51. https://doi.org/10.1093/nar/gkac920.
5. Alemnew M, Gelaw A, Nibret K, Getu, A, Berhane N. Resistance gene detection database for antimicrobial resistance investigations emphasizing on genomics and metagenomics techniques. J Life Sci Biomed. 2024. https://doi.org/10.54203/jlsb.2024.8
6. Ali M, Park IH, Kim J, Kim G, Oh J, You JS, Kim J, Shin J-S, Yoon SS. How deep learning in antiviral molecular profiling identified anti-SARS-CoV-2 inhibitors. Biomedicines. 2023;11. https://doi.org/10.3390/biomedicines11123134.
7. Alkan C, O'Brien T, Kenyon V, Ikegami T. Computer-selected antiviral compounds: assessing in vitro efficacies against rift valley fever virus. Viruses. 2024;16. https://doi.org/10.3390/v16010088.
8. Andrianov A, Shuldau M, Furs K, Yushkevich A, Tuzikov A. AI-driven De Novo design and molecular modeling for discovery of small-molecule compounds as potential drug candidates targeting SARS-CoV-2 main protease. Int J Mol Sci. 2023;24. https://doi.org/10.3390/ijms24098083.
9. Anstine D, Isayev O. Generative models as an emerging paradigm in the chemical sciences. J Am Chem Soc. 2023;145:8736–50. https://doi.org/10.1021/jacs.2c13467.
10. Awan R, Zainab S, Yousuf F, Mughal S. AI-driven drug discovery: exploring Abaucin as a promising treatment against multidrug-resistant *Acinetobacter baumannii*. Health Sci Rep. 2024;7. https://doi.org/10.1002/hsr2.2150.
11. Ayon N. High-throughput screening of natural product and synthetic molecule libraries for antibacterial drug discovery. Metabolites. 2023;13. https://doi.org/10.3390/metabo13050625.
12. Batool M, Ahmad B, Choi S. A structure-based drug discovery paradigm. Int J Mol Sci. 2019;20. https://doi.org/10.3390/ijms20112783.
13. Bess A, Berglind F, Mukhopadhyay S, Brylinski M, Alvin C, Fattah F, Wasan K. Identification of oral therapeutics using an AI platform against the virus responsible for COVID-19, SARS-CoV-2. Front Pharmacol. 2023;14. https://doi.org/10.3389/fphar.2023.1297924.
14. Bilodeau C, Jin W, Jaakkola T, Barzilay R, Jensen K. Generative models for molecular discovery: recent advances and challenges. Wiley Interdiscipl Rev Comput Molecul Sci. 2022;12. https://doi.org/10.1002/wcms.1608.
15. Bist PS, Tayara H, Chong K. Generative AI in the advancement of viral therapeutics for predicting and targeting immune-evasive SARS-CoV-2 mutations. IEEE J Biomed Health Inform. 2024;28:6974–82. https://doi.org/10.1109/JBHI.2024.3432649.
16. Blanco-González A, Cabezon A, Seco-González A, Conde-Torres D, Antelo-Riveiro P, Piñeiro Á, García-Fandiño R. The role of AI in drug discovery: challenges, opportunities, and strategies. Pharmaceuticals. 2022;16. https://doi.org/10.3390/ph16060891.
17. Brizuela C, Liu G, Stokes J, De La Fuente-Nunez C. AI methods for antimicrobial peptides: progress and challenges. Microb Biotechnol. 2025;18. https://doi.org/10.1111/1751-7915.70072.
18. Bucataru C, Ciobanasu C. Antimicrobial peptides: opportunities and challenges in overcoming resistance. Microbiol Res. 2024;286:127822. https://doi.org/10.1016/j.micres.2024.127822.
19. Byun H-E, Kim B, Lee J. Multi-step lookahead Bayesian optimization with active learning using reinforcement learning and its application to data-driven batch-to-batch optimization. Comput Chem Eng. 2022;167:107987. https://doi.org/10.1016/j.compchemeng.2022.107987.
20. Cai H, Zhang H, Zhao D, Wu J, Wang L (2022). FP-GNN: a versatile deep learning architecture for enhanced molecular property prediction. Brief Bioinform. https://doi.org/10.48550/arXiv.2205.03834

21. Capecchi A, Cai X, Personne H, Köhler T, Van Delden C, Reymond J. Machine learning designs non-hemolytic antimicrobial peptides. Chem Sci. 2021;12:9221–32. https://doi.org/10.1039/d1sc01713f.
22. Cerchia C, Lavecchia A. New avenues in artificial-intelligence-assisted drug discovery. Drug Discovery Today. 2023;103516. https://doi.org/10.1016/j.drudis.2023.103516.
23. Cesaro A, Bagheri M, Torres M, Wan F, De La Fuente-Nunez C. Deep learning tools to accelerate antibiotic discovery. Expert Opin Drug Discov. 2023;18:1245–57. https://doi.org/10.1080/17460441.2023.2250721.
24. Cesaro A, Hoffman S, Das P, De La Fuente-Nunez C. Challenges and applications of artificial intelligence in infectious diseases and antimicrobial resistance. Npj Antimicrobials Resist. 2025;3. https://doi.org/10.1038/s44259-024-00068-x.
25. Chen B, Pan Z, Mou M, Zhou Y, Fu W. Is fragment-based graph a better graph-based molecular representation for drug design? A comparison study of graph-based models. Comput Biol Med. 2023;169:107811. https://doi.org/10.1016/j.compbiomed.2023.107811.
26. Chen S, Wulamu A, Zou Q, Zheng H, Wen L, Guo X, Chen H, Zhang T, Zhang Y. MD-GNN: a mechanism-data-driven graph neural network for molecular properties prediction and new material discovery. J Mol Graph Model. 2023;123:108506. https://doi.org/10.1016/j.jmgm.2023.108506.
27. Chenthamarakshan V, Das P, Padhi I, Strobelt H, Lim KW, Hoover B, Hoffman S, Mojsilovic A. Target-specific and selective drug design for COVID-19 using deep generative models. 2020. ArXiv, abs/2004.01215
28. Chikhale R, Choudhary R, Eldesoky G, Kolpe MS, Shinde O, Hossain D. Generative AI, molecular docking and molecular dynamics simulations assisted identification of novel transcriptional repressor EthR inhibitors to target *Mycobacterium tuberculosis*. Heliyon. 2025;11:4. https://doi.org/10.1016/j.heliyon.2025.e42593.
29. Chrobak D. Revolutionizing new drug research: the role of AI and machine learning in the discovery of new antibiotics. In: Education and science of today: intersectoral issues and development of sciences. 2023. https://doi.org/10.36074/logos-18.08.2023.27
30. Church N, McKillip J. Antibiotic resistance crisis: challenges and imperatives. Biologia. 2021;76:1535–50. https://doi.org/10.1007/s11756-021-00697-x.
31. Das P, Sercu T, Wadhawan K, Padhi I, Gehrmann S, Cipcigan F, Chenthamarakshan V, Strobelt H, Santos CD, Chen P-Y, Yang YY, Tan J, Hedrick J, Crain J, Mojsilovic A. Accelerated antimicrobial discovery via deep generative models and molecular dynamics simulations. Nat Biomed Eng. 2021;5:613–23. https://doi.org/10.1038/s41551-021-00689-x.
32. Davis ML, Ricke SC, Donaldson JR. Establishment of Listeria monocytogenes in the gastrointestinal tract. Microorganisms. 2019;7(3). https://doi.org/10.3390/MICROORGANISMS7030075
33. Di Fiore F, Nardelli M, Mainini L. Active learning and Bayesian optimization: a unified perspective to learn with a goal. ArXiv, abs/2303.01560. 2023. https://doi.org/10.1007/s11831-024-10064-z
34. Diakou I, Papakonstantinou E, Papageorgiou L, Pierouli K, Dragoumani K, Spandidos D, Bacopoulou F, Chrousos G, Eliopoulos E, Vlachakis D. Novel computational pipelines in antiviral structure-based drug design (review). Biomed Rep. 2022;17. https://doi.org/10.3892/br.2022.1580.
35. Dietz E, Pritchard E, Pouwels K, Ehsaan M, Blake J, Gaughan C, Haduli E, Boothe H, Vihta KD, Peto T, Stoesser N, Matthews P, Taylor N, Diamond I, Studley R, Rourke E, Birrell P, De Angelis D, Fowler T, Walker AS et al. SARS-CoV-2, influenza A/B and respiratory syncytial virus positivity and association with influenza-like illness and self-reported symptoms, over the 2022/23 winter season in the UK: a longitudinal surveillance cohort. BMC Med. 2024;22(1):1–13. https://doi.org/10.1186/S12916-024-03351-W/TABLES/3.
36. Dong R, Liu R, Liu Z, Liu Y, Zhao G, Li H, Hou S, Xiaohan KH, Liu J, Guo F, Zhao P, Wang J, Wang C, Wu X, Ye S, Zhu C (2025) Exploring the repository of de novo-designed bifunctional antimicrobial peptides through deep learning. eLife 13. https://doi.org/10.7554/eLife.97330

37. Du S, Hu X, Menéndez-Arias L, Zhan P, Liu X. Target-based drug design strategies to overcome resistance to antiviral agents: opportunities and challenges. Drug Resist Updates Rev Comment Antimicrobial Anticancer Chemother. 2024;73:101053. https://doi.org/10.1016/j.drup.2024.101053.
38. Dunn I, Koes D (2023) Accelerating inference in molecular diffusion models with latent representations of protein structure. ArXiv. https://doi.org/10.48550/arXiv.2311.13466
39. Fairlamb A, Gow N, Matthews K, Waters A. Drug resistance in eukaryotic microorganisms. Nat Microbiol. 2016;1. https://doi.org/10.1038/nmicrobiol.2016.92.
40. Fang X, Liu L, Lei J, He D, Zhang S, Zhou J, Wang F, Wu H, Wang H. Geometry-enhanced molecular representation learning for property prediction. Nat Mach Intell. 2021;4:127–34. https://doi.org/10.1038/s42256-021-00438-4.
41. Feldgarden M, Brover V, González-Escalona N, Frye J, Haendiges J, Haft D, Hoffmann M, Pettengill J, Prasad A, Tillman G, Tyson G, Klimke W. AMRFinderPlus and the reference gene catalog facilitate examination of the genomic links among antimicrobial resistance, stress response, and virulence. Sci Rep. 2021;11. https://doi.org/10.1038/s41598-021-91456-0.
42. Florensa A, Kaas R, Clausen P, Aytan-Aktug D, Aarestrup F. ResFinder—an open online resource for identification of antimicrobial resistance genes in next-generation sequencing data and prediction of phenotypes from genotypes. Microbial Genom. 2022;8. https://doi.org/10.1099/mgen.0.000748.
43. Franceschelli G, Musolesi M. Reinforcement learning for generative AI: state of the art, opportunities and open research challenges. J Artif Intell Res. 2023;79:417–46. https://doi.org/10.1613/jair.1.15278.
44. Gangwal A, Ansari A, Ahmad I, Azad AK, Sulaiman W. Current strategies to address data scarcity in artificial intelligence-based drug discovery: a comprehensive review. Comput Biol Med. 2024;179:108734. https://doi.org/10.1016/j.compbiomed.2024.108734.
45. Gangwal A, Lavecchia A. Artificial intelligence in natural product drug discovery: current applications and future perspectives. J Med Chem. 2025;68:3948–69. https://doi.org/10.1021/acs.jmedchem.4c01257.
46. Han R, Yoon H, Kim G, Lee H, Lee Y. Revolutionizing medicinal chemistry: the application of artificial intelligence (AI) in early drug discovery. Pharmaceuticals. 2023;16. https://doi.org/10.3390/ph16091259.
47. Hehuan, Bian Y, Rong Y, Huang W, Xu T, Xie W-Y, Ye G, Huang J (2020) Multi-view graph neural networks for molecular property prediction. In: Quantitative methods
48. Hendriksen R, Bortolaia V, Tate H, Tyson G, Aarestrup F, McDermott P. Using genomics to track global antimicrobial resistance. Front Public Health. 2019;7. https://doi.org/10.3389/fpubh.2019.00242.
49. Hu K, Meyer F, Deng Z-l, Asgari E, Kuo T-H, Münch P, Mchardy A. Assessing computational predictions of antimicrobial resistance phenotypes from microbial genomes. Brief Bioinf. 2024;25. https://doi.org/10.1101/2024.01.31.578169
50. Huang D, Yang M, Wen X, Xia S, Yuan B. AI-driven drug discovery: accelerating the development of novel therapeutics in biopharmaceuticals. Int Med Sci Res J. 2024. https://doi.org/10.51594/imsrj.v4i8.1458
51. Huang J-J, Xu Y, Xue Y, Huang Y, Li X, Chen X, Xu Y, Zhang D, Zhang P, Zhao J, Ji J. Identification of potent antimicrobial peptides via a machine-learning pipeline that mines the entire space of peptide sequences. Nat Biomed Eng. 2023;7:797–810. https://doi.org/10.1038/s41551-022-00991-2.
52. Imrie F, Bradley A, Schaar M, Deane C. Deep generative models for 3D linker design. J Chem Inf Model. 2020;60:1983–95. https://doi.org/10.1021/acs.jcim.9b01120.
53. Jang S, Yoo S, Kang N. Generative design by reinforcement learning: enhancing the diversity of topology optimization designs. Comput Aided Des. 2020;146:103225. https://doi.org/10.1016/j.cad.2022.103225.
54. Jia B, Raphenya A, Alcock B, Waglechner N, Guo P, Tsang K, Lago B, Dave B, Pereira S, Sharma A, Doshi S, Courtot M, Lo R, Williams L, Frye J, Elsayegh T, Sardar D, Westman E, Pawlowski A, McArthur A. CARD 2017: expansion and model-centric curation of the

comprehensive antibiotic resistance database. Nucleic Acids Res. 2016;45. https://doi.org/10.1093/nar/gkw1004.
55. Jiang Z, Feng J, Wang F, Wang J, Wang N, Zhang M, Hsieh C-Y, Hou T, Cui W, Limin. AI-guided design of antimicrobial peptide hydrogels for precise treatment of drug-resistant bacterial infections. Adv Mater. 2025;37. https://doi.org/10.1002/adma.202500043
56. Kadan A, Ryczko K, Roitberg A, Yamazaki T. Guided multi-objective generative AI to enhance structure-based drug design. Chem Sci. 2024. https://doi.org/10.48550/arXiv.2405.11785
57. Kamal A. Generative AI in pharmaceutical research: accelerating drug discovery through predictive analytics and big data integration. J Front Multidiscipl Res. 2023. https://doi.org/10.54660/.ijfmr.2023.4.2.27-33
58. Khaledi A, Weimann A, Schniederjans M, Asgari E, Kuo T-H, Oliver A, Cabot G, Kola A, Gastmeier P, Hogardt M, Jonas D, Mofrad M, Bremges A, Mchardy A, Haeussler S. Predicting antimicrobial resistance in *Pseudomonas aeruginosa* with machine learning-enabled molecular diagnostics. EMBO Molecul Med. 2020;12. https://doi.org/10.15252/emmm.201910264
59. Khare P, Sahu U, Pandey SC, Samant M. Current approaches for target-specific drug discovery using natural compounds against SARS-CoV-2 infection. Virus Res. 2020;290. https://doi.org/10.1016/j.virusres.2020.198169.
60. Korshunova M, Huang N, Capuzzi S, Radchenko D, Savych O, Moroz Y, Wells C, Willson T, Tropsha A, Isayev O. Generative and reinforcement learning approaches for the automated de novo design of bioactive compounds. Commun Chem. 2022;5. https://doi.org/10.1038/s42004-022-00733-0.
61. Li C, Zhang T, Du X, Zhang Y, Xie H. Generative AI models for different steps in architectural design: a literature review. Front Architect Res. 2024. https://doi.org/10.1016/j.foar.2024.10.001.
62. Li P, Wang J, Qiao Y, Chen H, Yu Y, Yao X, Gao P, Xie G, Song S. An effective self-supervised framework for learning expressive molecular global representations to drug discovery. Brief Bioinform. 2021. https://doi.org/10.1093/bib/bbab109.
63. Lin B, Yan S, Zhen B. A machine learning method for predicting molecular antimicrobial activity. Sci Rep. 2025;15. https://doi.org/10.1038/s41598-025-91190-x.
64. Lin H-H, Zhang Q, Kong X, Zhang L, Zhang Y, Tang Y, Xu H. Machine learning prediction of antiviral-HPV protein interactions for anti-HPV pharmacotherapy. Sci Rep. 2021;11. https://doi.org/10.1038/s41598-021-03000-9.
65. Lipková J, Chen R, Chen B, Lu M, Barbieri M, Shao D, Vaidya A, Chen C, Zhuang L, Williamson D, Shaban M, Chen T, Mahmood F. Artificial intelligence for multimodal data integration in oncology. Cancer Cell. 2022;40(10):1095–110. https://doi.org/10.1016/j.ccell.2022.09.012.
66. Liu G, Catacutan D, Rathod K, Swanson K, Jin W, Mohammed J, Chiappino-Pepe A, Syed S, Fragis M, Rachwalski K, Magolan J, Surette M, Coombes B, Jaakkola T, Barzilay R, Collins J, Stokes J. Deep learning-guided discovery of an antibiotic targeting Acinetobacter baumannii. Nat Chem Biol. 2023;19:1342–50. https://doi.org/10.1038/s41589-023-01349-8.
67. Liu X, Zhang W, Tong X, Zhong F, Li Z, Xiong Z, Xiong J, Wu X, Fu Z, Tan X, Liu Z, Zhang S, Jiang H, Li X, Zheng M. MolFilterGAN: a progressively augmented generative adversarial network for triaging AI-designed molecules. J Cheminform. 2023;15. https://doi.org/10.1186/s13321-023-00711-1.
68. Liu Y, Fan Q, Xu C, Ning X, Wang Y, Liu Y, Zhang Y, Chen Y, Liu H. GDMol: generative double-masking self-supervised learning for molecular property prediction. Mol Inf. 2024;44. https://doi.org/10.1002/minf.202400146.
69. Lyu S, Sowlati-Hashjin S, Garton M. Variational autoencoder for design of synthetic viral vector serotypes. Nat Mac Intell. 2024;6:147–60. https://doi.org/10.1038/s42256-023-00787-2.
70. Macedo B, Vaz IR, Gomes TT. MedGAN: optimized generative adversarial network with graph convolutional networks for novel molecule design. Sci Rep. 2024;14. https://doi.org/10.1038/s41598-023-50834-6.

71. Maheswari P, Verma P, Soni G. Exploring reinforcement learning for optimizing generative AI systems. J Artif Neural Netw Learn Syst. 2025. https://doi.org/10.46610/joannls.2025.v02i01.003
72. Majidifar S, Zabihian A, Hooshmand M. Combination therapy synergism prediction for virus treatment using machine learning models. PLoS ONE. 2024;19. https://doi.org/10.1371/journal.pone.0309733.
73. Mao J, Wang J, Zeb A, Cho K-H, Jin H, Kim J, Lee O-C, Wang Y, No KT. Transformer-based molecular generative model for antiviral drug design. J Chem Inf Model. 2023;64:2733–45. https://doi.org/10.1021/acs.jcim.3c00536.
74. Mba I, Nweze E. Antimicrobial peptides therapy: an emerging alternative for treating drug-resistant bacteria. Yale J Biol Med. 2022;95:445–63.
75. Ocaña A, Pandiella A, Privat C, Bravo I, Luengo-Oroz M, Amir E, Győrffy B. Integrating artificial intelligence in drug discovery and early drug development: a transformative approach. Biomark Res. 2025;13. https://doi.org/10.1186/s40364-025-00758-2.
76. Ozaki R, Ishikawa K, Kanzaki Y, Suzuki S, Takeno S, Takeuchi I, Karasuyama M. Multi-objective Bayesian optimization with active preference learning. 2023. ArXiv, abs/2311.13460. https://doi.org/10.48550/arXiv.2311.13460
77. Pang Y, Yao L, Jhong J-H, Wang Z, Lee T-Y. AVPIden: a new scheme for identification and functional prediction of antiviral peptides based on machine learning approaches. Brief Bioinform. 2021. https://doi.org/10.1093/bib/bbab263.
78. Periyasamy M. AI-driven multi-omics integration for enhanced drug discovery pipelines. Int Conf Multi-Agent Syst Collab Intell (ICMSCI). 2025;2025:1553–8. https://doi.org/10.1109/ICMSCI62561.2025.10894291.
79. Potineni B. Generative AI in drug discovery: accelerating the search for new therapeutics. World J Adv Eng Technol Sci. 2025. https://doi.org/10.30574/wjaets.2025.15.1.0298
80. Potlitz F, Link A, Schulig L. Advances in the discovery of new chemotypes through ultra-large library docking. Expert Opin Drug Discov. 2023;18:303–13. https://doi.org/10.1080/17460441.2023.2171984.
81. Prasad K, Kumar V. Artificial intelligence-driven drug repurposing and structural biology for SARS-CoV-2. Current Res Pharmacol Drug Disc. 2021;2. https://doi.org/10.1016/j.crphar.2021.100042.
82. Qadri H, Shah A, Mir M. Novel strategiesto combat the emerging drug resistance in human pathogenic microbes. Curr Drug Targets. 2020. https://doi.org/10.2174/1389450121666201228123212.
83. Quesado P, Torres L, Ribeiro B, Arrais J. A Hybrid GNN approach for improved molecular property prediction. J Comput Biol J Comput Molecul Cell Biol. 2024. https://doi.org/10.1089/cmb.2023.0452
84. Qureshi A, Kaur G, Kumar M. AVCpred: an integrated web server for prediction and design of antiviral compounds. Chem Biol Drug Des. 2016;89:74–83. https://doi.org/10.1111/cbdd.12834.
85. Qureshi R, Irfan M, Gondal TM, Khan S, Wu J, Hadi MU, Heymach J, Le X, Yan H, Alam T. AI in drug discovery and its clinical relevance. Heliyon. 2023;9. https://doi.org/10.1016/j.heliyon.2023.e17575.
86. Raman K, Yeturu K, Chandra N. TargetTB: a target identification pipeline for Mycobacterium tuberculosis through an interactome, reactome and genome-scale structural analysis. BMC Syst Biol. 2008;2:109. https://doi.org/10.1186/1752-0509-2-109.
87. Rasool A, Rahman J, Iqbal Q. Optimizing GNN architectures through nonlinear activation functions for potent molecular property prediction. Computation. 2024. https://doi.org/10.3390/computation12110212.
88. Reddy S. Generative AI in healthcare: an implementation science informed translational path on application, integration and governance. Implement Sci IS. 2024;19. https://doi.org/10.1186/s13012-024-01357-9.
89. Ren G-P, Wu K, He Y. Enhancing molecular representations via graph transformation layers. J Chem Inf Model. 2023. https://doi.org/10.1021/acs.jcim.3c00059.

90. Rong Y, Bian Y, Xu T, Xie W-Y, Wei Y, Huang W-B, Huang J. Self-supervised graph transformer on large-scale molecular data. 2020. arXiv: Biomolecules.
91. Rudramurthy G, Swamy M, Sinniah U, Ghasemzadeh A. Nanoparticles: alternatives against drug-resistant pathogenic microbes. Molecules. 2016;21. https://doi.org/10.3390/molecules21070836.
92. Saramago L, Santana M, Gomes B, Dantas R, Senger M, Borges PHO, Ferreira V, Rosa A, Tucci A, Miranda M, Lukacik P, Strain-Damerell C, Owen C, Walsh M, Ferreira S, Silva-Junior F. AI-driven discovery of SARS-CoV-2 main protease fragment-like inhibitors with antiviral activity in vitro. J Chem Inf Model. 2023. https://doi.org/10.1021/acs.jcim.3c00409.
93. Schuh M, Hesse J, Sieber S. AI-guided antibiotic discovery pipeline from target selection to compound identification. 2025. ArXiv, abs/2504.11091. https://doi.org/10.48550/arXiv.2504.11091
94. Segler M, Kogej T, Tyrchan C, Waller M. Generating focused molecule libraries for drug discovery with recurrent neural networks. ACS Cent Sci. 2017;4:120–31. https://doi.org/10.1021/acscentsci.7b00512.
95. Sharma S, Chauhan A, Ranjan A, Mathkor D, Haque S, Ramniwas S, Tuli H, Jindal T, Yadav V. Emerging challenges in antimicrobial resistance: implications for pathogenic microorganisms, novel antibiotics, and their impact on sustainability. Front Microbiol. 2024;15. https://doi.org/10.3389/fmicb.2024.1403168.
96. Sherry N, Lee J, Giulieri S, Connor C, Horan K, Lacey J, Lane C, Carter G, Seemann T, Egli A, Stinear T, Howden B. Genomics for antimicrobial resistance—progress and future directions. Antimicrob Agents Chemother. 2025;69. https://doi.org/10.1128/aac.01082-24.
97. Shimizu Y, Ohta M, Ishida S, Terayama K, Osawa M, Honma T, Ikeda K. AI-driven molecular generation of not-patented pharmaceutical compounds using world open patent data. J Cheminform. 2023;15. https://doi.org/10.1186/s13321-023-00791-z.
98. Singh K, Kapania R. ALMO: active learning-based multi-objective optimization for accelerating constrained evolutionary algorithms. Appl Sci. 2024. https://doi.org/10.3390/app14219975.
99. Škalič M, Jiménez J, Sabbadin D, Fabritiis G. Shape-based generative modeling for de novo drug design. J Chem Inf Model. 2019;59(3):1205–14. https://doi.org/10.1021/acs.jcim.8b00706.
100. Song Y, Zheng S, Niu Z, Fu Z-H, Lu Y, Yang Y. Communicative representation learning on attributed molecular graphs. 2020;2831–2838. https://doi.org/10.24963/ijcai.2020/392
101. Sousa T, Correia J, Pereira V, Rocha M. Generative deep learning for targeted compound design. J Chem Inf Model. 2021. https://doi.org/10.1021/acs.jcim.0c01496.
102. Speck-Planche A, Cordeiro M. Enabling virtual screening of potent and safer antimicrobial agents against Noma: Mtk-QSBER model for simultaneous prediction of antibacterial activities and ADMET properties. Mini Rev Med Chem. 2015;15(3):194–202. https://doi.org/10.2174/1389557515031503l2120519.
103. Srivastava R. Advancing precision oncology with AI-powered genomic analysis. Front Pharmacol. 2025;16. https://doi.org/10.3389/fphar.2025.1591696.
104. Stewart I, Buehler M. Molecular analysis and design using generative artificial intelligence via multi-agent modeling. Molecul Syst Des Eng. 2025;10:314–37. https://doi.org/10.1039/d4me00174e.
105. Swanson K, Liu G, Catacutan D, Arnold A, Zou J, Stokes J. Generative AI for designing and validating easily synthesizable and structurally novel antibiotics. Nat Mac Intell. 2024;6:338–53. https://doi.org/10.1038/s42256-024-00809-7.
106. Talib AM, Al-Hgaish A, Atan RB, Alshammari A, Alomary FO, Yaakob R, Alsahli A, Osman MH. Evaluating critical success factors in AI-driven drug discovery using AHP: a strategic framework for optimization. IEEE Access. 2025;13:42045–63. https://doi.org/10.1109/ACCESS.2025.3546925.
107. Tang B, Ewalt J, Ng H. Generative AI models for drug discovery. Biophys Comput Tools Drug Discov. 2021. https://doi.org/10.1007/7355_2021_124.

108. Tang M, Li B, Chen H. Application of message passing neural networks for molecular property prediction. Curr Opin Struct Biol. 2023;81:102616. https://doi.org/10.1016/j.sbi.2023.102616.
109. Tang X, Dai H, Knight E, Wu F, Li Y, Li T, Gerstein M. A survey of generative AI for de novo drug design: new frontiers in molecule and protein generation. Brief Bioinform. 2024;25. https://doi.org/10.1093/bib/bbae338.
110. Tarasova O, Poroikov V. Machine learning in discovery of new antivirals and optimization of viral infections therapy. Curr Med Chem. 2021. https://doi.org/10.2174/0929867328666210504114351.
111. Thakur V, Gupta V, Sharma P, Gupta A, Capalash N. A novel AI-designed antimicrobial peptide synergistically potentiates aminoglycosides against colistin- and carbapenem-resistant *Acinetobacter baumannii.* 2023. https://doi.org/10.1101/2023.11.23.568446
112. Tong X, Liu X, Tan X, Li X, Jiang J, Xiong Z, Xu T, Jiang H, Qiao N, Zheng M. Generative models for de novo drug design. J Med Chem. 2021. https://doi.org/10.1021/acs.jmedchem.1c00927.
113. Torres L, Ribeiro B, Arrais J. Few-shot learning with transformers via graph embeddings for molecular property prediction. Expert Syst Appl. 2023;225:120005. https://doi.org/10.1016/j.eswa.2023.120005.
114. Torres M, Zeng Y, Wan F, Maus N, Gardner J, De La Fuente-Nunez C. A generative artificial intelligence approach for antibiotic optimization. 2024. https://doi.org/10.1101/2024.11.27.625757
115. Tripathi M, Nath A, Singh T, Ethayathulla A, Kaur P. Evolving scenario of big data and Artificial Intelligence (AI) in drug discovery. Mol Diversity. 2021;25:1439–60. https://doi.org/10.1007/s11030-021-10256-w.
116. Turchetta M, Krause A, Trimpe S. Robust model-free reinforcement learning with multi-objective Bayesian optimization. In: 2020 IEEE international conference on robotics and automation (ICRA). 2019. 10702–10708. https://doi.org/10.1109/ICRA40945.2020.9197000
117. Uddin TM, Chakraborty AJ, Khusro A, Zidan BRM, Mitra S, Emran TB, Dhama K, Ripon MKH, Gajdács M, Sahibzada MUK, Hossain MJ, Koirala N. Antibiotic resistance in microbes: History, mechanisms, therapeutic strategies and future prospects. J Infect Public Health. 2021;14(12):1750–66. https://doi.org/10.1016/J.JIPH.2021.10.020.
118. Ullah M, Akbar S, Raza A, Zou Q. DeepAVP-TPPred: identification of antiviral peptides using transformed image-based localized descriptors and binary tree growth algorithm. Bioinformatics. 2024;40. https://doi.org/10.1093/bioinformatics/btae305.
119. Upadhayay A, Ling J, Pal D, Xie Y, Ping F-F, Kumar A. Resistance-proof antimicrobial drug discovery to combat global antimicrobial resistance threat. Drug Resist Updates Rev Comment Antimicrob Anticancer Chemother. 2022;66:100890. https://doi.org/10.1016/j.drup.2022.100890.
120. Van Der Horst D, Carter-Timofte M, Danneels A, Da Costa LS, Kurmasheva N, Thielke A, Hansen A, Chorošajev V, Holm C, Belouzard S, De Weber I, Beny C, Olagnier D. Large-scale deep learning identifies the antiviral potential of PKI-179 and MTI-31 against coronaviruses. Antiviral Res. 2024;106012. https://doi.org/10.1016/j.antiviral.2024.106012.
121. Wang C, Garlick S, Zloh M. Deep learning for novel antimicrobial peptide design. Biomolecules. 2021;11. https://doi.org/10.3390/biom11030471.
122. Wang G, Vaisman I, Van Hoek M. Machine learning prediction of antimicrobial peptides. Methods Mol Biol. 2022;2405:1–37. https://doi.org/10.1007/978-1-0716-1855-4_1.
123. Wang J, Hsieh C-Y, Wang M, Wang X, Wu Z, Jiang D, Liao B, Zhang X, Yang B, He Q, Cao D, Chen X, Hou T. Multi-constraint molecular generation based on conditional transformer, knowledge distillation and reinforcement learning. Nat Mach Intell. 2021;3:914–22. https://doi.org/10.1038/s42256-021-00403-1.
124. Wang Y, Gong H, Li X, Li L, Zhao Y, Bao P, Kong Q, Wan B, Zhang Y, Zhang J, Ni J, Han Z, Nan X, Ju K, Sun L, Chang H, Zheng M, Yu Y, Yang X, Li Y, et al. De novo multi-mechanism antimicrobial peptide design via multimodal deep learning. 2024. https://doi.org/10.1101/2024.01.02.573846

125. Wang Y, Li Z, Farimani A. Graph neural networks for molecules. 2022. ArXiv, abs/2209.05582. https://doi.org/10.1007/978-3-031-37196-7_2
126. Wang Z, Zhu G, Li S. Mapping knowledge landscapes and emerging trends in artificial intelligence for antimicrobial resistance: bibliometric and visualization analysis. Front Med. 2025;12. https://doi.org/10.3389/fmed.2025.1492709.
127. Wu K, Xia Y, Deng P, Liu R, Zhang Y, Guo H, Cui Y, Pei Q, Wu L, Xie S, Chen S, Lu X, Hu S, Wu J, Chan C, Chen S, Zhou L, Yu N, Chen E, Liu T-Y, et al. TamGen: drug design with target-aware molecule generation through a chemical language model. Nat Commun. 2024;15. https://doi.org/10.1038/s41467-024-53632-4.
128. Wu X, Wang H, Gong Y, Fan D, Ding P, Li Q, Qian Q. Graph neural networks for molecular and materials representation. J Mater Inf. 2023. https://doi.org/10.20517/jmi.2023.10
129. Xie F, Wang L, Li S, Hu L, Wen Y, Li X, Ye K, Duan Z, Wang Q, Guan Y, Zhang Y, Shi Q, Yang J, Xia H, Xie L. Large-scale genomic analysis reveals significant role of insertion sequences in antimicrobial resistance of *Acinetobacter baumannii.* mBio. 2025;16. https://doi.org/10.1128/mbio.02852-24
130. Xiong J, Xiong Z, Chen K, Jiang H, Zheng M. Graph neural networks for automated de novo drug design. Drug Disc Today. 2021. https://doi.org/10.1016/j.drudis.2021.02.011.
131. Xu L, Xia L, Pan S, Li Z. Triple generative self-supervised learning method for molecular property prediction. Int J Mol Sci. 2024;25. https://doi.org/10.3390/ijms25073794.
132. Ye J, Chen X. Current promising strategies against antibiotic-resistant bacterial infections. Antibiotics. 2022;12. https://doi.org/10.3390/antibiotics12010067.
133. Yu Z, Gao H. Molecular representation learning via heterogeneous motif graph neural networks. 2022;25581–25594.
134. Yue, Frutos-Beltrán E, Kang D, Pannecouque C, De Clercq E, Menéndez-Arias L, Liu X, Zhan P. Medicinal chemistry strategies for discovering antivirals effective against drug-resistant viruses. Chem Soc Rev. 2021. https://doi.org/10.1039/d0cs01084g
135. Zaccagnino R, Benevento G, Laurenzano G, Malandrino D, Petescia A, Zaccagnino G. Navigating the explainable molecular graph: best practices for representation learning in bioinformatics. In: 2024 IEEE 24th international conference on bioinformatics and bioengineering (BIBE). 2024;1–8. https://doi.org/10.1109/BIBE63649.2024.10820484
136. Zang X, Zhao X, Tang B. Hierarchical molecular graph self-supervised learning for property prediction. Commun Chem. 2023;6. https://doi.org/10.1038/s42004-023-00825-5.
137. Zankari E, Hasman H, Cosentino S, Vestergaard M, Rasmussen S, Lund O, Aarestrup F, Larsen M. Identification of acquired antimicrobial resistance genes. J Antimicrob Chemother. 2012;67:2640–4. https://doi.org/10.1093/jac/dks261.
138. Zeng X, Wang F, Luo Y-Z, Kang S-G, Tang J, Lightstone F, Fang E, Cornell W, Nussinov R, Cheng F. Deep generative molecular design reshapes drug discovery. Cell Rep Med. 2022;3. https://doi.org/10.1016/j.xcrm.2022.100794.
139. Zervou M, Doutsi E, Pantazis Y, Tsakalides P. De Novo antimicrobial peptide design with feedback generative adversarial networks. Int J Mol Sci. 2024;25. https://doi.org/10.3390/ijms25105506.
140. Zhang F. The mechanism of bacterial resistance and potential bacteriostatic strategies. Antibiotics. 2022;11. https://doi.org/10.3390/antibiotics11091215.
141. Zhang X-C, Wu C, Yang Z-J, Wu Z, Yi J, Hsieh C-Y, Hou T, Cao D. MG-BERT: leveraging unsupervised atomic representation learning for molecular property prediction. Brief Bioinform. 2021. https://doi.org/10.1093/bib/bbab152.
142. Zhang Z, Liu Q, Wang H, Lu C, Lee C-K. Motif-based graph self-supervised learning for molecular property prediction. 2021;15870–15882.

Chapter 10
Integration of AI-Driven Drug Design in Industry and Academia

José Antonio García Gordillo, Luis Javier Hernández García, and Carlos Arturo González Núñez

Abstract Recent advances in computational chemistry, artificial intelligence, and structural biology have changed how de novo drug design works. Researchers can now develop new molecules with better pharmacological properties much faster. For example, a newly designed CDK4/6 inhibitor has already reached pre-clinical trials, showing how these technologies are making a real difference in cancer research. This review covers current methods that bring together generative deep learning, molecular modeling, virtual screening, and accurate protein structure prediction in drug discovery. We will focus mainly on generative deep learning models and molecular modeling, with shorter sections on virtual screening and protein structure prediction. This way, readers can easily find the topics most relevant to their interests, generative adversarial networks, reinforcement learning, and diffusion-based methods enable researchers to explore chemical spaces that traditional screening cannot reach. By quantifying the "chemical space unreachable" by classic high-throughput screening (HTS), we can provide a sharper contrast, highlighting the unique capability of these generative models. For example, while traditional HTS might be limited to billions of compounds, generative models can theoretically explore chemical spaces on the order of trillions, offering an advantage of at least three orders of magnitude. Molecular dynamics simulations help predict the stability of ligand-target interactions, identify allosteric sites, and improve docking results. Faster GPUs and better sampling methods have made these advances possible. Hybrid virtual screening methods that mix ligand-based and structure-based approaches often use machine learning to improve results. These techniques help prioritize hits and assess ADMET properties more efficiently. Better protein structure prediction has also expanded the number of druggable targets, especially for proteins without known structures. By combining these computational tools, early drug discovery is faster, less expensive, and produces higher-quality candidates. Together, these changes are reshaping medicinal chemistry.

J. A. G. Gordillo (✉) · L. J. H. García · C. A. G. Núñez
Instituto Nacional de Cancerología, Ciudad de México, México
e-mail: joseantonio.garciagordillo@gmail.com

A. A. Parikesit and A. N. M. Ansori (eds.), *Recent Computational Techniques in De Novo Drug Design*, Bio-IT and AI, https://doi.org/10.1007/978-3-032-20428-8_10

Keywords Generative deep learning · De novo drug design · Chemical space exploration · Molecular dynamics simulations · Hybrid virtual screening

10.1 Introduction

Developing new drugs is essential for modern medicine, but the process has traditionally been slow, expensive, and uncertain. In oncology, drug discovery often depended on time-consuming trial-and-error methods, particularly during the target validation and lead optimization phases. AI significantly shortens these steps by up to 50%, enabling researchers to validate targets and optimize leads more efficiently than traditional methods. Now, artificial intelligence is changing the way drugs are designed, making the process more efficient and innovative.

This shift is especially significant in translational oncology and precision medicine, where the objective is to efficiently connect scientific discovery with clinical application. Traditional challenges in drug development, including the identification of novel molecular targets and the synthesis of compounds with optimal properties and minimal toxicity, are now addressed by AI with greater speed and accuracy.

Computational methods in de novo drug design now go beyond traditional high-throughput screening. Instead of just testing existing molecules, AI can create entirely new chemical compounds designed for specific treatments [4–6]. While repurposing existing drugs has proven effective and quick for assessing new therapeutic potentials, de novo design expands these efforts by enabling the discovery of novel compounds that may offer additional benefits or alternative mechanisms for treatment. Advanced algorithms and deep learning help explore a wide range of chemical space to identify new compounds with strong anticancer activity. This approach offers real clinical benefits, speeds up the development of personalized therapies, and could bring better treatments to patients more quickly.

This chapter examines recent progress in de novo drug design, focusing on how these advances affect molecular medicine and on their growing role in clinical practice. It also discusses current challenges in this rapidly changing field.

10.2 De Novo Drug Design: Concept and Clinical Relevance

De novo drug design integrates mathematical models, artificial intelligence, and biological data to develop new molecules that meet specific requirements: high affinity for the therapeutic target, metabolic stability, adequate absorption, and minimal toxicity.

It's important to understand the difference between de novo drug design and high-throughput screening. High-throughput screening tests large libraries of existing compounds, sometimes millions at a time. De novo drug design, on the other hand, explores almost limitless possible chemical structures, going beyond what traditional methods can achieve.

In clinical practice, de novo drug design allows scientists to imagine and create molecules that have never been made before. This helps develop compounds tailored for specific diseases. For example, in cancer research, the goal is to design selective inhibitors of proteins such as CDK4/6 and PI3K. In infectious disease, algorithms have enabled the design of compounds targeting viral proteins for influenza, HIV, and SARS-CoV-2, even before their full structures were known.

10.3 Generative AI in Drug Design

10.3.1 Generative Deep Learning Model

Generative deep learning is a type of AI that learns patterns from complex data and uses this knowledge to create new, realistic examples. In drug discovery, methods such as diffusion models, variational autoencoders, and generative adversarial networks can be used to design new chemical structures that meet specific criteria, such as strong binding, favorable chemical properties, and suitable absorption and metabolism. To guide the evaluation of these generative models, specific metrics will be utilized, including validity, novelty, and synthesizability of the generated compounds. Declaring these criteria upfront will facilitate a structured assessment later in the review [1–3, 6].

These models capture associations between molecular features and biological activity, facilitating the identification of candidates with enhanced potency, specificity, and safety. Generative deep learning is a subset of artificial intelligence architectures that learn the joint probability distribution of complex data to produce new, valid instances [7].

10.3.2 Core Models of Generative AI in Drug Design

AI now allows researchers not only to analyze data but also to generate new ideas for molecules. Some of the most important generative models include:

Variational Autoencoders (VAEs): These are machine learning algorithms that learn simplified versions of chemical structures. From these simplified forms, they can create new versions of molecules. This process helps researchers gradually optimize the desired properties of a drug [11, 12].

Researchers have developed a series of tyrosine kinase inhibitors by modifying existing drugs to generate derivatives with enhanced selectivity and reduced hepatotoxicity. Utilizing variational autoencoders (VAEs), these modified inhibitors have shown a selectivity improvement of approximately 30% over traditional methodologies. This significant enhancement highlights the practical benefit of VAE-driven drug design in achieving targeted therapeutic effects with fewer side effects.

Generative Adversarial Networks (GANs): These are two-part machine learning systems. One part creates new molecules; the other part evaluates if they meet the required standards. Through this back-and-forth competition, the model generates more effective drug candidates [13, 14].

Diffusion Models: These algorithms start with random noise and gradually adjust it to form realistic molecular structures. By reversing noise into structure, they can generate molecules even when the structure of the protein target is not known [8–10].

Reinforcement Learning: In this method, the computer program gets rewards for designing compounds with good properties. These properties might include strong attraction to the target protein (binding affinity), the ability to dissolve in solution (solubility), and low likelihood of causing harm (low toxicity). The reward system guides the search for better candidate molecules.

10.3.3 Principles of Molecular Dynamics

These simulations often use mathematical descriptions, called force fields, such as AMBER (Assisted Model Building with Energy Refinement), CHARMM (Chemistry at HARvard Macromolecular Mechanics), and GROMOS (Groningen Molecular Simulation). These force fields provide details about how atoms interact. Proteins, ligands (drug molecules), water molecules, and ions are treated as particles that interact, allowing detailed study of protein movement, shape changes (allosteric transitions), and effects of the surrounding solution [16].

10.3.4 Applications in Drug Discovery

Ligand-receptor stability assessment: After molecules are placed in the protein's binding site (docking), the stability of these interactions—such as hydrogen bonds (connections involving hydrogen atoms) and hydrophobic interactions (connections between water-repellent regions)—is measured over extremely short time periods, like nanoseconds or microseconds [16].

Allosteric binding site identification: Molecular dynamics (MD), which simulates how molecules move, can reveal temporary cavities in proteins. These findings help design drugs that bind to parts of the protein other than the main active site (non-orthosteric binding sites) [16–18].

Binding stability assessment: Running molecular dynamics (MD) simulations after initial docking checks if a drug candidate maintains important contacts, such as hydrogen bonds (weak connections involving hydrogen atoms) or hydrophobic interactions (water-repelling areas), inside the protein's binding site over time [16].

Thermodynamic and kinetic profiling: Using methods like metadynamics (a way of sampling rare events in molecular simulations) or free energy calculations (measuring how strongly a molecule binds), researchers can accurately predict how well molecules bind to their targets [17, 20].

Protein flexibility in virtual screening: Molecular dynamics simulations (computer-based movement simulations) produce many versions of a protein's shape. This approach considers that proteins are flexible, which improves the process of finding drug molecules that can bind well [16, 19].

Graphics processing units (GPUs), which are specialized computer chips for fast, complex calculations, have made molecular dynamics simulations much faster. As a result, researchers can now study many molecules at the same time and use these simulations more often in everyday drug design [18].

10.3.5 Technical Advances in Molecular Dynamics

Molecular dynamics simulations are valuable for studying how molecules behave over time and for understanding protein–ligand interactions. Recent improvements have made these simulations faster and more accurate, so they are now more useful in drug discovery and development. A crucial aspect of utilizing molecular dynamics outputs is connecting them to practical applications in medicinal chemistry. For example, analyzing Root Mean Square Fluctuation (RMSF) can reveal specific hotspots of flexibility within the protein structure. These hotspots can guide chemists in suggesting scaffold modifications to design better drug candidates. By explicitly linking these dynamic insights to potential chemical adjustments, the bridge between computational findings and medicinal applications is strengthened, enhancing interdisciplinary collaboration.

Graphics processing unit (GPU) acceleration substantially reduces simulation times. This approach enables microsecond-scale molecular dynamics simulations on workstations with moderate computational resources [18, 23]. Crucially, enhanced sampling techniques like accelerated Gaussian molecular dynamics (MD) and replica exchange address the issue of force-field bias by expanding the range of protein structures and movements that can be studied, an aspect that will be important when discussing limitations later [18, 23].

Enhanced sampling techniques, like accelerated Gaussian molecular dynamics (MD) and replica exchange, help simulations avoid getting stuck in a single structure. These methods allow researchers to study a wider range of protein shapes and movements that are relevant to drug discovery [20].

Hydrogen mass redistribution (HMR): This is a technique that slightly shifts mass from heavy atoms to their bonded hydrogen atoms in a simulation. By doing so, scientists can use longer time steps when running molecular dynamics, making simulations both faster and still accurate.

Coarse-grained modeling strategies: These approaches simplify molecular systems for simulation by grouping individual atoms into larger particles. This method makes it easier to study large molecular complexes or longer biological events than with detailed atom-by-atom simulations [23, 24].

10.3.6 Post-simulation Analyses

Post-simulation analyses of MD trajectories provide valuable insights into the stability and behavior of a protein–ligand complex. Common metrics include:

Root Mean Square Deviation (RMSD) quantifies the overall structural stability of the protein–ligand complex over the course of the simulation [16].

Root Mean Square Fluctuation (RMSF) characterizes the flexibility or rigidity of specific protein regions, with emphasis on the active site and surrounding residues [16].

The radius of Gyration (Rg) quantifies the overall compactness of the protein–ligand complex during molecular dynamics simulation [16].

Hydrogen Bond Analysis examines the persistence of hydrogen bonds throughout the simulation. Greater persistence indicates stronger ligand–protein interactions [21].

10.3.7 Limitations: Despite These Advancements, MD Still Faces Challenges

Timescale limitations: Numerous biologically relevant molecular motions occur on timescales inaccessible to conventional molecular dynamics (MD) simulations unless specialized computational resources are employed [16, 20].

Force-field dependence: The accuracy of the results strongly depends on the chosen force field and its handling of solvation and entropy [16].

Sampling depth: In the absence of enhanced sampling techniques, molecular dynamics simulations frequently fail to identify relevant alternative conformational states [18, 20].

10.4 Virtual Screening and Hybrid Approaches

Virtual screening (VS) has evolved from simple ligand-based similarity searches to sophisticated hybrid methods that combine multiple layers of information. Virtual screening is the computational evaluation of large compound libraries to prioritize the most promising compounds [25–27]. There are two main approaches:

Ligand-based virtual screening (LBVS) compares candidate compounds with known active molecules using descriptors such as molecular fingerprints or pharmacophores. LBVS is especially valuable when the three-dimensional structure of the target protein is unavailable but sufficient bioactivity data on active ligands are available [28, 29].

Structure-based VS (SBVS): This uses the three-dimensional structure of the target protein to simulate the binding of each candidate using docking [30].

Hybrid approaches that integrate ligand-based and structure-based strategies with machine learning currently demonstrate the highest efficiency. These methods facilitate compound classification based on simulated affinity and enable early assessment of pharmacokinetic and toxicological properties, including absorption, distribution, metabolism, excretion, and toxicity (ADMET), prior to experimental synthesis [25–27]. For example, studies suggest that hybrid virtual screening can lead to a 2-threefold increase in hit identification rates compared to ligand-based virtual screening (LBVS) alone. This significant improvement underscores the value of hybrid methods in enhancing the decision-making process in drug development.

Platforms such as AutoDock Vina, Glide, SwissADME, and pkCSM have established themselves as standard tools in the pharmaceutical industry and academia.

10.4.1 Applications of Virtual Screening in Modern Drug Design

The applications of hybrid virtual screening (VS) extend beyond antiviral research. This approach has been implemented in the following areas:

Identify kinase inhibitors in oncology (e.g., novel EGFR, ALK, and CDK4/6 inhibitors) [32, 33].

Fragment-based drug discovery is supported by combining fragment hits with pharmacophore constraints [34].

Commonly used platforms and tools include LigandScout for pharmacophore modeling, AutoDock Vina or Glide for docking simulations, and pkCSM or SwissADME for ADMET estimation. Combining VS with molecular dynamics and binding free energy calculations (e.g., MM-GBSA) supports the validation of whether candidate molecules remain stably bound under near-physiological conditions [34].

By combining computational methods in hybrid virtual screening, researchers can identify high-quality lead compounds faster and at lower cost for experimental testing.

10.5 Predictive Modeling and AI-Enhanced Design

Accurately predicting protein structures, almost as well as experiments, is a major step forward. Machine learning also helps predict important drug properties, such as absorption, distribution, metabolism, excretion, and toxicity, early in the process. Using these predictive models in de novo drug design lets researchers focus on compounds with the best safety and pharmacokinetic profiles before they are made.

Reinforcement learning (RL) approaches, such as the Activity Cliff-Aware Reinforcement Learning (ACARL) framework, are particularly noteworthy. ACARL explicitly models abrupt activity changes with minor structural modifications, a phenomenon known as activity cliffs, thus enabling the generation of compounds that effectively navigate these complex structure–activity landscapes [35].

10.6 Protein Structure Prediction

The ability to model protein structures with high precision has significantly expanded the capabilities of structure-based drug design (SBDD). Computational predictions now enable the investigation of target molecules without experimental structures, thus opening new avenues for therapeutic interventions. Methods such as AlphaFold2, RoseTTAFold, and trRosetta, as well as complementary modeling approaches, have made reliable templates available for many previously inaccessible proteins [36–38].

Even with recent progress, prediction accuracy still has limits. Major challenges include modeling side-chain positions within binding sites and determining the appropriate conformations for ligand interactions. Combining predicted models with physics-based simulations, medicinal chemistry, and experimental testing helps ensure that the structures used in drug design are truly relevant [36–39].

10.6.1 Predictive Modeling and AI-Enhanced Design

Machine learning enables the early prediction of ADMET properties. Reinforcement learning frameworks such as the Activity Cliff-Aware RL (ACARL) model abrupt activity changes and guide connection generation through complex structure–activity landscapes [35].

10.6.2 Clinical Applications and Industry Trends

The pharmaceutical industry has quickly started using these technologies. Some startups say their drug discovery success rates are up to 100 times higher than with older methods. In cancer research, AI-designed drug candidates are already being tested against targets such as CDK4/6, EGFR, and proteins in the tumor microenvironment.

These technological advances reduce costs and development time and facilitate the implementation of computer-assisted personalized medicine. In this approach, molecular design is tailored to the specific genomic and proteomic profiles of individual patients.

10.7 Limitations and the Path Forward

Although computational drug design offers significant potential, the field faces substantial challenges that necessitate a balanced approach. The most critical issue is data bias. Artificial intelligence (AI) demonstrates strong pattern recognition capabilities, but its effectiveness depends on the quality and diversity of training data. Incomplete or biased datasets result in algorithms that produce similarly biased predictions. Consequently, the generated molecules may be effective only against limited biological targets or specific patient populations. This issue is especially pertinent in personalized medicine, which aims to develop drugs tailored to individual genomic profiles.

It's also important to remember that computer-based drug design still needs real-world testing. Even the best models can't capture all the complexity of living systems. Turning a promising design into a safe and effective drug requires thorough lab and animal testing to check how well it works, its toxicity, and how the body processes it. AI's role is to speed up and guide these experiments, not to replace them.

Scalability and ethical or regulatory issues are also major concerns in modern drug discovery. Many advanced simulations require substantial computing power, limiting their widespread use. As AI-based therapies get closer to clinical use, they raise questions about data transparency, intellectual property, and the rules needed to keep patients safe and maintain public trust.

10.8 Conclusion: A Paradigm Shift in Molecular Medicine

Combining artificial intelligence, molecular modeling, and accurate protein structure prediction is a major step forward for molecular medicine. The field is shifting from slow, traditional drug discovery to smarter and more efficient approaches. Processes that were once costly and time-consuming are now much quicker and more flexible.

Recent progress includes designing new molecules from scratch, exploring huge chemical spaces, and cutting the time and cost of early drug discovery. AI can now predict protein structures almost as accurately as experiments, opening new targets for diseases once thought untreatable.

The primary promise of this technological revolution lies in its potential to. The real promise of these new technologies is their ability to advance personalized medicine and translational pharmacology, not just to make things faster. The main goal is to create treatments that are both effective and highly specific, matched to each patient's unique genetic and protein profile. This approach connects lab research with clinical care, allowing computational design to directly benefit patients. The computational power of algorithms with the rigor of experimental validation. The success of the next generation of innovative therapies will rely on effectively harnessing this convergence and addressing challenges with vision and a strong commitment to scientific integrity.

References

1. Tang X, Dai H, Knight E, Wu F, Li Y, Li T, et al. A survey of generative AI for de novo drug design: new frontiers in molecule and protein generation. Brief Bioinform. 2024;25(4).
2. Lima AN, Philot EA, Trossini GHG, Scott LPB, Maltarollo VG, Honorio KM. Use of machine learning approaches for novel drug discovery. Expert Opin Drug Discov. 2016;11(3):225–39.
3. Wang M, Wang Z, Sun H, Wang J, Shen C, Weng G, et al. Deep learning approaches for de novo drug design: an overview. Curr Opin Struct Biol. 2022;72:135–44.
4. Kuenzi BM, Park J, Fong SH, Sanchez KS, Lee J, Kreisberg JF, et al. Predicting drug response and synergy using a deep learning model of human cancer cells. Cancer Cell [Internet]. 2020;38(5):672–684.e6. Available from: https://www.cell.com/cancer-cell/pdfExtended/S1535-6108(20)30488-8.
5. Cai L, Gao J, Zhao D. A review of the application of deep learning in medical image classification and segmentation. Ann Trans Med. 2020;8(11):713–23.
6. Polykovskiy D, Zhebrak A, Sanchez-Lengeling B, Golovanov S, Tatanov O, Belyaev S, et al. Molecular sets (MOSES): a benchmarking platform for molecular generation models. Front Pharmacol. 2020;11.
7. Zhavoronkov A, Li R, Ma C, Mamoshina P. Deep biomarkers of aging and longevity: from research to applications. Aging. 2019;11(22):10771–80.
8. Kazerouni A, Aghdam EK, Heidari M, Azad R, Fayyaz M, Hacihaliloglu I, et al. Diffusion models in medical imaging: a comprehensive survey. Med Image Anal [Internet]. 2023;88:102846. Available from: https://www.sciencedirect.com/science/article/pii/S1361841523001068?via%3Dihub.
9. Luo C. Understanding diffusion models: a unified perspective. arXiv Cornell University; 2022. Available from: http://arxiv.org/abs/2208.11970
10. Khader F, Müller-Franzes G, Tayebi Arasteh S, Han T, Haarburger C, Schulze-Hagen M, et al. Denoising diffusion probabilistic models for 3D medical image generation. Sci Rep. 2023;13(1):7303–3.
11. Rais K, Amroune M, Benmachiche A, Haouam MY. Exploring variational autoencoders for medical image generation: a comprehensive study. arXiv Cornell University; 2024.
12. Gómez-Bombarelli R, Wei JN, Duvenaud D, Hernández-Lobato JM, Sánchez-Lengeling B, Sheberla D, et al. Automatic chemical design using a data-driven continuous representation of molecules. ACS Cent Sci. 2018;4(2):268–76.
13. Hussain J, Båth M, Ivarsson J. Generative adversarial networks in medical image reconstruction: a systematic literature review. Comput Biol Med [Internet]. 2025;191:110094. Available from: https://www.sciencedirect.com/science/article/pii/S0010482525004457.
14. Kadurin A, Nikolenko S, Khrabrov K, Aliper A, Zhavoronkov A. DruGAN: an advanced generative adversarial autoencoder model for de novo generation of new molecules with desired molecular properties in silico. Mol Pharm. 2017;14(9):3098–104.
15. Olivecrona M, Blaschke T, Engkvist O, Chen H. Molecular de-novo design through deep reinforcement learning. J Cheminform. 2017;9(1).

16. Adelusi TI, Oyedele AQK, Boyenle ID, Ogunlana AT, Adeyemi RO, Ukachi CD, et al. Molecular modeling in drug discovery. Inform Med Unlocked [Internet]. 2022;29:100880. Available from: https://www.sciencedirect.com/science/article/pii/S235291482200034X#fig.
17. Perez C, Faust B, Mehdipour AR, Francesconi KA, Forrest LR, Ziegler C. Substrate-bound outward-open state of the betaine transporter BetP provides insights into Na+ coupling. Nat Commun. 2014;5(1).
18. Strecker C, Meyer B. Plasticity of the binding site of renin: optimized selection of protein structures for ensemble docking. J Chem Inf Model. 2018;58(5):1121–31.
19. Ivetac A, Andrew MJ. Mapping the druggable allosteric space of G-protein coupled receptors: a fragment-based molecular dynamics approach. Chem Biol Drug Des. 2010;5:1.
20. Lan NT, Vu KB, Dao Ngoc MK, Tran PT, Hiep DM, Tung NT, et al. Prediction of AChE-ligand affinity using the umbrella sampling simulation. J Mol Graph Model. 2019;93:107441.
21. Kandeel M, Kitade Y. Molecular dynamics and binding selectivity of nucleotides and polynucleotide substrates with EIF2C2/Ago2 PAZ domain. Int J Biol Macromol. 2018;107:2566–73.
22. Bhattarai A, Wang J, Miao Y. Retrospective ensemble docking of allosteric modulators in an adenosine G-protein-coupled receptor. Biochimica et Biophysica Acta (BBA)—General Subjects. 2020;1864(8):129615.
23. Souza PCT, Thallmair S, Conflitti P, Ramírez-Palacios C, Alessandri R, Raniolo S, et al. Protein–ligand binding with the coarse-grained Martini model. Nat Commun. 2020;11(1).
24. Shinoda W, DeVane R, Klein ML. Multi-property fitting and parameterization of a coarse grained model for aqueous surfactants. Mol Simul. 2007;33(1–2):27–36.
25. Eid AM, Selim A, Khaled M, Elfiky AA. Hybrid virtual screening approach to predict novel natural compounds against HIV-1 CCR5. J Phys Chem B. 2024.
26. Lionta E, Spyrou G, Vassilatis D, Cournia Z. Structure-based virtual screening for drug discovery: principles, applications and recent advances. Curr Top Med Chem. 2014;14(16):1923–38.
27. Walters WPatrick, Murcko MA. Prediction of "drug-likeness." Adv Drug Deliv Rev. 2002;54(3):255–71.
28. Willett P. Similarity-based virtual screening using 2D fingerprints. Drug Discovery Today. 2006;11(23–24):1046–53.
29. Schneider G. Virtual screening: an endless staircase? Nat Rev Drug Discovery. 2010;9(4):273–6.
30. Meng XY, Zhang HX, Mezei M, Cui M. Molecular docking: a powerful approach for structure-based drug discovery. Current Comput Aided-Drug Des [Internet]. 2011;7(2):146–57. Available from: https://pmc.ncbi.nlm.nih.gov/articles/PMC3151162/.
31. Sliwoski G, Kothiwale S, Meiler J, Lowe EW. Computational methods in drug discovery. pharmacological reviews [Internet]. 2013;66(1):334–95. Available from: https://www.ncbi.nlm.nih.gov/pmc/articles/PMC3880464/.
32. Xia S, Chen E, Zhang Y. Integrated molecular modeling and machine learning for drug design. J Chem Theory Comput Am Chem Soc. 2023;19:7478–95.
33. Reddy AT, Lakshmi SP, Reddy RC. PPARγas a novel therapeutic target in lung cancer. PPAR Res. 2016;2016:1–7.
34. Erlanson DA, Fesik SW, Hubbard RE, Jahnke W, Jhoti H. Twenty years on: the impact of fragments on drug discovery. Nat Rev Drug Discovery. 2016;15(9):605–19.
35. Moingeon P, Kuenemann M, Guedj M. Artificial intelligence-enhanced drug design and development: toward a computational precision medicine. Drug Discovery Today. 2021.
36. Jumper J, Evans R, Pritzel A, Green T, Figurnov M, Ronneberger O, et al. Highly accurate protein structure prediction with Alphafold. Nature. 2021;596(7873):583–9.
37. Baek M, DiMaio F, Anishchenko I, Dauparas J, Ovchinnikov S, Lee GR, et al. Accurate prediction of protein structures and interactions using a three-track neural network. Science. 2021;373(6557):871–6.
38. Paiardini A. Protein structure prediction in drug discovery. Biomolecules [Internet]. 2023;13(8):1258–8. Available from: https://www.ncbi.nlm.nih.gov/pmc/articles/PMC10452104/#:~:text=They%20have%20advanced%20our%20understanding.

39. Naceri S, Marc D, Blot R, Flatters D, Anne-Claude Camproux. Druggable pockets at the RNA interface region of influenza A virus NS1 protein are conserved across sequence variants from distinct subtypes. Biomolecules. 2022;13(1):64–4.
40. Bhowmick S, Jing T, Wang W, Zhang EY, Zhang F, Yang Y. In silico protein folding prediction of COVID-19 mutations and variants. Biomolecules. 2022;12(11):1665.
41. Dick A, Meuser ME, Cocklin S. Clade-specific alterations within the HIV-1 capsid protein with implications for nuclear translocation. Biomolecules [Internet]. 2022 [cited 2022 Oct 3];12(5):695. Available from: https://www.ncbi.nlm.nih.gov/pmc/articles/PMC9138599/
42. Rosignoli S, Paiardini A. Boosting the full potential of PyMOL with structural biology plugins. Biomolecules. 2022;12(12):1764.
43. Ferreira L, dos Santos R, Oliva G, Andricopulo A. Molecular docking and structure-based drug design strategies. Molecules. 2015;20(7):13384–421.

Chapter 11
Trends of Open-Source Tools and Frameworks for AI Drug Design

Saurav Kumar Mishra, Priya Mukhia, Kusum Gurung, Pragya Rai, Apurba Roy, Tabsum Chhetri, and John J. Georrge

Abstract Emerging technology, along with various algorithms and artificial intelligence (AI), has accelerated the drug design process. The growing concern of emerging pathogens and their associated complications has created severe health challenges. To address this, research now utilises various tools, databases, and associated resources to streamline the drug discovery process. Conventional drug design methods are time-consuming, costly, and involve multiple stages. In response to the urgent need for safer and faster therapeutic solutions, the field is shifting toward rational, tool-based approaches. To support this transformation, several tools and database servers have been developed to aid and speed up the process. Open-source resources are particularly valuable, as they are not only effective and accessible but also help reduce overall development costs. Moreover, AI has become a key driver in developing more accurate and efficient tools for drug design. Several AI-integrated resources have been developed to support therapeutic design and enhance outcomes. This study discusses the role of open-access tools, databases, and packages, as well as AI-assisted resources in drug design. It provides insights into their applications, accessibility, and relevance in pharmaceutical research. Additionally, the study includes AI-based navigation within the pharmaceutical sector, highlighting how AI contributes to innovation and efficiency. Finally, the advantages and limitations of these resources are discussed to offer a comprehensive understanding of their current stage, implications, and potential in designing novel drugs through AI-driven approaches.

Keywords Artificial intelligence · Drug design · Open resource · Database · Tools · Pharmaceutical sector

---

S. K. Mishra · P. Mukhia · K. Gurung · P. Rai · T. Chhetri · J. J. Georrge (✉)
Department of Bioinformatics, University of North Bengal, Darjeeling 734013, West Bengal, India
e-mail: johnjgeorrge@gmail.com

A. Roy
Department of Pharmaceutical Technology, University of North Bengal, Darjeeling 734013, West Bengal, India

A. A. Parikesit and A. N. M. Ansori (eds.), *Recent Computational Techniques in De Novo Drug Design*, Bio-IT and AI, https://doi.org/10.1007/978-3-032-20428-8_11

11.1 Introduction

Traditional drug discovery relies heavily on manual target discovery and trial-and-error methods, where chemical compounds are tested on animal cells to assess their apparent effect on treatment, followed by clinical trials and failures, which can take approximately 10–15 years and incur significant costs [1]. The incorporation of Artificial Intelligence (AI) and computational approaches has revolutionised the traditional drug design process, transforming it into rational drug design and accelerating. It enables high-throughput screening to streamline the drug discovery process by screening millions of compounds to quickly identify and select the most promising drug candidates. AI-based algorithms, such as Machine Learning (ML), are utilised to identify disease-related target proteins for targeted therapies. Additionally, AI supports individual analyses of genetic profiles, enabling personalised treatments, increasing efficiency, and minimising side effects [2]. Many ML models, like QSAR and virtual screening, are used to anticipate the effectiveness of potential drugs and their binding affinity, accelerating the target selection procedure and reducing the laborious laboratory screening efforts. Also, AI-based combined CNN/RNN approaches are utilised to predict the toxicity and safety of the potential drug at the early stage of processing, lowering the risk of last-stage failure and can also predict graph-based models for the drug target interactions to predict any possible side effects of the compounds, overall accelerating drug processing and ensuring interpretability and regulatory alignment [3]. Since commercial drug design industries are led by profit-oriented pharmaceutical companies that utilise closed-source, expensive commercial software, involving costly clinical trials that require substantial capital, this ultimately limits access. It prioritises illnesses with high market value potential, neglecting rare or non-profitable diseases. As a result, this lowers the rate of innovation in drug development due to cost barriers and market risks [4, 5]. The integration of open-source tools and platforms provides the fundamental backbone for AI-based drug design, enabling the faster, more efficient, accessible, and cost-effective discovery of therapeutic compounds, while offering professional-level functionality at no cost.

As a part of the computational approaches, the application was also involved in the therapeutic finding in terms of promoting inhibitors. Open-source computational tools are essential in modern drug design as they facilitate researchers to study molecular structures, behaviour, and thermodynamic properties, accelerating and improving novel drug developments, at the same time encouraging collaboration among diverse researchers, boosting innovation and problem-solving skills across the globe, and promoting non-profit research and creating a common knowledge base [6–10]. Open-source platforms stipulate that all research outputs be publicly accessible, providing low-cost innovations that build transparency, promote scrutiny, consistency, and trust among the scientific community and patients. This approach eliminates redundancy, enabling advancements on existing work rather than duplicating it. The World Health Organisation (WHO) is increasingly promoting Open-source platform-based innovation as it encourages expanding access to treatments, especially to underdeveloped nations [11]. The collaboration of open-source tools and platforms in drug development is significant in managing costs. They make drug discovery more accessible by eliminating the need for expensive commercial software, also minimising costly wet lab experiments. Since pre-clinical screening with open-source tools enables researchers to eliminate compounds at early stages, thereby preventing costly failures in later stages, this approach saves both time and money [6]. Since there is a lot of data sharing, it is crucial for Open-source technologies to ensure the security and privacy of sensitive data. Therefore, privacy legislation, such as GDPR and CCPA, is implemented to prevent unnecessary access, while also enabling real-time monitoring and open data utilisation, promoting accountability and fairness [12]. Open-source technologies often prioritise combating neglected illnesses by emphasising the development of inexpensive diagnostics, which also allows open dataset models to detect and identify local infection variations with great accuracy [13]. GitHub, a web-based platform popular among researchers, is widely used to share code for drug design workflows, providing accessibility, reproducibility, transparency, and quick sharing of data with the open-source community, also enabling collaboration globally, making it easier for researchers to repeat and validate results, improving the workflow and computational approach for open-source drug development [14]. Subsequently, the incorporation of user-friendly modern technology like cloud computing facilitates open-source drug designing as it allows researchers to access very complex, powerful software and huge storage without the need to have supercomputers and advanced devices, providing on-demand resources, dynamic flexibility according to the required need and also providing VMs to run multiple programs simultaneously [15]. Overall, the integration of open-source tools and platforms in the drug development industry has accelerated the process, enhancing accessibility and innovation, minimising cost, and paving the way for cost-effective personalised treatments [16]. As their involvement occurs frequently, this study aims to provide precise information about the available resources, including tools, frameworks, databases, and insightful information to guide their effective application in modern drug design, along with an overview of advancements and challenges. An overview of this study is illustrated in Fig. 11.1.

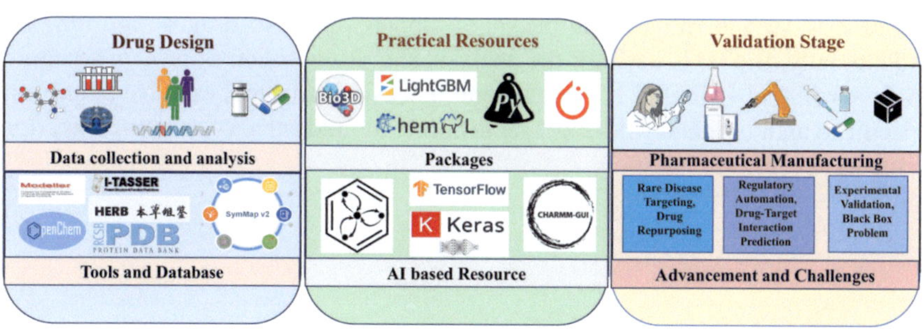

Fig. 11.1 Overview of the involved steps in drug design and their information included in this study

11.2 Insight from a Database Perspective

Databases refer to collections of data that are organized in various servers and systems, allowing individuals to access, upload, download, or manipulate the information as needed [17]. In drug discovery, the introduction of databases has significantly accelerated the process, as they facilitate the acquisition of a wide range of data related to potential drug candidates and disease-specific information [18, 19]. Databases such as GEO, GeneCards, and OMIM contain information about diseases and the genes associated with them [20–22]. Additionally, some databases store data on the use of phytochemicals as potential drugs, such as TCMSP and SymMap [23, 24]. With new drugs constantly being developed, investigated, and approved, databases such as DrugBank provide valuable information on these compounds [25]. Furthermore, some databases store protein structures and related information, such as PDB and UniProt [26, 27]. It has simplified data storage and also enhanced data retrieval and analysis. A comprehensive list of the databases, along with their use, is highlighted in Table 11.1.

11.3 Insights from a Package Perspective

A set of programming modules and functions used in languages like Python or R is called a package. It is primarily used to analyse and manage large datasets [75]. The drug discovery process is tedious, requiring substantial manpower and money. The introduction of such packages has made it simpler to handle a large volume of data. The analysis of drugs or any result during drug discovery was prone to human errors that could compromise the reliability of the results. These packages not only reduce human error associated with manual data analysis but also enhance accuracy, reproducibility, and scalability in processing biological data [76]. Various packages have been developed, each serving specialised functions in drug discovery, like Scikit-learn, which is a package used for protein model generation [77]. The R package ChemmineR facilitates cheminformatics analysis of small molecules, while DrugEx, a deep learning-based package, supports de novo drug design [78, 79]. Additionally, Bio3D is used for the analysis and visualisation of protein structures [80]. Additionally, Table 11.2 lists several other programs that each have functions specific to their type and work together to accelerate and enhance contemporary drug discovery.

Table 11.1 List of databases used in various phases of drug discovery

Sl.No.	Name	Description	URL	References
1	TCMSP	A comprehensive data repository and analysis platform for studying and understanding traditional Chinese medicine	https://tcmsp-e.com/	[24]
2	ETCM	ETCM contains information on TCM small plants and their phytochemicals	http://www.tcmip.cn/ETCM/	[28]
3	SymMap (Symptom mapping)	SymMap provides information on herbs, targets, clinical symptoms and diseases combining both Chinese and modern medicine	http://www.symmap.org/	[23]
4	BATMAN-TCM (A bioinformatics analysis tool for molecular mechanism of traditional Chinese medicine)	BATMAN-TCM is designed to analyse the TCM biological procedure	http://bionet.ncpsb.org.cn/batman-tcm/index.php/Home/Index/index	[29]
5	TCMID	A comprehensive database that builds a bridge between traditional and modern science	https://bidd.group/TCMID/	[30]
6	HERB (A high-throughput experiment and reference-guided database of traditional Chinese medicine)	HERB is used to analyse gene expression profiles and supports modernization for TCM	http://herb.ac.cn/	[31]
7	TCM Database@TaiWan	TCM Database@TaiWan is the largest non-commercial database that contains data from Chinese medical texts and publications	http://tcm.cmu.edu.tw/	[32]
8	TCM-ID (Traditional Chinese medicine information database)	TCM-ID enables clinical studies, herb verification, and big data linkage	https://www.bidd.group/TCMID/	[33]
9	HIT 2.0	HIT contains herbal targets, enabling target mining and also developing a personal target profile online	http://www.badd-cao.net:2345/	[34]
10	RCSB PDB	PDB is a database that provides access to 3-D protein structures for analysis and research	https://www.rcsb.org/	[27]
11	Proteomics identification database (PRIDE)	A comprehensive repository with certain criteria and open access to the mass spectrometry proteomics data	https://www.ebi.ac.uk/pride/	[35]

(continued)

Table 11.1 (continued)

Sl.No.	Name	Description	URL	References
12	UniProt	UniProt contains information related to protein sequence and functional analysis	https://www.uniprot.org/	[26]
13	InterPro	InterPro can functionally analyze proteins through family classification, domain and significant site prediction, and other methods	https://www.ebi.ac.uk/interpro/	[36]
14	VARIDT	Comprehensive information on every facet of drug transporters' variability is available through VARIDT	https://varidt.idrblab.net/	[37]
15	Ensembl	Centralised resources provide robust functions, including regulatory function predictions and gene annotation, alongside centralised genomic data	https://useast.ensembl.org/index.html	[38]
16	Gene expression omnibus (GEO)	GEO is an open access for high-throughput functional genome data	https://www.ncbi.nlm.nih.gov/geo/	[21]
17	University of California at Santa Cruz (UCSC) Genome	The UCSC is utilised for visualization and analysis of genetic data	https://genome.ucsc.edu/	[39]
18	DrugBank	The database provides information on drugs and drug targets along with their classification, reaction, pathways, and many more	https://go.drugbank.com/	[25]
19	ChEMBL	A database that provides access to the bioactive molecules with drug-like characteristics, along with the comprehensive data for the identification of novel drugs	https://www.ebi.ac.uk/chembl/	[40]
20	PubChem	PubChem contains chemical, physical and biological information for molecules especially small molecules	https://pubchem.ncbi.nlm.nih.gov/	[41]
21	DrugMAP	A thorough list of interacting molecules for medications and therapeutic prospects is provided by DrugMAP, along with details on patterns of differential expression	https://drugmap.idrblab.net/	[42]
22	GenBank	A sequence database that provides open access to the available nucleic acid sequences	https://www.ncbi.nlm.nih.gov/genbank/	[43]

(continued)

Table 11.1 (continued)

Sl.No.	Name	Description	URL	References
23	RefSeq	An annotated set of reference sequences, such as genomic, protein, and transcripts, is provided by a comprehensive RefSeq database	https://www.ncbi.nlm.nih.gov/refseq/	[44]
24	Expression Atlas	The EMBL-EBI Expression Atlas provides information on the expression of genes and proteins across various species	https://www.ebi.ac.uk/gxa/home	[45]
25	TTD	TTD contains detailed information on drug targets and helps predict the drugability rate	https://idrblab.net/ttd/	[46]
26	ChemSpider	ChemSpider provides chemical and structural information for compounds, along with their identifiers	https://www.chemspider.com/	[47]
27	PHAROS	PHAROS offers an all-inclusive, cohesive information repository for the druggable genome	https://pharos.nih.gov/	[48]
28	TCGA	It provides comprehensive data on cancer, along with computational tools and molecular classification platforms	https://www.cancer.gov/ccg/research/genome-sequencing/tcga	[49]
29	DisGenNET	DisGeNET is a comprehensive database focused on information related to gene-associated diseases	https://disgenet.com/	[50]
30	GeneCards	GeneCards provides functional, disease pathway, proteomic, and genetic information of predicted genes	https://www.genecards.org/	[20]
31	ClinVar	ClinVar provides information on the relationships between human genetic variations and their clinical significance	https://www.ncbi.nlm.nih.gov/clinvar/	[51]
32	HPRD (Human protein reference database)	HPRD provides detailed information on a gene's function, expression, and related disease pathways	http://www.hprd.org/	[52]
33	OMIM	OMIM databases contain thousands of genes, phenotypic and genotypic data mostly related to Mendelian diseases	https://www.omim.org/	[22]

(continued)

Table 11.1 (continued)

Sl.No.	Name	Description	URL	References
34	BioCarta	Online maps of metabolic and signalling pathways	https://maayanlab.cloud/Harmonizome/dataset/Biocarta+Pathways	[53]
35	STITCH (Search tool for interactions of chemicals)	Aids in compound-associated interactive target analysis	http://stitch.embl.de/	[54]
36	PharmGKB	A helpful database to navigate genetic alterations for designing therapeutics	https://www.pharmgkb.org/	[55]
37	KEGG	Knowledge base that systematically analyses gene functions	https://www.kegg.jp/	[56]
38	RxNorm	It provides distinct identifiers and standard names by relating to the drug terminologies	https://www.nlm.nih.gov/research/umls/rxnorm/index.html	[57]
39	BindingDB	Database focused on experimentally determined binding affinities between small molecules and proteins, particularly those considered drug targets	https://www.bindingdb.org/	[58]
40	ZINC	It contains a vast number of compounds in 3D formats, readily usable for molecular docking and other computational studies	https://zinc.docking.org/	[59]
41	SuperPred	Web server for ATC target prediction, along with other information	http://bioinformatics.charite.de/superpred	[60]
42	CMap	Includes functional relationships between drugs, genes, and diseases	https://www.broadinstitute.org/connectivity-map-cmap	[61]
43	Reactome	A pathway database that provides open access to analyse, visualize, and interpret the biological pathways and molecular interactions	https://reactome.org/	[62]

(continued)

Table 11.1 (continued)

Sl.No.	Name	Description	URL	References
44	PDBbind	Helpful to navigate and experimentally classify the binding affinity of a biological complex	https://www.pdbbind-plus.org.cn/	[63]
45	IntAct	A public, open-source database focused on molecular interactions, particularly protein–protein interactions	https://www.ebi.ac.uk/intact/home	[64]
46	BioGRID	Helpful for interaction datasets of genomic, proteomic, and chemical molecules	https://thebiogrid.org/	[65]
47	ChemProt	A database for chemical-protein interactions, along with their association with different diseases	http://www.cbs.dtu.dk/	[66]
48	Clusters of orthologous gene (COGs)	A database including genomes from bacteria and archaeal species, along with the annotation, references, and PDB links	https://www.ncbi.nlm.nih.gov/research/cog	[67]
49	Consensus Path DataBase (CPDB)	The database provides integration information of genetic, proteomic, and drug target interactions	http://cpdb.molgen.mpg.de/	[68]
50	Database for annotation, visualization and integrated discovery (DAVID)	It contains tools to predict ontological assessment, ID conversion, an ortholog tool, gene classification, gene name, and a batch viewer	https://davidbioinformatics.nih.gov/	[69]
51	Database of interacting proteins (DIP)	It provides information on protein–protein interactions	https://dip.doe-mbi.ucla.edu/dip/Main.cgi	[70]
52	MMsINC	A database that provides queries on chemical structures based on structure and their similarity to PDB ligands	http://mms.dsfarm.unipd.it/MMsINC/search/	[71]
53	Integrated interactions database (IID)	It contains interactions between proteins in 18 species that are detected and predicted	https://iid.ophid.utoronto.ca/	[72]
54	SignaLink	Signalling pathway, regulatory enzymes, transcription factors, and miRNA analysis resource	http://signalink.org/	[73]
55	T3DB	Helps to analyze the information of compounds to predict their toxicity for drug safety purposes	www.t3db.ca	[74]

Table 11.2 List of various Python and R-based packages used for various stages of drug discovery

Sl.No.	Name	Description	URLs/GitHub	References
1	Scikit-learn	Scikit-learn is an open-source Python library used for developing predictive models for drug design	https://github.com/scikit-learn/scikit-learn	[77]
2	PyTorch	It is a built-in Python DL library that provides accelerated GPU support and dynamic graphs	https://pytorch.org/	[81]
3	LightGBM	A package offering an R interface to a tree-based learning algorithm for handling large datasets and high-resolution features	https://github.com/Microsoft/LightGBM	[82]
4	RScore	RScore is an R package used to explore single-cell RNA-seq data along with molecular networks	https://github.com/wycwycpku/RSCORE	[83]
5	Guacamol	A Python package designed for comparing classical and DL based algorithms for novel molecule building	https://github.com/BenevolentAI/guacamol	[84]
6	DrugEx	A free-access online package for novel drug development is also used for multi-objective reinforcement learning	https://github.com/CDDLeiden/DrugEx	[79]
7	ChemML	An open-source software package for making ML and data science more accessible	https://github.com/hachmannlab/chemml	[85]
8	Pybel	It is a Python wrapper for OpenBabel, helping with molecule conversion and other functions	https://openbabel.org/docs/UseTheLibrary/Python_Pybel.html	[86]
9	ChemmineR	It is a chemoinformatics package used to study and evaluate drug-like small molecules	https://www.bioconductor.org/packages/devel/bioc/vignettes/ChemmineR/inst/doc/ChemmineR.html	[78]
10	Bio3d	Useful package to interpret model structure in various ways	http://thegrantlab.org/bio3d/	[80]

11.4 Insights from a Tool Perspective

Apart from packages, numerous software platforms and tools have been developed and introduced, designed for specific tasks across various requirements [87, 88]. The tools are also utilised in various phases of drug development and have contributed significantly by reducing costs and labour, as well as optimising the process. With the introduction and advancement of these new tools, one can now predict the molecular mechanism of a compound as well as perform target identification, virtual screening, and docking, among other tasks [88, 89]. The morphology of the target is extremely significant for the drug discovery process; therefore, tools such as MODELLER and SWISS-MODEL are used for structure prediction and homology modelling of the compounds [90, 91]. To understand how ligands bind to targets, tools such as AutoDock Vina and GROMACS are employed for molecular docking, which helps predict the potential interaction between a drug and its target [92, 93]. Furthermore, different tools have various input formats; therefore, tools like Open Babel are used to convert file formats accordingly [94]. These tools contribute significantly to the drug discovery process, and Table 11.3 presents a list of widely used tools applied at various stages of drug development.

11.5 Insights from AI and DL Integrated Resources

The implementation of AI and DL in a variety of domains, including drug research and discovery, has increased in recent years. This has not only accelerated but also automated a large portion of the drug discovery process [89, 110]. In addition to enhancing the software tools and packages already in place, many AI and ML integrated frameworks and libraries have been developed, offering fresh viewpoints and insights into pre-existing models and improving the functionality of software and tools to produce more accurate results [111, 112]. To evaluate the compounds and create models that can be used further for drug development, libraries such as TensorFlow and Chainer Chemistry are employed [113, 114], followed by protein modelling and protein structure prediction via AI-integrated platforms, including MolAICal and AlphaFold [115, 116]. Furthermore, several frameworks that utilise the DL method to aid in drug discovery and molecular modelling have been established, such as DeepChem [117]. Additionally, the training and comparison of generated generative models in de novo molecular modeling has been made possible by the introduction of Molecular Sets (MOSES) [118]. A list of AI and DL-integrated resources that can be utilised for multiple phases of modern drug development is highlighted in Table 11.4.

11.6 AI-Based Navigation in the Pharmaceutical Area

The introduction of AI in the medical sector has truly revolutionised the area by enhancing productivity, accelerating the manufacturing process, and producing better quality goods, along with increased security. Moreover, an overview of this section is presented in Fig. 11.1. Artificial Intelligence has enhanced traditional methods that previously required time, expense, and considerable patience, as they were susceptible to failure. The various challenges faced in the pharmaceutical industry, such as

Table 11.3 Overview of various tools utilized in the drug discovery process

Sl.No.	Name	Description	URLs/GitHub	References
1	PaDEL-Descriptor	It is a Java-based software that is used to compute chemical descriptions and fingerprints in drug design	http://padel.nus.edu.sg/software/padeldescriptor	[95]
2	AiZynthFinder	An open-source tool for retrosynthetic planning	http://www.github.com/MolecularAI/aizynthfinder	[96]
3	Open Drug Discovery Toolkit (ODDT)	It's used for computer-aided drug discovery pipelines, integrating machine learning scoring functions	https://github.com/oddt/oddt	[97]
4	Open Bable	Open Bable is a chemical toolbox specifically designed to understand and communicate various languages of chemical data	https://www.cheminfo.org/Chemistry/Cheminformatics/FormatConverter/index.html	[94]
5	Chemistry Development Kit (CDK)	Useful Java-assisted library for chemoinformatics analysis	https://cdk.github.io/	[98]
6	OpenChem	A Python-based tool used to train models for drug discovery	https://github.com/Mariewelt/OpenChem	[99]
7	MODELLER	Aids in de novo modeling, optimization, multiple alignment, clustering, and database searching	https://salilab.org/modeller/	[91]
8	SWISS-MODEL	It is a web tool used for homology modelling as well as visualisation of protein structures	https://swissmodel.expasy.org	[90]
9	Phyre2	It is utilised for protein-structure prediction, analysis, and visualisation	https://www.sbg.bio.ic.ac.uk/~phyre2/html/page.cgi?id=index	[100]
10	I-TASSER	It is used for predicting 3D structures of proteins and their function, and also combines threading, assembly, and refinement to generate accurate models	https://zhanggroup.org/I-TASSER/	[101]
11	Rosetta	It is used for protein structure modelling, designing as well as docking	https://rosettacommons.org/	[102]

(continued)

Table 11.3 (continued)

Sl.No.	Name	Description	URLs/GitHub	References
12	HHPred	It is a tool for homology modelling and structure prediction and uses hidden Markov models	https://toolkit.tuebingen.mpg.de/tools/hhpred	[103]
13	CABS-flex3.0	It is used for protein structure flexibility prediction and de novo structure prediction	https://lcbio.pl/cabsflex3/	[104]
14	GalaxyHonomer	It is a template-based web server used to predict protein homo-oligomer structures	http://galaxy.seoklab.org/homomer	[105]
15	GROMACS	It is open-source software used for molecular dynamics simulation	https://www.gromacs.org/	[93]
16	ProSMoS	It is an open-source tool used for protein secondary structure prediction	http://prodata.swmed.edu/ProSMoS/	[106]
17	QUARK	It is used to predict 3-D models of small proteins in the absence of homologous template	https://zhanggroup.org/QUARK/	[107]
18	CONFOLD2	It is a tool that predicts protein folding using the ab initio method	https://github.com/multicom-toolbox/CONFOLD2	[108]
19	Psipred	It is utilized to predict protein secondary structure	https://github.com/psipred/psipred	[109]
20	AutoDock Vina	It is a molecular docking software used to predict the binding of small compounds with the receptor	https://vina.scripps.edu/	[92]

drug research and development, its reutilization, clinical trials, and the overall broadening of pharmaceutical outcomes, have been simplified by the incorporation of AI and its tools [132, 133]. Thus, the implementation of AI has served as a quicker, less expensive, and more efficient alternative to traditional procedures [134]. The process of improving workflow, which ultimately reduces costs and boosts precision, is facilitated by the incorporation of a predictive system trained using information, machine learning, and deep learning [133]. However, the potential of AI comes with certain limitations, such as a lack of interpretability, the need for comprehensive and extensive datasets, and ethical concerns. Hence, extensive research in this sector, followed by the optimisation of the technology, needs to be carried out to address these issues [135].

Table 11.4 AI and DL-integrated resources for various stages of modern drug discovery

Sl.No.	Name	Description	URLs/GitHub	References
1	Keras	Keras is an API that helps in simplifying the process of building deep neural networks	https://github.com/fchollet/keras	[119]
2	DeepChem	It is a Python-based library used for multiple applications in drug development and quantum physics	https://github.com/deepchem/deepchem	[117]
3	TensorFlow	TensorFlow is a Python library used for developing predictive models for drug design	https://www.tensorflow.org/	[113]
4	TLlib	TLlib an open-source library for transfer learning	https://github.com/thuml/Transfer-Learning-Library	[120]
5	ANI-1	ANI-1 is a dataset made for training and evaluating ML potentials for molecules, and neural network-based models especially	https://github.com/isayev/ANI1_dataset	[121]
6	TorchDrug	It is an ML platform used in multiple drug development processes through deep generative models	https://torchdrug.ai/	[122]
7	Chemistry42	A software platform for designing and optimizing drug-like small molecules with the right properties	https://pharma.ai/chemistry42	[123]
8	AlphaFold	Prediction of the underlying physical structure of the protein	https://alphafold.ebi.ac.uk/	[116]
9	MolAICal	A software tool combined with AI and programming to predict 3-D drug design of the target protein	https://molaical.github.io	[115]
10	HydraScreen	A DL tool that speeds up and improves the structure-based drug discovery	https://github.com/Ro5-ai/hydrascreen	[124]
11	AMPL	An open-source software pipeline that makes ML in drug discovery more collaborative	https://github.com/ATOMconsortium/AMPL	[125]
12	Chainer Chemistry	A DL framework that is applicable to biology and chemistry	https://github.com/chainer/chainer-chemistry	[114]

(continued)

Table 11.4 (continued)

Sl.No.	Name	Description	URLs/GitHub	References
13	Spaya-API	AI-driven retrosynthesis platform for converting target compounds into commercially available materials	https://spaya.ai	[83]
14	RSPred	A hidden Markov model using tool that is used to identify malaria surface antigens	https://mybiosoftware.com/rspred-rifinstevor-prediction-tool.html#google_vignette	[126]
15	DeepTOX	A tool for predicting chemical toxicity	www.bioinf.jku.at/research/DeepTox	[127]
16	RaptorX	A DL based tool utilized to predict structural and behavioural characteristics of proteins	http://raptorx6.uchicago.edu/	[128]
17	LOMETS	A meta server that predicts protein structures using a threading or fold recognition method	https://zhanggroup.org/LOMETS/	[129]
18	CHARMM	It is a GUI used for molecular modelling for the molecular simulation of macromolecules	https://www.charmm-gui.org/	[130]
19	Chemprop	It is a DL-based package using the D-MPNN architecture for the prediction of the chemical property of a structure	https://chemprop.readthedocs.io/en/latest/index.html	[131]
20	Molecular Sets (MOSES)	It is a platform used for training and comparison of predictive models in entirely novel molecular modelling	https://github.com/molecularsets/moses	[118]

11.6.1 *Insight into Drug Design*

The major problem faced by the department of drug formulation is the high cost and the slow productivity. However, the integration of AI has accelerated the procedure by utilising diverse approaches for drug discovery, formulation, and design frameworks, handling various aspects such as target identification, virtual screening, molecular engineering, drug repurposing, and many more, leading to more affordable, personalised treatments [3]. The system identifies the potential target associated with a certain disease as it evaluates the metabolic, genomic, and proteomic data of the target. For categorisation of the target, potential tools like TargetNet, DeepTarget, and HitPick are used. The subclass of AI, specifically the ML (machine learning) algorithm, also assists in calculating the binding affinity and predicting toxicity with the aid of various tools designed to simplify the procedure [136]. The Artificial

Neural Network (ANN) facilitates drug target interactions, whether with the molecular target or the cellular membrane, subsequently evaluating features at the target sites, such as solubility, speed, distribution, and execution [137, 138]. Furthermore, the implementation of multi-omics data (including genomics, proteomics, transcriptomics, etc.) into models created via machine learning and deep learning methods successfully predicts the relationship between genes and disease. The discovery of biomarkers via tools like DeepSEA, along with the classification of diseases, has established the path for personalised treatment. AI has streamlined such a framework [139, 140]. Regarding the risks associated with personalised therapy, the decision support system can help identify potential risks, such as allergies, by utilising patient-level data, thereby customising the treatment safely. Some of the AI-powered tools, like GENTRL and REINVENT, assist in de novo drug designing purposes, while AlphaFold and DeepTox can be utilised in protein structure and toxicity prediction in addition to that, they facilitate drug target interaction via DL models like Convolution Neural Networks (CNN) and MultiLayer Perceptron (MLPs) [141–143]. Approaches like SHAP and LIME, which utilise Explainable AI (XAI), aid in toxicity prediction, target ranking, and many other complex decisions [144]. Furthermore, fresh generative models, including flow-based, transformer-based, and multi-agent reinforcement learning, are shifting the entire paradigm of drug design. Notably, ChemBERTa and MolBERT employ SMILES strings to predict molecular properties. When combined, it holds the potential to enhance factors such as bioavailability, toxicology, and synthetic accessibility of a molecule. The framework facilitates easy incorporation of other procedures, making it into a more innovative and efficient system [145].

11.6.2 Insight into Clinical Trials

The clinical trials are a significant and more expensive procedure, and the incorporation of AI in this component may ensure a lower cost and a lower failure rate, making them more accurate and efficient. Prior to the sample collection for the clinical trial, the AI system facilitates the evaluation of the patient and the previous trial information by doing so, the system enhances the selection standard along with the design of the protocol that would quicken the procedure, additionally uncovering the structure and links by the assessment of the trial information [146, 147]. The research protocol for selecting the heterogeneous reduced sample, along with the dosage and specific outcome, is improved by AI using Bayesian nonparametric models. Based on the design provided by the Bayesian logistic regression model (BLRM), the optimisation of the dosage is considered while taking into account data diversity from several populations [148, 149]. It can further help streamline enrollment, reduce expenses, and improve the quality of data in half the time, using tools like NLP/OCR and LLM monitors to enable continuous health assessment of the patient, ensuring compliance and continuous testing of the biomarker [150–152]. In order to optimise the pre-clinical and clinical trials the ML algorithms were developed to predict factors like toxicity, bioavailability, and the drug candidate's binding affinity, also designing ideal dosage for administration which helped in making the process a lot more efficient with less failure rates with a significantly lower mortality rates seen in the second and third phase of the trial [76]. Further, the Natural Language Processing (NLP) and ML have helped facilitate the FDA approval with post-market analysis by identification of the side effects of drugs [153]. In order to improve regulatory compliance and accurate reporting, the concern for accessibility, transparency, and data tracking while using AI procedures can be complemented with TRIPOD AI, CONSORT AI, and DECIDE AI. Regarding the quality assurance of data, bias recognition, privacy, and security controls, the FDA or ISO protocols require a sophisticated regulatory framework for the approval of AI-based technology [154].

11.6.3 Insight into Pharmaceutical Manufacturing

Through the use of AI-powered technologies guided by data, the pharmaceutical sector has evolved into a standard, reliable, and cost-efficient development process. The pharmaceutical manufacturing process encompasses the entire spectrum, from drug discovery to approval and delivery to end-users. At the same time, the AI has not only streamlined the drug designing and its clinical trials but has also extended to its manufacturing procedure [155]. The primary purpose of its involvement in the sector is to reduce expenses and increase productivity, which it achieves by enabling certain features. Through the use of predictive maintenance, equipment failure can be detected and defects in products and packaging identified via computer vision-based quality control. To boost batch yield while minimising waste, the AI analyses the parameters using a digital twin [156]. The AI further aids in the optimisation of functions such as vendor evaluation, asset control, and customer demand forecasting, thereby easing distribution difficulties [135]. An analytical E-VAI platform, generated via Eularis, utilises a machine learning (ML) algorithm to estimate significant factors, such as major partners and the current market position, in order to facilitate a statistical framework that maximises revenue and facilitates switching investment areas for declining sales [157]. By ensuring the continuous evaluation of critical process parameters (CPPs) according to the FDA's Process Analytical Technology pipeline, product quality is ensured [158]. AI and ML models are trained to detect adverse drug reactions (ADRs) from data collected daily to ensure drug safety evaluation. Such procedures can be analysed via NLP and deep learning algorithms [159, 160]. To achieve a human-AI collaborative scheme, human operators were integrated into the workflow that optimises functionality, operator reliability, and instructional method development and manufacturing of medicinal products. A healthcare-specific technology, such as Industry 4.0 or Pharma 4.0, improves the manufacturing process, making it more efficient and achieving more accurate outcomes [161]. The distribution pipeline estimation of demand has also been made simple through the incorporation of AI. The system analyses the prescription history, client conduct, and the changing seasons to successfully predict future demands and prevent the probability of overproduction or stockouts [137]. The integration of blockchain technologies with AI has also shown potential in detecting counterfeits and thereby enhancing the authenticity of products. The technology requires explicit regulations, accessibility, technical surveillance, and an extensive record, which further enhances security and reliability, making it a secure and productive space for healthcare data sharing [162].

11.6.4 Insight into Ethical, Regulatory, and Future Perspectives

It is essential to consider the moral risk of involving AI in the healthcare sector. Several factors must be considered from an ethical perspective, including biased data. When training an AI model, the training data must be free of bias, as it can lead to incorrect treatment suggestions [163]. The privacy of the health data being used, the consent from the patients, and their confidentiality must be taken into account. According to the GDPR and HIPAA laws, the anonymity and privacy of the data must be validated before it is used as training data for AI models. Federated data that facilitates the use of anonymous datasets from independent datasets for the AI training model can be the solution for safeguarding privacy [164, 165]. For the future, the utilisation of quantum technology for molecular modelling, combined with AI-supported artificial biology infrastructure, may facilitate the on-demand design of drugs, followed by their production [166]. Despite the limitations, establishing faith through strong confidentiality protection and compliance is necessary, as it aids in discovering the true potential of AI in the pharmaceutical industry, which is efficient, accurate, and secure (Fig. 11.2).

11.7 Open-Source Frameworks, Tools, and Databases: Current Trends

In the current digital landscape, pipelines, databases, and tools integrated with AI have become integral to research and development. The resource is openly accessible with a user-friendly interface, placing a significant emphasis on procedures such as data collection and evaluation, outcome prediction, process automation, and more [167]. The adaptation of these resources has led to significant improvements in cost-effectiveness, productivity, development, and efficiency in the area [168]. The contributions made by these communities yielded significant results with strong potential for further optimisation and provided a platform for further innovation.

The AI-based tool has various applications, as it serves as a diagnostic, probability-based decision-making tool, and many more [169]. The tools require a standard training dataset and a particular pipeline to function efficiently, as the models are based on machine learning and deep learning algorithms. While most current tools

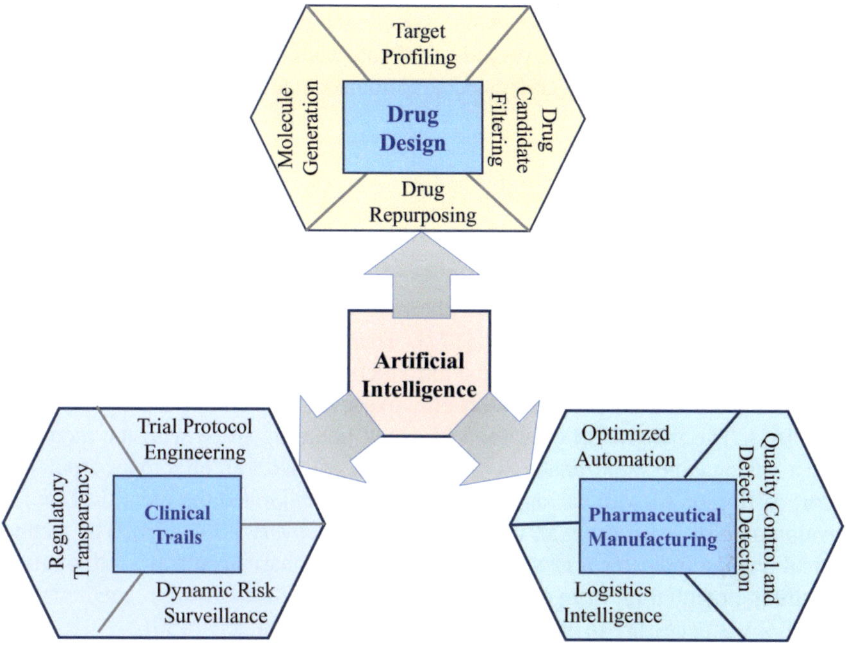

Fig. 11.2 Illustration of AI-assisted insight in the pharmaceutical and its related sector

have been optimised by incorporating cloud-based storage and containerization, their integrated approach demonstrates interoperability, but a lack of decision-making may be their major shortcoming [170]. The incorporation of various tools in several areas of research highlights the variability and its potential. The continuous validation and refinement of tools' functionality, followed by their integration with more optimised resources, can lead to the possible extension of their practicality [171]. The AI framework offers a structured and comprehensive application that includes pre-structured components, a data library, and an architecture that enables models to function efficiently. The objective functionality of the framework appears to be influenced by factors such as local impact, social and moral standards, and the need for computational and architectural structure within the system [172]. The current framework only supports the computational base and operational setting and hence lacks the function of interpretability and accountability [170, 172]. In some sectors, such as clinical ones, the interpretation of the regulatory framework may require continuous validation for the sake of safety [173]. An open-source database with AI integration is transforming the standard information platform by enhancing search, automating management tasks such as lineage monitoring and targeting, improving efficiency, and facilitating abnormality detection through natural language queries and a user-friendly interface [174]. Focus to learn more. However, concerns still arise regarding the bias, interpretability, and transparency of AI algorithms. The incorporation of AI into the database has improved its efficiency in handling extensive and comprehensive databases, which was a significant problem in the past. Additionally, further optimisation is displayed in the integration of reinforcement learning, quantum computing, and AI-navigated abnormality identification, which allows for future growth [175, 176].

11.8 Advantages and Limitations

AI has revolutionised the field of the pharmaceutical industry, serving as a bridge between technology and medicine [177]. Traditional workflows often relied on experimental procedures; however, the introduction of AI has streamlined numerous processes, focusing on precision and effectiveness, and reduced the workload by half [164]. The integration of AI and Machine Learning (ML) with the medicinal aspect has enabled rapid and cost-effective analysis with an effective output. The incorporation of AI with various potent tools has accelerated the identification and development of novel drugs in the field of drug discovery [134, 178]. The technological change indicates a new era of innovation in pharmaceuticals. Although AI has strong potential for drug discovery, understanding its application, methods, and limitations is necessary to avoid any eventual propaganda [179, 180].

A huge margin reduces the timeline for identifying novel drug candidates, as AI can efficiently evaluate massive datasets consisting of years of initial discovery work [181]. The models created through ML algorithms can improve the pre-clinical and clinical success rates by predicting significant variables like toxicity, bioavailability, and binding affinity of the drug candidate [76]. The implementation of AI reduces the cost of experimentation by streamlining the processes involved, making the framework both cost-effective and efficient [164]. The AI system, introduced as AlphaFold for protein structure modelling, has drastically improved accuracy, as well as design and target interaction analysis [134]. To evaluate protein-drug interactions through virtual screening and estimate the bioactivity of compounds, as well as their development in an efficient manner, the implementation of AI was a major factor [153]. The critical step of designing the drug design pipeline can be conveniently executed via machine learning methods [179]. Machine learning algorithms can assist in drug repurposing by identifying new potential targets, thus discovering novel therapeutic uses for approved medications [182]. Soon, identification and assessment of rare diseases can be executed via the utilisation of AI methods with multi-omics data, predicting disease-gene associations and identifying biological targets [76]. With the study of integrated and vast omics data, biomarker discovery and categorisation of patients, leading to disease classification and personalised therapy, can be simplified through AI [139]. The implementation of ML and NLP facilitates the automation of reports for FDA approval or patent applications, as well as post-market analysis, by detecting side effects and analysing their sentiment [153]. Deep Learning models, such as CNNs and multilayer perceptrons, can predict drug target interactions using expression data. Additionally, WideDTA and DeepDTA can aid in the precise prediction of drug target affinity for binding [134, 183]. An AI-generated molecule used for drug development can be authentically generated via machine learning tools like RScore and RSPred. The incorporation of a cloud-based platform facilitates data exchange, multinational collaboration, and the adaptation of AI tools, directly promoting the drug design framework in both research and the industrial sector [83].

Although the AI-driven model may possess high performance ability that provides accurate outcomes, it only works when provided with extensive and diverse datasets. However, in drug discovery, data are often rare, restricted, or even biased [136]. The majority of models, notably DNN, function as "black boxes" with a lack of transparency on how they operate, thus making their interpretation difficult [184]. An appropriate specification must be introduced for the use of AI in clinical research, considering the ethical issues of individual information, including anonymity, informed consent, and equity, to ensure the algorithm's accuracy and personal safety [76, 164]. An excessive reliance on algorithmic predictions may result in misguided pursuits. Despite Computer-Aided Drug Discovery (CADD) being a resilient tool, a balance between experimental findings and technological estimations is necessary [16]. Although AI-generated molecules provide a breakthrough, the feasibility of the product remains a question. Not all synthetically generated molecules are functional, thus making this method impractical in reality [185]. The involvement of AI in existing research sectors requires technical modifications, data consistency, and a user-friendly interface for ordinary people, as it faces cultural barriers and a lack of knowledge about AI [139]. There is a lack of standardised techniques for evaluating and reviewing AI models in the context of drug research, which ultimately reduces the consistency and comparability of the outcomes produced by these models [76]. Smaller laboratories are unable to utilise the high-performance models, such as those requiring deep learning technologies, as they require significant computational power and energy, along with a high associated cost [16, 76]. The gap between the skill and the training implemented in the AI system may present a drawback in the implementation of AI for the identification and optimisation of drugs. Hence, it requires the collaboration of the experts in the respective field to use and interpret the model [164]. The AI is required to comprehend the complex biological system of the human system to predict the effects of a drug effectively. Hence, a detailed understanding of the interaction between these biological networks for the AI model is necessary [76]. Many data-related limitations can be observed in the drug development field, including issues with data format, visualisation, label inconsistencies, confidentiality, sample sizes, model training methods, and model interpretation [153]. Although the AI holds an important and transformative role in modern drug development, it relies explicitly on collaborative efforts that integrate theoretical, computational, operational, and ethical aspects. Thus, AI can become a fundamental and reliable component of next-generation medical technology by resolving these issues.

11.9 Conclusion

The implementation of open-source tools and platforms in drug development has transformed the pharmaceutical industry by providing accessible, faster, and affordable solutions. Open-source technologies encourage innovation through collaboration, transparency, and the reduction of duplication, along with the integration of modern technologies that enhance accessibility to complex software and large-scale storage, eliminating the need for expensive, super-powered hardware. With a focus on open-source treatment, options for neglected and rare diseases are expanding, aiming for equal healthcare progress. The wide range of open-source tools and databases enables comprehensive analysis and drug design, also offering data-driven decisions through the integration of proteomics and genomics to predict the toxicity and efficacy of potential drugs at an early stage, thereby avoiding late-stage failure. Since open-source technologies are free and accessible to all, they bridge the gap between academic and commercial researchers, providing an opportunity to foster multidisciplinary drug development research. However, certain challenges, such as data quality issues, expertise shortages, interpretability problems, and concerns related to privacy and ethics, hinder high-performance ability. However, with the integration of multi-omics data and XAI tools, the future prioritises unbiased data quality and provides transparency in prediction, thereby shaping the future of personalised medicine.

Acknowledgements All authors would like to acknowledge their respective affiliations for the facility support to complete the work. The authors would also like to acknowledge the NIAID NIH BIOART Source (https://bioart.niaid.nih.gov/) from which some figures in this manuscript were designed.

Author Contribution **SKM**: Conceptualisation, Data curation, Original draft writing. **PM, KG, TC**: Original draft writing and image design. **AR** and **PR**: Data collection, image design. **JJG**: Formal Analysis, Supervision, Writing review and editing. All authors approved the final version of this manuscript.

Funding None.

Data Availability Statement Not applicable.

Declaration of Competing Interests The authors declare that they have no known competing financial interests or personal relationships that could have appeared to influence the work reported in this study.

References

1. Khan SR, Al Rijjal D, Piro A, Wheeler MB. Integration of AI and traditional medicine in drug discovery. Drug Discov Today. 2021;26(4):982–92.
2. Arabi AA. Artificial intelligence in drug design: algorithms, applications, challenges and ethics. Future Drug Disc. 2021;3(2):FDD59.

3. Blanco-Gonzalez A, Cabezon A, Seco-Gonzalez A, Conde-Torres D, Antelo-Riveiro P, Pineiro A, et al. The role of AI in drug discovery: challenges, opportunities, and strategies. Pharmaceuticals. 2023;16(6):891.
4. Doytchinova I. Drug design—past, present, future. Molecules. 2022;27(5):1496.
5. Urbina F, Lentzos F, Invernizzi C, Ekins S. Dual use of artificial-intelligence-powered drug discovery. Nat Mach Intell. 2022;4(3):189–91.
6. Patel JR, Joshi HV, Shah UA, Patel JK. A review on computational software tools for drug design and discovery. Indo Global J Pharm Sci. 2022;3(4):5. https://doi.org/10.35652/IGJPS.2022.12006.
7. Georrge JJ, Mishra SK, Chhetri T, Roy S, Gurung K. Computational tools and software in drug discovery. In: Molecular modeling and docking techniques for drug discovery and design. IGI Global Scientific Publishing; 2025. p. 183–216.
8. Mishra SK, Kumar N, Tóth Z, Jardan YAB, Akash S, Georrge JJ. Unveiling the molecular activity of HIV towards the CD4: a study based on subtype C via docking and dynamics approach. J Genetic Eng Biotechnol. 2025;23(1):100457.
9. Mishra SK, Jeba Praba J, Georrge JJ. An emerging trends of bioinformatics and big data analytics in healthcare. Germany: Walter de Gruyter GmbH & Co KG. Berlin; 2024.
10. Mishra SK, Roy S, Chhetri T, Georrge JJ. Computer-assisted methods and tools for structure- and ligand-based drug design. Comput Methods Rat Drug Des. 2025;69–95.
11. Årdal C, Røttingen J-A. Open source drug discovery in practice: a case study. PLoS Negl Trop Dis. 2012;e1827. https://doi.org/10.1371/journal.pntd.0001827.
12. Carvalho T, Antunes L, Costa Santos C, Moniz N. Empowering open data sharing for social good: a privacy-aware approach. Sci Data. 2025;12(1):248.
13. Ward P, Dahlberg P, Lagatie O, Larsson J, Tynong A, Vlaminck J, et al. Affordable artificial intelligence-based digital pathology for neglected tropical diseases: a proof-of-concept for the detection of soil-transmitted helminths and Schistosoma mansoni eggs in Kato-Katz stool thick smears. PLoS Negl Trop Dis. 2022;16(6):e0010500.
14. Schaduangrat N, Lampa S, Simeon S, Gleeson MP, Spjuth O, Nantasenamat C. Towards reproducible computational drug discovery. J Cheminformat. 2020;12(1):9.
15. Spjuth O, Frid J, Hellander A. The machine learning life cycle and the cloud: implications for drug discovery. Expert Opin Drug Discov. 2021;16(9):1071–9.
16. Niazi SK, Mariam Z. Computer-aided drug design and drug discovery: a prospective analysis. Pharmaceuticals. 2023;17(1):22.
17. Watson RT. Data management, databases and organizations. Wile; 2008.
18. Trifan A, Oliveira JL. Patient data discovery platforms as enablers of biomedical and translational research: a systematic review. J Biomed Inform. 2019;93:103154. https://doi.org/10.1016/j.jbi.2019.103154.
19. Wang RC, Wang Z. Precision medicine: disease subtyping and tailored treatment. Cancers (Basel). 2023;15(15). https://doi.org/10.3390/cancers15153837.
20. Safran M, Rosen N, Twik M, BarShir R, Stein TI, Dahary D, et al. The genecards suite. In: Practical guide to life science databases. Springer; 2022. p. 27–56.
21. Clough E, Barrett T, Wilhite SE, Ledoux P, Evangelista C, Kim IF, et al. NCBI GEO: archive for gene expression and epigenomics data sets: 23-year update. Nucleic Acids Res. 2024;52(D1):D138–44.
22. Amberger JS, Hamosh A. Searching online Mendelian inheritance in man (OMIM): a knowledgebase of human genes and genetic phenotypes. Curr Protoc Bioinformat. 2017;58(1):1.2.1–2.12.
23. Wu Y, Zhang F, Yang K, Fang S, Bu D, Li H, et al. SymMap: an integrative database of traditional Chinese medicine enhanced by symptom mapping. Nucleic Acids Res. 2019;47(D1):D1110–7.
24. Ru J, Li P, Wang J, Zhou W, Li B, Huang C, et al. TCMSP: a database of systems pharmacology for drug discovery from herbal medicines. J Cheminformat. 2014;6(1):13.
25. Wishart DS, Knox C, Guo AC, Cheng D, Shrivastava S, Tzur D, et al. DrugBank: a knowledgebase for drugs, drug actions and drug targets. Nucleic acids research. 2008;36(suppl_1):D901-D6.

26. UniProt: the universal protein knowledgebase in 2025. Nucleic Acids Res. 2025;53(D1):D609–D17.
27. Burley SK, Bhatt R, Bhikadiya C, Bi C, Biester A, Biswas P, et al. Updated resources for exploring experimentally-determined PDB structures and computed structure models at the RCSB protein data bank. Nucleic Acids Res. 2025;53(D1):D564–74.
28. Zhang Y, Li X, Shi Y, Chen T, Xu Z, Wang P, et al. ETCM v2. 0: an update with comprehensive resource and rich annotations for traditional Chinese medicine. Acta Pharmaceutica Sinica B. 2023;13(6):2559–71.
29. Liu Z, Guo F, Wang Y, Li C, Zhang X, Li H, et al. BATMAN-TCM: a bioinformatics analysis tool for molecular mechanism of traditional Chinese medicine. Sci Rep. 2016;6(1):21146.
30. Huang L, Xie D, Yu Y, Liu H, Shi Y, Shi T, et al. TCMID 2.0: a comprehensive resource for TCM. Nucleic acids research. 2018;46(D1):D1117–D20.
31. Fang S, Dong L, Liu L, Guo J, Zhao L, Zhang J, et al. HERB: a high-throughput experiment-and reference-guided database of traditional Chinese medicine. Nucleic Acids Res. 2021;49(D1):D1197–206.
32. Chen CY-C. TCM Database@ Taiwan: the world's largest traditional Chinese medicine database for drug screening in silico. PloS one. 2011;6(1):e15939.
33. Xue R, Fang Z, Zhang M, Yi Z, Wen C, Shi T. TCMID: traditional Chinese medicine integrative database for herb molecular mechanism analysis. Nucleic Acids Res. 2012;41(D1):D1089–95.
34. Yan D, Zheng G, Wang C, Chen Z, Mao T, Gao J, et al. HIT 2.0: an enhanced platform for herbal ingredients' targets. Nucleic Acids Res. 2022;50(D1):D1238–D43.
35. Perez-Riverol Y, Bandla C, Kundu DJ, Kamatchinathan S, Bai J, Hewapathirana S, et al. The PRIDE database at 20 years: 2025 update. Nucleic Acids Res. 2025;53(D1):D543–53.
36. Blum M, Andreeva A, Florentino LC, Chuguransky SR, Grego T, Hobbs E, et al. InterPro: the protein sequence classification resource in 2025. Nucleic Acids Res. 2025;53(D1):D444–56.
37. Yin J, Chen Z, You N, Li F, Zhang H, Xue J, et al. VARIDT 3.0: the phenotypic and regulatory variability of drug transporter. Nucleic Acids Res. 2024;52(D1):D1490–D502.
38. Zerbino DR, Achuthan P, Akanni W, Amode MR, Barrell D, Bhai J, et al. Ensembl 2018. Nucleic Acids Res. 2018;46(D1):D754–61.
39. Perez G, Barber GP, Benet-Pages A, Casper J, Clawson H, Diekhans M, et al. The UCSC genome browser database: 2025 update. Nucleic Acids Res. 2025;53(D1):D1243–9.
40. Zdrazil B. Fifteen years of ChEMBL and its role in cheminformatics and drug discovery. J Cheminformat. 2025;17(1):1–9.
41. Kim S, Chen J, Cheng T, Gindulyte A, He J, He S, et al. PubChem 2025 update. Nucleic Acids Res. 2025;53(D1):D1516–25.
42. Li F, Yin J, Lu M, Mou M, Li Z, Zeng Z, et al. DrugMAP: molecular atlas and pharma-information of all drugs. Nucleic Acids Res. 2023;51(D1):D1288–99.
43. Sayers EW, Cavanaugh M, Frisse L, Pruitt KD, Schneider VA, Underwood BA, et al. GenBank 2025 update. Nucleic Acids Res. 2025;53(D1):D56–61.
44. Goldfarb T, Kodali VK, Pujar S, Brover V, Robbertse B, Farrell CM, et al. NCBI RefSeq: reference sequence standards through 25 years of curation and annotation. Nucleic Acids Res. 2025;53(D1):D243–57.
45. Moreno P, Fexova S, George N, Manning JR, Miao Z, Mohammed S, et al. Expression Atlas update: gene and protein expression in multiple species. Nucleic Acids Res. 2022;50(D1):D129–40.
46. Zhou Y, Zhang Y, Zhao D, Yu X, Shen X, Zhou Y, et al. TTD: therapeutic target database describing target druggability information. Nucleic Acids Res. 2024;52(D1):D1465–77.
47. Pence HE, Williams A. ChemSpider: an online chemical information resource. ACS Publications; 2010.
48. Kelleher KJ, Sheils TK, Mathias SL, Yang JJ, Metzger VT, Siramshetty VB, et al. Pharos 2023: an integrated resource for the understudied human proteome. Nucleic Acids Res. 2023;51(D1):D1405–16.
49. Weinstein JN, Collisson EA, Mills GB, Shaw KR, Ozenberger BA, Ellrott K, et al. The cancer genome atlas pan-cancer analysis project. Nat Genet. 2013;45(10):1113–20.

50. Hu Y, Guo X, Yun Y, Lu L, Huang X, Jia S. DisGeNet: a disease-centric interaction database among diseases and various associated genes. Database. 2025;2025:baae122.
51. Landrum MJ, Chitipiralla S, Kaur K, Brown G, Chen C, Hart J, et al. ClinVar: updates to support classifications of both germline and somatic variants. Nucleic Acids Res. 2025;53(D1):D1313–21.
52. Goel R, Harsha H, Pandey A, Prasad TK. Human protein reference database and human proteinpedia as resources for phosphoproteome analysis. Mol BioSyst. 2012;8(2):453–63.
53. BioCarta ND. Biotech software & internet report. Comput Softw J Scient. 2001;2(3):117–20.
54. Szklarczyk D, Santos A, Von Mering C, Jensen LJ, Bork P, Kuhn M. STITCH 5: augmenting protein–chemical interaction networks with tissue and affinity data. Nucleic Acids Res. 2016;44(D1):D380–4.
55. Gong L, Whirl-Carrillo M, Klein TE. PharmGKB, an integrated resource of pharmacogenomic knowledge. Curr Protoc. 2021;1(8):e226.
56. Kanehisa M, Furumichi M, Tanabe M, Sato Y, Morishima K. KEGG: new perspectives on genomes, pathways, diseases and drugs. Nucleic Acids Res. 2017;45(D1):D353–61.
57. Liu S, Ma W, Moore R, Ganesan V, Nelson S. RxNorm: prescription for electronic drug information exchange. IT Prof. 2005;7(5):17–23.
58. Liu T, Hwang L, Burley SK, Nitsche CI, Southan C, Walters WP, et al. BindingDB in 2024: a FAIR knowledgebase of protein-small molecule binding data. Nucleic Acids Res. 2025;53(D1):D1633–44.
59. Irwin JJ, Shoichet BK. ZINC—a free database of commercially available compounds for virtual screening. J Chem Inf Model. 2005;45(1):177–82.
60. Nickel J, Gohlke B-O, Erehman J, Banerjee P, Rong WW, Goede A, et al. SuperPred: update on drug classification and target prediction. Nucleic Acids Res. 2014;42(W1):W26–31.
61. Gao Y, Kim S, Lee Y-I, Lee J. Cellular stress-modulating drugs can potentially be identified by in silico screening with connectivity map (CMap). Int J Mol Sci. 2019;20(22):5601.
62. Gillespie M, Jassal B, Stephan R, Milacic M, Rothfels K, Senff-Ribeiro A, et al. The reactome pathway knowledgebase 2022. Nucleic Acids Res. 2022;50(D1):D687–92.
63. Wang Y, Sun K, Li J, Guan X, Zhang O, Bagni D, et al. PDBBind optimization to create a high-quality protein-ligand binding dataset for binding affinity prediction. arXiv e-prints. 2024:arXiv: 2411.01223.
64. Del Toro N, Shrivastava A, Ragueneau E, Meldal B, Combe C, Barrera E, et al. The IntAct database: efficient access to fine-grained molecular interaction data. Nucleic Acids Res. 2022;50(D1):D648–53.
65. Oughtred R, Stark C, Breitkreutz B-J, Rust J, Boucher L, Chang C, et al. The BioGRID interaction database: 2019 update. Nucleic Acids Res. 2019;47(D1):D529–41.
66. Taboureau O, Nielsen SK, Audouze K, Weinhold N, Edsgärd D, Roque FS, et al. ChemProt: a disease chemical biology database. Nucleic Acids Res. 2010;39(suppl_1):D367–D72.
67. Koonin EV. The Clusters of Orthologous Groups (COGs) Database: phylogenetic classification of proteins from complete genomes. The NCBI handbook; 2002.
68. Lo W-C, Lee C-C, Lee C-Y, Lyu P-C. CPDB: a database of circular permutation in proteins. Nucleic Acids Res. 2009;37(suppl_1):D328–D32.
69. Huang DW, Sherman BT, Tan Q, Kir J, Liu D, Bryant D, et al. DAVID Bioinformatics Resources: expanded annotation database and novel algorithms to better extract biology from large gene lists. Nucleic Acids Res. 2007;35(suppl_2):W169–W75.
70. Xenarios I, Rice DW, Salwinski L, Baron MK, Marcotte EM, Eisenberg D. DIP: the database of interacting proteins. Nucleic Acids Res. 2000;28(1):289–91.
71. Masciocchi J, Fanton M, Floris M, Frau G, Sturlese M, Palla P, et al. MMsINC®: a new public large-scale chemoinformatics database system.
72. Kotlyar M, Pastrello C, Sheahan N, Jurisica I. Integrated interactions database: tissue-specific view of the human and model organism interactomes. Nucleic Acids Res. 2016;44(D1):D536–41.
73. Korcsmáros T, Farkas IJ, Szalay MS, Rovó P, Fazekas D, Spiró Z, et al. Uniformly curated signaling pathways reveal tissue-specific cross-talks and support drug target discovery. Bioinformatics. 2010;26(16):2042–50.

74. Wishart D, Arndt D, Pon A, Sajed T, Guo AC, Djoumbou Y, et al. T3DB: the toxic exposome database. Nucleic Acids Res. 2015;43(D1):D928–34.
75. Hill C, Du L, Johnson M, McCullough B. Comparing programming languages for data analytics: accuracy of estimation in Python and R. Wiley Interdisc Rev Data Mining Knowl Discov. 2024;14(3):e1531.
76. Tiwari PC, Pal R, Chaudhary MJ, Nath R. Artificial intelligence revolutionizing drug development: exploring opportunities and challenges. Drug Dev Res. 2023;84(8):1652–63.
77. Pedregosa F, Varoquaux G, Gramfort A, Michel V, Thirion B, Grisel O, et al. Scikit-learn: Machine learning in Python. J Mach Learn Res. 2011;12:2825–30.
78. Cao Y, Charisi A, Cheng L-C, Jiang T, Girke T. ChemmineR: a compound mining framework for R. Bioinformatics. 2008;24(15):1733–4.
79. Sicho M, Luukkonen S, van den Maagdenberg HW, Schoenmaker L, Béquignon OJ, van Westen GJ. DrugEx: deep learning models and tools for exploration of drug-like chemical space. J Chem Inf Model. 2023;63(12):3629–36.
80. Grant BJ, Rodrigues AP, ElSawy KM, McCammon JA, Caves LS. Bio3d: an R package for the comparative analysis of protein structures. Bioinformatics. 2006;22(21):2695–6.
81. Imambi S, Prakash KB, Kanagachidambaresan G. PyTorch. Programming with TensorFlow: solution for edge computing applications. Springer; 2021. p. 87–104.
82. Ke G, Meng Q, Finley T, Wang T, Chen W, Ma W, et al. Lightgbm: A highly efficient gradient boosting decision tree. Adv Neural Inf Process Syst. 2017;30.
83. Dong J, Zhou P, Wu Y, Chen Y, Xie H, Gao Y, et al. Integrating single-cell datasets with ambiguous batch information by incorporating molecular network features. Briefings in Bioinformatics. 2022;23(1):bbab366.
84. Brown N, Fiscato M, Segler MH, Vaucher AC. GuacaMol: benchmarking models for de novo molecular design. J Chem Inf Model. 2019;59(3):1096–108.
85. Haghighatlari M, Vishwakarma G, Altarawy D, Subramanian R, Kota BU, Sonpal A, et al. ChemML: a machine learning and informatics program package for the analysis, mining, and modeling of chemical and materials data. Wiley Interdisc Rev Comput Molec Sci. 2020;10(4):e1458.
86. O'Boyle NM, Morley C, Hutchison GR. Pybel: a Python wrapper for the OpenBabel cheminformatics toolkit. Chem Cent J. 2008;2(1):5.
87. Scacchi W. Understanding the requirements for developing open source software systems. IEE Proc Softw. 2002;149(1):24–39.
88. Marion TJ, Fixson SK. The transformation of the innovation process: how digital tools are changing work, collaboration, and organizations in new product development. J Prod Innov Manag. 2021;38(1):192–215.
89. Nag S, Baidya AT, Mandal A, Mathew AT, Das B, Devi B, et al. Deep learning tools for advancing drug discovery and development. 3 Biotech. 2022;12(5):110.
90. Waterhouse A, Bertoni M, Bienert S, Studer G, Tauriello G, Gumienny R, et al. SWISS-MODEL: homology modelling of protein structures and complexes. Nucleic Acids Res. 2018;46(W1):W296–303.
91. Webb B, Sali A. Protein structure modeling with MODELLER. Functional genomics: Methods and protocols. Springer; 2017. p. 39–54.
92. Trott O, Olson AJ. AutoDock Vina: improving the speed and accuracy of docking with a new scoring function, efficient optimization, and multithreading. J Comput Chem. 2010;31(2):455–61.
93. Abraham MJ, Murtola T, Schulz R, Páll S, Smith JC, Hess B, et al. GROMACS: High performance molecular simulations through multi-level parallelism from laptops to supercomputers. SoftwareX. 2015;1:19–25.
94. O'Boyle NM, Banck M, James CA, Morley C, Vandermeersch T, Hutchison GR. Open Babel: an open chemical toolbox. J Cheminf. 2011;3(1):33.
95. Yap CW. PaDEL-descriptor: an open source software to calculate molecular descriptors and fingerprints. J Comput Chem. 2011;32(7):1466–74.

96. Genheden S, Thakkar A, Chadimová V, Reymond J-L, Engkvist O, Bjerrum E. AiZynthFinder: a fast, robust and flexible open-source software for retrosynthetic planning. J Cheminform. 2020;12(1):70.
97. Wójcikowski M, Zielenkiewicz P, Siedlecki P. Open drug discovery toolkit (ODDT): a new open-source player in the drug discovery field. J Cheminform. 2015;7(1):26.
98. Willighagen EL, Mayfield JW, Alvarsson J, Berg A, Carlsson L, Jeliazkova N, et al. The Chemistry Development Kit (CDK) v2. 0: atom typing, depiction, molecular formulas, and substructure searching. J Cheminform. 2017;9(1):33.
99. Korshunova M, Ginsburg B, Tropsha A, Isayev O. OpenChem: a deep learning toolkit for computational chemistry and drug design. J Chem Inf Model. 2021;61(1):7–13.
100. Kelley LA, Mezulis S, Yates CM, Wass MN, Sternberg MJ. The Phyre2 web portal for protein modeling, prediction and analysis. Nat Protoc. 2015;10(6):845–58.
101. Yang J, Yan R, Roy A, Xu D, Poisson J, Zhang Y. The I-TASSER Suite: protein structure and function prediction. Nat Methods. 2015;12(1):7–8.
102. Leman JK, Weitzner BD, Lewis SM, Adolf-Bryfogle J, Alam N, Alford RF, et al. Macromolecular modeling and design in Rosetta: recent methods and frameworks. Nat Methods. 2020;17(7):665–80.
103. Söding J, Biegert A, Lupas AN. The HHpred interactive server for protein homology detection and structure prediction. Nucleic Acids Res. 2005;33(suppl_2):W244–W8.
104. Wróblewski K, Zalewski M, Kuriata A, Kmiecik S. CABS-flex 3.0: an online tool for simulating protein structural flexibility and peptide modeling. Nucleic Acids Res. 2025:gkaf412.
105. Baek M, Park T, Heo L, Park C, Seok C. GalaxyHomomer: a web server for protein homo-oligomer structure prediction from a monomer sequence or structure. Nucleic Acids Res. 2017;45(W1):W320–4.
106. Shi S, Chitturi B, Grishin NV. ProSMoS server: a pattern-based search using interaction matrix representation of protein structures. Nucleic Acids Res. 2009;37(suppl_2):W526–W31.
107. Xu D, Zhang Y. Ab initio protein structure assembly using continuous structure fragments and optimized knowledge-based force field. Proteins Struct Func Bioinform. 2012;80(7):1715–35.
108. Adhikari B, Cheng J. CONFOLD2: improved contact-driven ab initio protein structure modeling. BMC Bioinformatics. 2018;19(1):22.
109. Buchan DW, Jones DT. The PSIPRED protein analysis workbench: 20 years on. Nucleic Acids Res. 2019;47(W1):W402–7.
110. Gupta R, Srivastava D, Sahu M, Tiwari S, Ambasta RK, Kumar P. Artificial intelligence to deep learning: machine intelligence approach for drug discovery. Mol Diversity. 2021;25(3):1315–60.
111. Didas M, Mutajwaa SA. Roadmap to systematic understanding of artificial intelligence technology development and implementation: a narrative review of frameworks, platforms, scenarios of use and empirical guide. J ICT Syst. 2025;3(1):34–62.
112. Rane NL, Paramesha M, Choudhary SP, Rane J. Machine learning and deep learning for big data analytics: a review of methods and applications. Partners Univ Int Innov J. 2024;2(3):172–97.
113. Developers T. TensorFlow. Zenodo; 2022. https://doi.org/10.5281/zenodo.18569529
114. Nakago K. Release chainer chemistry: a library for deep learning in biology and chemistry; 2017.
115. Bai Q, Tan S, Xu T, Liu H, Huang J, Yao X. MolAICal: a soft tool for 3D drug design of protein targets by artificial intelligence and classical algorithm. Brief Bioinform. 2021;22(3).
116. Varadi M, Bertoni D, Magana P, Paramval U, Pidruchna I, Radhakrishnan M, et al. AlphaFold protein structure database in 2024: providing structure coverage for over 214 million protein sequences. Nucleic Acids Res. 2024;52(D1):D368–75.
117. Ramsundar B, Eastman P, Walters P, Pande V. Deep learning for the life sciences: applying deep learning to genomics, microscopy, drug discovery, and more. O'Reilly Media; 2019.
118. Polykovskiy D, Zhebrak A, Sanchez-Lengeling B, Golovanov S, Tatanov O, Belyaev S, et al. Molecular sets (MOSES): a benchmarking platform for molecular generation models. Front Pharmacol. 2020;11:565644.

119. Gulli A, Pal S. Deep learning with Keras. Packt Publishing Ltd; 2017.
120. Junguang Jiang B, Fu B. Transfer-learning-library; 2020.
121. Smith JS, Isayev O, Roitberg AE. ANI-1, A data set of 20 million calculated off-equilibrium conformations for organic molecules. Sci Data. 2017;4(1):1–8.
122. Zhu Z, Shi C, Zhang Z, Liu S, Xu M, Yuan X, et al. Torchdrug: A powerful and flexible machine learning platform for drug discovery. arXiv preprint arXiv:220208320. 2022.
123. Ivanenkov YA, Polykovskiy D, Bezrukov D, Zagribelnyy B, Aladinskiy V, Kamya P, et al. Chemistry42: an AI-driven platform for molecular design and optimization. J Chem Inf Model. 2023;63(3):695–701.
124. Prat A, Abdel Aty H, Bastas O, Kamuntavicius G, Paquet T, Norvaisas P, et al. HydraScreen: a generalizable structure-based deep learning approach to drug discovery. J Chem Inf Model. 2024;64(15):5817–31.
125. Minnich AJ, McLoughlin K, Tse M, Deng J, Weber A, Murad N, et al. AMPL: a data-driven modeling pipeline for drug discovery. J Chem Inf Model. 2020;60(4):1955–68.
126. Joannin N, Kallberg Y, Wahlgren M, Persson B. RSpred, a set of Hidden Markov Models to detect and classify the RIFIN and STEVOR proteins of Plasmodium falciparum. BMC Genomics. 2011;12(1):119.
127. Mayr A, Klambauer G, Unterthiner T, Hochreiter S. DeepTox: toxicity prediction using deep learning. Front Environ Sci. 2016;3:80.
128. Wang S, Li W, Liu S, Xu J. RaptorX-Property: a web server for protein structure property prediction. Nucleic Acids Res. 2016;44(W1):W430–5.
129. Wu S, Zhang Y. LOMETS: a local meta-threading-server for protein structure prediction. Nucleic Acids Res. 2007;35(10):3375–82.
130. Jo S, Cheng X, Lee J, Kim S, Park SJ, Patel DS, et al. CHARMM-GUI 10 years for biomolecular modeling and simulation. J Comput Chem. 2017;38(15):1114–24.
131. Heid E, Greenman KP, Chung Y, Li S-C, Graff DE, Vermeire FH, et al. Chemprop: a machine learning package for chemical property prediction. J Chem Inf Model. 2023;64(1):9–17.
132. Ali KA, Mohin S, Mondal P, Goswami S, Ghosh S, Choudhuri S. Influence of artificial intelligence in modern pharmaceutical formulation and drug development. Future J Pharm Sci. 2024;10(1):53.
133. Kalyane D, Sanap G, Paul D, Shenoy S, Anup N, Polaka S, et al. Artificial intelligence in the pharmaceutical sector: current scene and future prospect. In: The future of pharmaceutical product development and research. Elsevier; 2020. p. 73–107.
134. Agu PC, Obulose CN. Piquing artificial intelligence towards drug discovery: tools, techniques, and applications. Drug Dev Res. 2024;85(2):e22159.
135. Vora LK, Gholap AD, Jetha K, Thakur RRS, Solanki HK, Chavda VP. Artificial intelligence in pharmaceutical technology and drug delivery design. Pharmaceutics. 2023;15(7):1916.
136. Walters WP, Murcko MA. Prediction of 'drug-likeness.' Adv Drug Deliv Rev. 2002;54(3):255–71.
137. Shahiwala A. AI approaches for the development of drug delivery systems. In: A handbook of artificial intelligence in drug delivery. Elsevier; 2023. p. 83–96.
138. Sable P, Khanvilkar VV. Pharmaceutical applications of artificial intelligence. Int J Pharma Res Health Sci. 2018;6:2342–5.
139. Mansur MA. Artificial intelligence in drug discovery towards strategic applications challenges and implementation frameworks for accelerated pharmaceutical innovation. Eur J Artif Intell Mach Learn. 2025;4(3):16–23.
140. Zhou J, Troyanskaya OG. Predicting effects of noncoding variants with deep learning-based sequence model. Nat Methods. 2015;12(10):931–4.
141. Bharadwaj S, Deepika K, Kumar A, Jaiswal S, Miglani S, Singh D, et al. Exploring the artificial intelligence and its impact in pharmaceutical sciences: insights toward the horizons where technology meets tradition. Chem Biol Drug Des. 2024;104(4):e14639.
142. Mak K-K, Wong Y-H, Pichika MR. Artificial intelligence in drug discovery and development. Drug Discov Eval Safety Pharmacokinet Assays. 2024:1461–98.

143. Zhavoronkov A, Ivanenkov YA, Aliper A, Veselov MS, Aladinskiy VA, Aladinskaya AV, et al. Deep learning enables rapid identification of potent DDR1 kinase inhibitors. Nat Biotechnol. 2019;37(9):1038–40.
144. Rane N, Choudhary S, Rane J. Explainable artificial intelligence (XAI) in healthcare: interpretable models for clinical decision support. Available at SSRN 4637897. 2023.
145. Mswahili ME, Jeong Y-S. Transformer-based models for chemical SMILES representation: a comprehensive literature review. Heliyon. 2024;10(20).
146. Shaki F, Amirkhanloo M, Jahani D, Chahardori M. Artificial intelligence in pharmaceuticals: exploring applications and legal challenges. Pharm Biomed Res. 2024;10(1):1–10.
147. Tarle S, Kakad A, Shaikh M. Overview: Embracing tools of artificial intelligence in pharmaceuticals. Int J Sci Acad Res. 2023;4(6):5749–55.
148. Kolluri S, Lin J, Liu R, Zhang Y, Zhang W. Machine learning and artificial intelligence in pharmaceutical research and development: a review. AAPS J. 2022;24(1):19.
149. Liu R, Lin J, Li P. Design considerations for phase I/II dose finding clinical trials in Immuno-oncology and cell therapy. Contemp Clin Trials. 2020;96:106083.
150. Ganesan P. LLM-powered observability enhancing monitoring and diagnostics. J Artif Intell Mach Learn Data Sci. 2024;2(2):1329–36.
151. LaBoone PA, Marques O. Overview of the future impact of wearables and artificial intelligence in healthcare workflows and technology. Int J Inf Manag Data Insights. 2024;4(2):100294.
152. Lu X, Yang C, Liang L, Hu G, Zhong Z, Jiang Z. Artificial intelligence for optimizing recruitment and retention in clinical trials: a scoping review. J Am Med Inform Assoc. 2024;31(11):2749–59.
153. Qureshi R, Irfan M, Gondal TM, Khan S, Wu J, Hadi MU, et al. AI in drug discovery and its clinical relevance. Heliyon. 2023;9(7).
154. Shiferaw KB, Roloff M, Balaur I, Welter D, Waltemath D, Zeleke AA. Guidelines and standard frameworks for artificial intelligence in medicine: a systematic review. JAMIA open. 2025;8(1):ooae155.
155. Arden NS, Fisher AC, Tyner K, Yu LX, Lee SL, Kopcha M. Industry 4.0 for pharmaceutical manufacturing: preparing for the smart factories of the future. Int J Pharm. 2021;602:120554.
156. Figueiredo J, Coito T, Martins M, Firme B, Vieira S, Sousa J. Assessing the impact of automation in pharmaceutical quality control labs using a digital twin. J Manuf Syst. 2022:270–85. https://doi.org/10.1016/j.jmsy.2021.11.014.
157. Agrawal P, Kushwaha V, Khan NF. Artificial intelligence: past, present and future. World J Pharm Pharm Sci. 2023;12(6):865–82. https://doi.org/10.20959/wjpps20236-25026.
158. Latif WB, Yasin IM, Ali MJ, Islam MN, Forid MS. Transforming applied medical sciences: the impact of AI, VR, and AR on research, education technology, and clinical practices. Integ Biomed Res. 2024;8(9):1–8.
159. Panda P, Mohapatra R. Revolutionizing patient safety: machine learning and AI for the early detection of adverse drug reactions and drug-induced toxicity. Curr Artif Intell. 2024.
160. Khemani B, Malave S, Shinde S, Shukla M, Shikalgar R, Talwar H. AI-driven pharmacovigilance: enhancing adverse drug reaction detection with deep learning and NLP. MethodsX. 2025:103460.
161. Mastrantonas A, Kokkas P, Chatzopoulos A, Papoutsidakis M, Stergiou C, Vairis A, et al. Identifying the effects of Industry 4.0 in the pharmaceutical sector: achieving the sustainable development goals. Discov Sustain. 2024;5(1):460.
162. Ettaloui N, Arezki S, Gadi T. An overview of blockchain-based electronic health records and compliance with GDPR and HIPAA. Data Metadata. 2023;2:166.
163. Mak K-K, Pichika MR. Artificial intelligence in drug development: present status and future prospects. Drug Discov Today. 2019;24(3):773–80.
164. Ferreira FJ, Carneiro AS. AI-Driven Drug Discovery: A Comprehensive Review. ACS omega. 2025.
165. Sangaraju VV. AI and data privacy in healthcare: compliance with HIPAA, GDPR, and emerging regulations. Int J Emerg Trends Comput Sci Inf Technol. 2025;67–74.

166. Jeyaraman N, Jeyaraman M, Yadav S, Ramasubramanian S, Balaji S. Revolutionizing healthcare: the emerging role of quantum computing in enhancing medical technology and treatment. Cureus. 2024;16(8).
167. Mishra D, Mishra RK, Agarwal R. Recent trends in artificial intelligence and its applications. Artif Intell Trends Appl. 2024;1:73–106.
168. Boza P, Evgeniou T. Implementing AI principles: frameworks, processes, and tools. INSEAD Working Paper No. 2021/04/DSC/TOM; 2021.
169. Lin Z, Ma W, Lin T, Zheng Y, Ge J, Wang J, et al. Open source AI-based SE tools: opportunities and challenges of collaborative software learning. ACM Trans Softw Eng Methodol. 2025;34(5):1–24.
170. Mora-Cantallops M, Sánchez-Alonso S, García-Barriocanal E, Sicilia M-A. Traceability for trustworthy AI: a review of models and tools. Big Data Cogn Comput. 2021;5(2):20.
171. Pimenov DY, Bustillo A, Wojciechowski S, Sharma VS, Gupta MK, Kuntoğlu M. Artificial intelligence systems for tool condition monitoring in machining: analysis and critical review. J Intell Manuf. 2023;34(5):2079–121.
172. Wirtz BW, Müller WM. An integrated artificial intelligence framework for public management. Public Manag Rev. 2019;21(7):1076–100.
173. Celsa C, Quartararo A, Maida M, Giusino G, Gaudioso V, Sparacino A, et al. The application of artificial intelligence-based tools in the management of hepatocellular carcinoma: current status and future perspectives. Hepatoma Research. 2025;11:N/A-N/A.
174. Parashar K, Dev A, Kumar A, Khatri D. Enhancing Productivity in database management through AI: A three-phase approach for database. arXiv preprint arXiv:250409288. 2025.
175. Zhou X, Chai C, Li G, Sun J. Database meets artificial intelligence: a survey. IEEE Trans Knowl Data Eng. 2020;34(3):1096–116.
176. Oloruntoba O. AI-Driven autonomous database management: self-tuning, predictive query optimization, and intelligent indexing in enterprise it environments. World J Adv Res Rev. 2025;25(2):1558–80.
177. Deng J, Yang Z, Ojima I, Samaras D, Wang F. Artificial intelligence in drug discovery: applications and techniques. Brief Bioinform. 2022;23(1).
178. Muller C, Rabal O, Diaz Gonzalez C. Artificial intelligence, machine learning, and deep learning in real-life drug design cases. Artif Intell Drug Des. 2021;383–407.
179. Álvarez-Machancoses Ó, Fernández-Martínez JL. Using artificial intelligence methods to speed up drug discovery. Expert Opin Drug Discov. 2019;14(8):769–77.
180. Bhattacharya S, Czejdo B, Agrawal R, Erdemir E, Gokaraju B. Open source platforms and frameworks for artificial intelligence and machine learning. St. Petersburg, FL, USA: IEEE; 2018.
181. Hsu Y-Y, Clyne M, Wei C-H, Khoury MJ, Lu Z. Using deep learning to identify translational research in genomic medicine beyond bench to bedside. Database. 2019;2019:baz010.
182. Aliper A, Plis S, Artemov A, Ulloa A, Mamoshina P, Zhavoronkov A. Deep learning applications for predicting pharmacological properties of drugs and drug repurposing using transcriptomic data. Mol Pharm. 2016;13(7):2524–30.
183. Hamdi A, Colas P. Yeast two-hybrid methods and their applications in drug discovery. Trends Pharmacol Sci. 2012;33(2):109–18.
184. Askr H, Elgeldawi E, Aboul Ella H, Elshaier YA, Gomaa MM, Hassanien AE. Deep learning in drug discovery: an integrative review and future challenges. Artif Intell Rev. 2023;56(7):5975–6037.
185. Gangwal A, Lavecchia A. Artificial intelligence in natural product drug discovery: current applications and future perspectives. J Med Chem. 2025;68(4):3948–69.

Chapter 12
Future Horizons: Collaboration of Advanced Pharmaceutical Labs and Computational Sciences

Julia Werle, Bozena Hosnedlova, Ondrej Mitrovsky, Hoai Viet Nguyen, Arli Aditya Parikesit, Geir Bjorklund, Agnes Pholosi, Narayanan Vedha Hari, Warawan Eiamphungporn, Carlos Fernandez, Karel Kotaska, Eva Klapkova, Richard Prusa, and Rene Kizek

Abstract This chapter explores the accelerating transformation of pharmaceutical laboratories into intelligent analytical systems, where experimental science and computational innovation converge. Manual techniques, such as polyacrylamide gel electrophoresis, continue to provide accessible and versatile tools for example probing nanoparticle modifications and protein–drug interactions at the exploratory

---

SCOPUS ID of the principal investigator: https://www.scopus.com/authid/detail.uri?authorId=24376674300

J. Werle · O. Mitrovsky · K. Kotaska · E. Klapkova · R. Prusa · R. Kizek (✉)
2nd Faculty of Medicine, Department of Medical Chemistry and Clinical Biochemistry, Charles University, Motol and Homolka University Hospital, Prague, Czechia
e-mail: kizekrene@gmail.com

B. Hosnedlova · R. Kizek
Department of Research and Development, EcoNanoLife, Prague, Czech Republic

H. V. Nguyen
Research Center for Environmental Monitoring and Modeling, University of Science, Vietnam National University, Hanoi, Vietnam

A. A. Parikesit
Department of Biotechnology, School of Health and Life Sciences, i3L University, Jakarta Timur, Indonesia

G. Bjorklund
Council for Nutritional and Environmental Medicine, Mo i Rana, Norway

A. Pholosi
Department of Natural Sciences, Vaal University of Technology, Vanderbijlpark, South Africa

N. V. Hari
Pharmaceutical Technology Lab, School of Chemical and Biotechnology, SASTRA Deemed University, Thanjavur, India

W. Eiamphungporn
Faculty of Medical Technology, Department of Clinical Microbiology and Applied Technology, Mahidol University, Salaya, Nakhon Pathom, Thailand

A. A. Parikesit and A. N. M. Ansori (eds.), *Recent Computational Techniques in De Novo Drug Design*, Bio-IT and AI, https://doi.org/10.1007/978-3-032-20428-8_12

stage. Their limitations, however, have been surpassed by automated capillary electrophoresis, which offers high-throughput capacity, reproducibility, and clinical applicability, exemplified by its central role in large-scale hemoglobin variant screening and monitoring. The latest developments combine automation with artificial intelligence, spectral analysis, and multi-omics data, creating tools that can predict outcomes, optimize experiments in real time, and reveal complex biological patterns. These advances depend on the close integration of experimental and computational sciences, improving efficiency, reproducibility, and the overall pace of discovery. Beyond simple data analysis, modern systems now enable automated optimization, context-aware modeling, and real-time feedback within closed-loop laboratory environments. This convergence is redefining pharmaceutical innovation ensuring rigorous standards while accelerating the development of new therapies. Looking ahead, intelligent laboratories that unite analytical instruments and artificial intelligence are expected to become central to pharmaceutical research, delivering precision-guided and personalized treatments and marking a decisive shift toward truly data-driven, patient-centered medicine.

Keywords Laboratory automation · Drug discovery · New biomarkers · Cancer treatment · Nanomedicine

12.1 Introduction

Over the past two decades, biochemical laboratories have experienced a fundamental transformation, marked by the progressive integration of automation technologies into core experimental and analytical workflows [1]. This evolution reflects not only a shift in conventional analytical techniques and processes but a strategic reconfiguration of the research environment to smart analytical chemistry in response to increasing molecular complexity, regulatory rigor, and the imperative for accelerated innovation of portable and powerful platforms for modern scientific and industrial needs [2, 3]. Historically, drug discovery and development have predominantly relied on labor-intensive procedures, subject to variability, limited throughput, and susceptibility to human error. These constraints are increasingly incompatible with the contemporary demands of biochemical laboratories, particularly in the context of high-throughput screening, combinatorial chemistry, and the growing emphasis on data-driven decision-making. In this context, laboratory automation has emerged as an essential enabler of scalability, reproducibility, and efficiency. Modern automation systems now encompass a wide spectrum of technologies including robotic liquid handlers, integrated analytical platforms, microfluidic devices, and advanced process analytical technologies [4]. These systems are capable of executing highly

C. Fernandez
School of Pharmacy, Applied Sciences and Public Health, Robert Gordon University, Aberdeen, UK

complex, multi-step experimental protocols with exceptional precision. Results can be directly interfaced with laboratory information management systems and real-time monitoring tools, ensuring enhanced traceability and regulatory compliance. Automated laboratories generate standardized, high-quality datasets that serve as the foundation for computational modeling, machine learning, and in silico simulation. Moreover, by liberating laboratory technicians from repetitive manual tasks, automation promotes a shift toward higher-order problem-solving, hypothesis generation, and interdisciplinary collaboration. In this chapter, the integration of advanced automation technologies over the years and computational frameworks is examined. Furthermore, it redefines the work in biochemical laboratories through research and development and how it redefines the future work in biochemical laboratories and research and development [5].

12.2 Manual Analysis—Gel Electrophoresis Showing Nanoparticles Modified with Albumin and Doxorubicin

Manual electrophoretic techniques, particularly polyacrylamide gel electrophoresis (PAGE), continue to play a vital role in pharmaceutical and nanobiotechnological research. Their enduring relevance can be attributed to their methodological flexibility, affordability, and suitability for visually assessing molecular interactions and particle modifications. These methods are especially valuable in early-stage experimental design, where rapid screening and qualitative analysis are prioritized over automation. As an illustrative example of manual electrophoretic techniques in pharmaceutical research, this study presents the use of PAGE in conjunction with the colortest analysis (Fig. 12.2) to investigate the stepwise modification of silver nanoparticles (AgNPs) synthesized from *Artemisia absinthium* extract. Other details will be published elsewhere. Manual methods such as PAGE remain highly relevant in nanobiotechnological workflows due to their accessibility, low operational costs, and capacity for visually assessing molecular interactions and nanoparticle behavior in a laboratory setting without the need for automation. Nanotechnology has been widely used in medicine, biology, physics and chemistry [6–9]. The application of silver nanoparticles (AgNPs) in nanomedicine and other sciences have a growing tendency due to their desirable properties, including their high surface to volume ratio, catalytic activity and antimicrobial effects [10–16]. Serum albumin, the most abundant blood plasma protein is produced in the liver and plays an important role in regulating blood volume by maintaining the oncotic pressure of the blood compartment. Furthermore, it is a non-toxic and biodegrable drug-carrier, which can be used as a cytostatic carrier and photosensitizer [17, 18]. Doxorubicin (DOX) is an anthracycline antibiotic derived from *Streptomyces peucetius* and is widely used as a chemotherapeutic agent due to its potent antitumor activity [9, 19, 20]. It serves as a cornerstone drug in the treatment of numerous cancers, including leukemia, lymphoma, breast cancer, sarcoma, and other solid tumors [19]. Different possibilities of using gel electrophoresis for nanoparticle separation are presented in several articles [21–23] and the separation of AgNPs in terms of their shape and size has been elaborated in detail in the work of Hanauer et al. [24]. In this example, AgNPs were progressively functionalized with biologically active macromolecules as shown in Fig. 12.1. Bovine serum albumin (BSA), a model protein carrier, and doxorubicin (DOX), a widely used chemotherapeutic agent are utilized to simulate a modular drug delivery system [25–28]. The 10% acrylamide gel was employed to resolve differences in particle mobility, which reflect surface charge alterations and size changes associated with each functionalization step. This was complemented by UV-Vis spectral analysis to monitor nanoparticle aggregation, binding events, and changes in optical properties during the modification process [29, 30]. Together, these techniques exemplify how manual analytical tools can be effectively applied to evaluate complex nanoparticle systems in the early phases of formulation development, prior to the implementation of automated or high-throughput methods [31, 32].

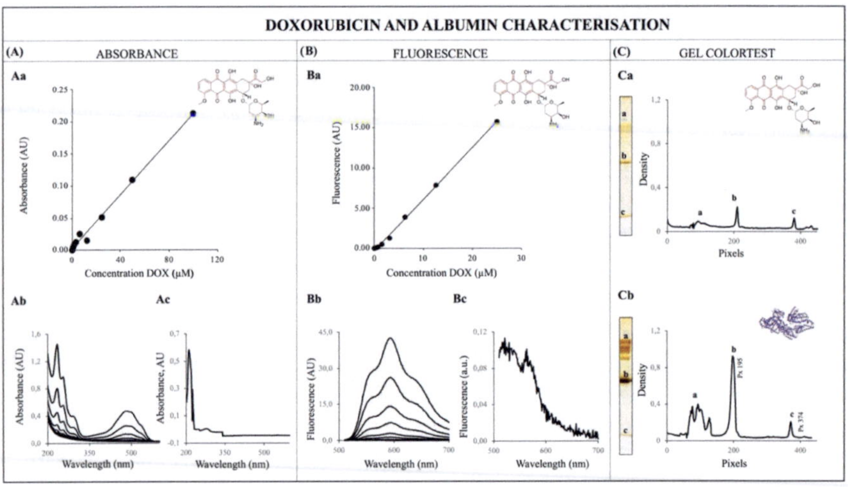

Fig. 12.1 Doxorubicin and albumin characterization. **A** (Aa) Absorbance calibration curve of doxorubicin. (Ab) Absorbance spectra of doxorubicin. (Ac) Absorbance spectra of albumin 50 μg/ml. **B** (Ba) Fluorescence calibration curve of doxorubicin. (Bb) Fluorescence spectra of doxorubicin. (Bc) Fluorescence spectra of albumin. **C** (Ca) Colortest analysis of doxorubicin in a 10% acrylamid gel. (Cb) Colortest analysis of gel alone. Gel (a) characteristic bands of BSA, (b) BSA band, similar to bands found in nanoparticles alone solution due to similar size of some proteins from the plants, (c) band indicating the end of the gel. Further details will be published

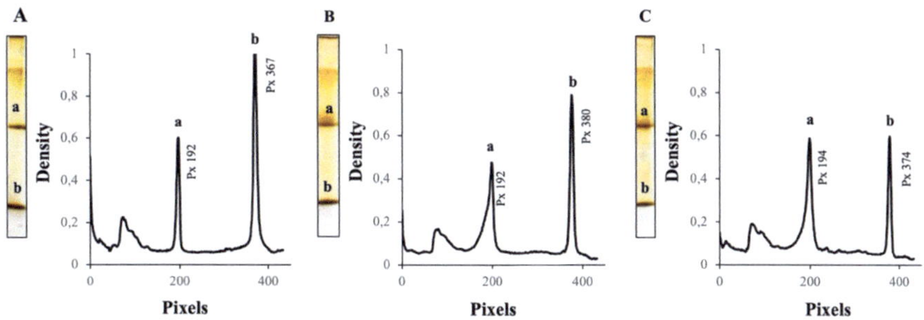

Fig. 12.2 Visualization of PAGE example electropherograms. Green synthesized silver nanoparticles from sagebrush. **A** AgNPs in extract dilution (1:16). **B** AgNPs treated 1 h with BSA in extract dilution (1:10). **C** AgNPs treated 24 h with BSA and 48 h with doxorubicin. Other details will be published elsewhere

12.3 From Manual to Automated: Capillary Electrophoresis in Modern Bioanalysis

While traditional gel electrophoresis remains a valuable tool in pharmaceutical and nanomaterials research, recent advances in automated capillary electrophoresis (CE) have transformed the analytical capabilities [33] available to modern laboratories. CE platforms offer high-resolution, high-throughput separation of charged biomolecules under strictly controlled conditions, with minimal sample volumes and superior reproducibility.

A particularly impactful application of automated CE is the detection and quantification of hemoglobin (Fig. 12.3) variants in clinical and translational settings [34, 35]. State-of-the-art instruments such as capillary zone electrophoresis systems enable the precise separation of hemoglobin A, A2, F, S, C, and other clinically relevant variants based on their charge-to-mass ratios [35–39]. These platforms are fully automated, incorporating sample injection, separation, detection (typically via UV absorbance at 415 nm), and integrated data analysis with standardized migration profiles. Compared to manual electrophoresis, automated CE systems minimize operator bias, reduce analysis time, and offer traceability compliant with clinical quality systems. Their application spans from neonatal screening and thalassemia diagnostics to monitoring of acquired hemoglobinopathies in patients undergoing drug treatment or transfusion therapy. Those advanced diagnostic tools are crucial for accurate diagnosis and may be relevant for patient management. As such, CE exemplifies the broader shift toward precision, scalability, and digital integration in pharmaceutical and biomedical analytics [33, 40].

Since the introduction of hemoglobin analysis in Motol Universtity Hospital in Prague, Czech Republic by CE in 2017, 139,002 blood samples have been analyzed between 2015 and 2025, including those targeted for HbF ($n = 172$) and HbA_2 ($n = 92$) as shown in Fig. 12.4. Between the years 2020 and 2025, a total of 88,010 blood

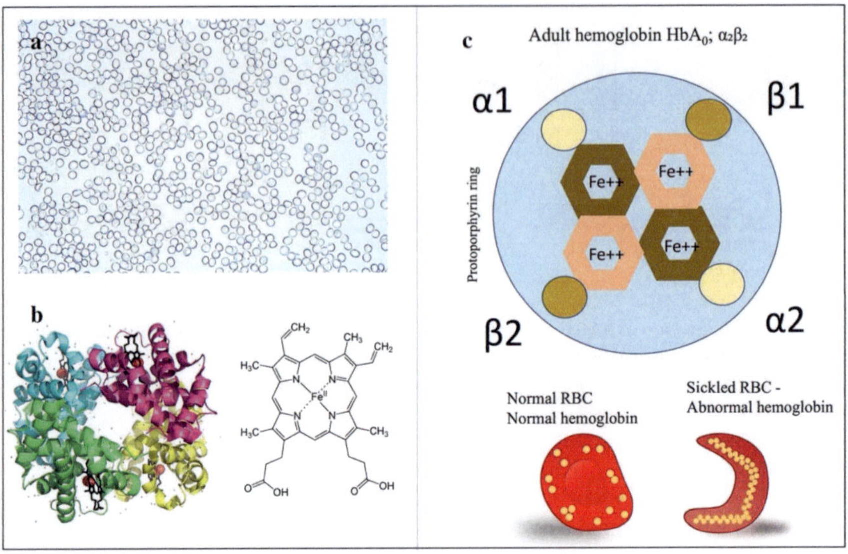

Fig. 12.3 Adult Human Hemoglobin and Red Blood Cells. Physiological function of erythrocytes linked to the primary and secondary structure of hemoglobin chains. **a** Microscopic image of a peripheral blood smear from a subject with abnormal hemoglobin (400×). **b** Structural model of hemoglobin highlighting the heme group and bound iron. **c** Schematic arrangement of globin chains and resulting abnormal erythrocyte morphology due to a defect in the primary structure

samples were subjected to routine analysis for HbA_{1c} using CE. As reported by Strickland et al. [41], the system Capillarys 2 Flex Piercing from Sebia enabled the detection of rare hemoglobin variants during HbA_{1c} measurement. Over a period of 14 months, 26,850 probands were evaluated, and 36 samples with atypical profiles were identified. Of these, 17 were evaluated as actual hemoglobinopathies, representing a prevalence of 0.06% [41]. This is consistent with the proportion of hemoglobin variants identified in our study through routine HbA_{1c} measurement.

Hemoglobin is a tetrameric protein facilitating the reversible binding and transport of molecular oxygen within erythrocytes as shown in Fig. 12.3. In neonates, the predominant isoform is fetal hemoglobin (HbF), composed of two α-globin and two γ-globin subunits ($\alpha_2\gamma_2$). Postnatally, it is gradually replaced by adult hemoglobin. In healthy adults, the major form is hemoglobin A0 (HbA_0; $\alpha_2\beta_2$), which constitutes approximately 97% of total hemoglobin. A smaller proportion is made up of hemoglobin A2 (HbA_2; $\alpha_2\delta_2$), typically accounting for 2.5–3.0%. Hemoglobinopathies are a heterogeneous group of genetic disorders affecting the structure or synthesis of globin chains, and are broadly classified into two categories: (i) structural variants, which arise from single-point mutations or small deletions leading to amino acid substitutions or truncated globin chains; and (ii) thalassemia syndromes, which are characterized by partial or complete suppression of α- or β-globin chain synthesis due to mutations in the corresponding globin genes, resulting in imbalanced chain production, ineffective erythropoiesis, and varying degrees

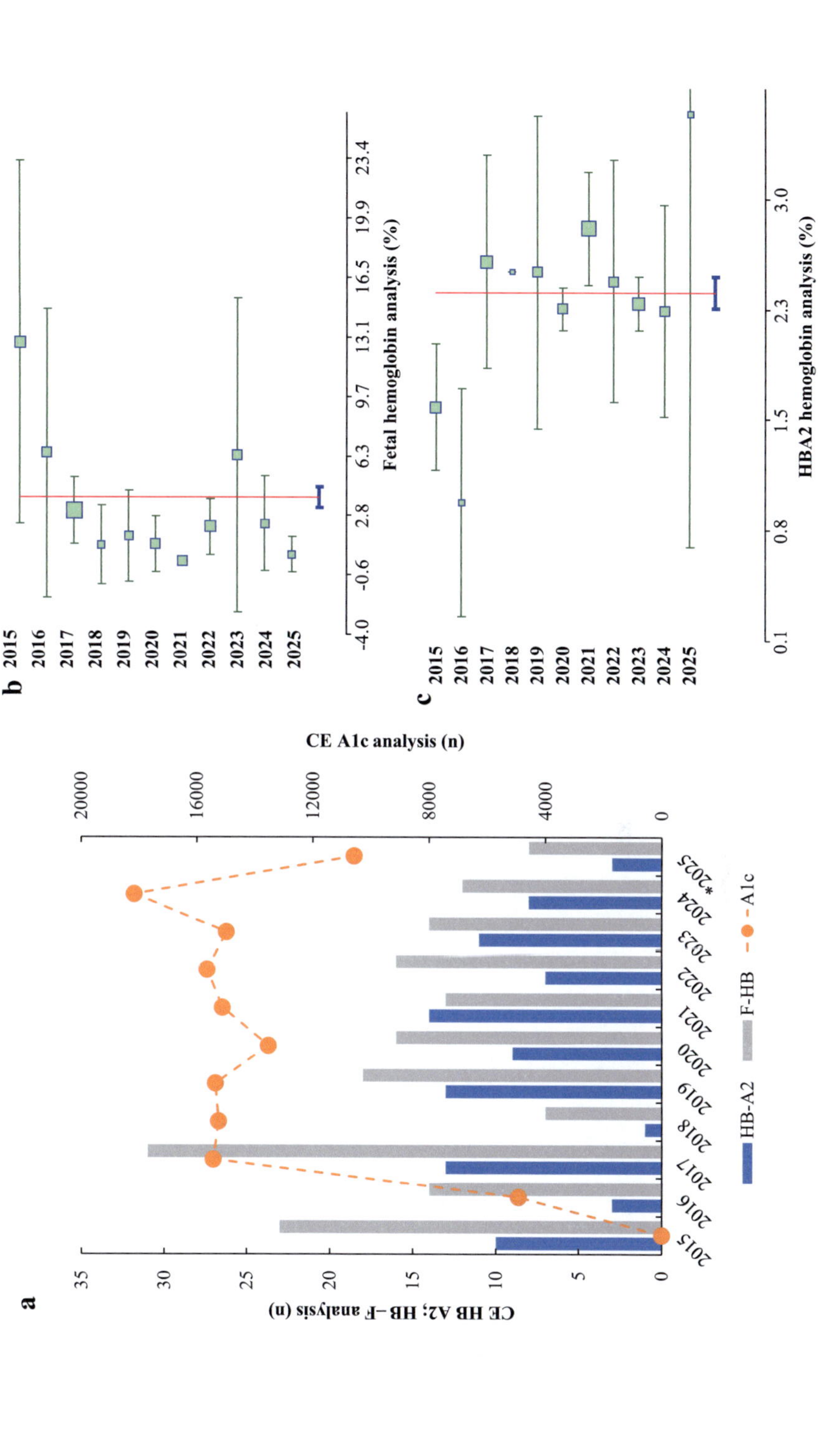

Fig. 12.4 Hemoglobin Examinations using CE at the Motol University Hospital, Prague. Presents the trends in the number of hemoglobin examinations using CE at the Motol University Hospital (2015–2025); HbA_2, F-Hb, and $A1_c$ (**a**). Forest plot analysis of fetal hemoglobin (**b**) and HbA_2 hemoglobin using CE (2015–2025) (**c**)

of anemia [42, 43]. Hemoglobinopathies constitute one of the most widespread groups of inherited disorders globally, with an estimated 7% of the population being affected [44, 45]. Among the most clinically consequential mutations of the β-globin gene is the point mutation resulting in altered amino acid sequences responsible for hemoglobin S (HbS). Under deoxygenated conditions, HbS molecules tend to aggregate resulting in distortion of erythrocyte morphology into a rigid, sickle-shaped conformation. These structurally compromised red blood cells exhibit impaired deformability, reduced lifespan, and a propensity to obstruct the microvasculature, thereby precipitating the hallmark features of sickle cell disease (SCD), including recurrent vaso-occlusive crises, chronic hemolytic anemia, and end-organ damage. Routinely and most commonly CE is employed for the quantitative analysis of glycated hemoglobin ($HbA1_c$), providing a reliable estimate of average blood glucose concentrations over the preceding 8–12 weeks.

For visualization of CE electropherogram examples of normal, glycated, and fetal and HbS hemoglobin samples are shown in Fig. 12.5. A typical electropherogram in a healthy proband is shown in Fig. 12.2A where HbA_{1c} (36 mmol/mol; 5.5%), HbA_0 (91.3%) and HbA_2 (2.0%) signals are very well resolved. Figure 12.5B shows a record of a diabetes patient with an increased HbA_{1c} value of 169 mmol/mol (17.6%) whereas other fractions are HbA_0 (74.8%) and HbA_2 (1.5%). In Fig. 12.5C fetal hemoglobin of a one-day old newborn with HbF of 72.6% is depicted.

To exclude falsified HbA_{1c} measurement possibly influencing patient treatment it is furthermore important to recognize hemoglobin variants [41]. For illustrative

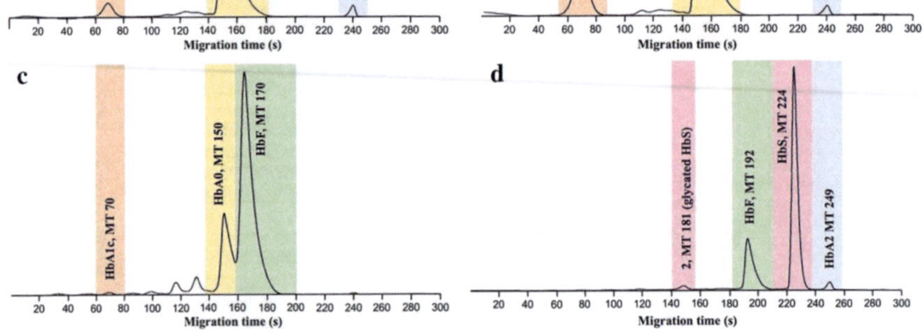

Fig. 12.5 Visualization of example electropherograms. Depicts typical electropherograms. **a** Shows a healthy proband, where $HbA1_c$, HbA, HbA_0 and HbA_2 signals are very well resolved. **b** Shows a record of a typical glycated hemoglobin with $HbA1_c$ of 169 mmol/mol. In **c** CE with visible fetal hemoglobin (72.6%, MT 170) is depicted from a one day-old newborn. **d** Shows HbS with HbS-β^0-thalassemia with RBC (2.45×10^{12}/L) Hb (102 g/L)

purposes, an electropherogram demonstrating the presence of the hemoglobin S (HbS) variant is shown in Fig. 12.5D. It presents the capillary electrophoresis profile of an 18-year-old female patient from Kenya with a confirmed diagnosis of sickle cell disease (SCD), who has been undergoing treatment with hydroxyurea since 2019. The CE shows a prominent HbF peak of 22.9% at MT 192s, HbS peak of 73.2% at MT 221s, and HbA_2 peak of 2.2% at MT 249s. This pattern is typical in the case of HbS-β^0-thalassemia [46, 47] which is also characterized by altered migration times. Hydroxyurea therapy is used in sickle cell patients to increase HbF since HbF is able to inhibit polymerization of HbS and thus prevents painful sickling crisis [48]. From an analytical perspective HbS does not interfere with the HbA_{1c} determination. Nevertheless, due to the shortened erythrocyte survival time, the HbA_{1c} value is lower resulting in an overestimation of the quality of diabetes control.

12.4 Future Outlook: Automatization, AI-Enhanced Spectral Evaluation, and Multi-omics Integration

The analytical sciences in pharmaceutical research are undergoing a progressive transformation, moving from manual techniques, through automated high-throughput systems, toward intelligent and self-optimizing analytical platforms. While the previously discussed example of manual gel electrophoresis demonstrates the adaptability and accessibility of traditional laboratory techniques, automated capillary electrophoresis (CE) systems already provide improved throughput, reproducibility, and standardization. The next decisive leap will be achieved through integration of automation, artificial intelligence (AI), and multi-omics data analysis [49–55].

The transition toward such systems is captured in Fig. 12.6, which illustrates the concept of a closed-loop automated laboratory. In this model, experimental workflows begin with data collection for example from electropherograms, UV-Vis spectra, fluorescence profiles, or other analytical modalities which is immediately processed by AI algorithms capable of pattern recognition, trend analysis, and predictive modeling [56–59]. This learn/design phase informs optimized protocols, which are executed by automated robotic platforms in the build phase. The generated products, such as nanoparticle formulations or purified biomolecules, undergo biological analysis to assess functional performance. The resulting data feed back into the AI system for evaluation in the test phase, completing the optimization loop. This cycle operates continuously, reducing human intervention, accelerating iteration rates, and ensuring consistent quality across experiments.

In electrophoresis and related spectroscopic applications spanning such as: ultraviolet-visible (UV-Vis), Raman, infrared (IR), nuclear magnetic resonance (NMR), and mass spectrometry, AI is increasingly capable of detecting subtle, multidimensional features in complex datasets that are often imperceptible through manual interpretation [58–60]. Machine learning algorithms not only enable faster

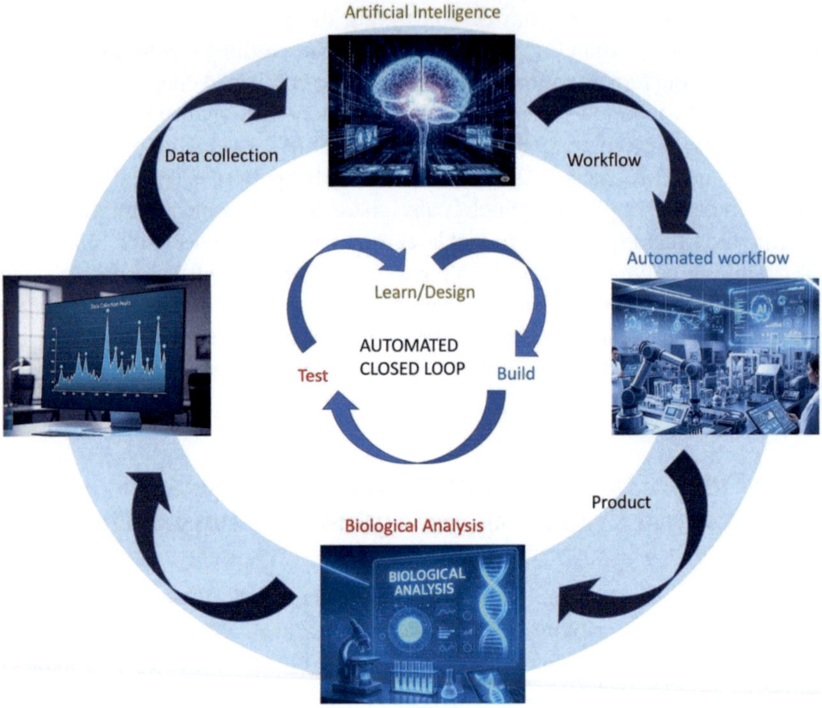

Fig. 12.6 Laboratory automation. A schematic representation of a fully automated and roboticized laboratory workflow utilizing optimization algorithms. Generated big data can be furthermore analyzed using artificial intelligence computational algorithms

classification and more accurate quantification but also allow forward prediction of molecular interactions and structure–function relationships, shifting analysis from a purely retrospective exercise to a proactive, design-driven process. The potential of AI expands further when paired with multi-omics integration linking genomics, transcriptomics, proteomics, and metabolomics data into unified analytical frameworks. This holistic approach connects electrophoretic and spectral signatures to underlying biological networks, enabling the discovery of biomarkers, mechanistic pathways, and therapeutic targets that single-modality analyses cannot reveal. The incorporation of explainable AI ensures that these models remain transparent and scientifically verifiable, which is essential for regulatory acceptance and clinical translation [61]. Emerging AI frameworks are also designed to operate across biological scales, linking genotype, environmental context, and phenotype to provide causal, rather than merely correlative, insights [56]. Such capabilities are already accelerating drug discovery pathways from target identification to drug repurposing by integrating high-content spectral data with multi-omics analyses [61, 62]. As these technologies mature, the laboratory will evolve into an intelligent analytical ecosystem: a setting where automated systems handle execution, AI algorithms guide decision-making, and human expertise focuses on innovation and interpretation. This progression, from manual electrophoresis, to automated CE, to AI-integrated, closed-loop automation, will redefine the pace and scope of pharmaceutical innovation, making laboratory workflows more predictive, adaptive, and capable of driving precision medicine forward see in Fig. 12.6.

12.5 Conclusion

The progression from manual electrophoresis, through automated capillary electrophoresis, to the integration of AI-driven spectral evaluation and multi-omics analysis reflects a continuous and accelerating transformation in analytical laboratory practice. Each stage represents a qualitative leap: manual techniques offering flexibility and foundational insight; automated systems providing reproducibility, standardization, and efficiency; and AI-enhanced approaches enabling predictive, adaptive, and context-aware interpretation of complex datasets. The next generation of analytical platforms will go beyond sample processing to actively inform experimental design, optimize workflows in real time, and reveal multidimensional biological relationships that were previously inaccessible. This evolution is not only technological marking a paradigm shift toward data-centric, precision-guided biomedical discovery, with the potential to redefine how pharmaceuticals are developed, validated, and brought to clinical application.

Acknowledgments The project is being implemented under the grant project LPR Prague. The authors would like to express their sincere gratitude to our collaborators, Dagmar Uhlířová, Dr. Branislav Ruttkay-Nedecký, Dr. Karel Sehnal, Luděk Melich and Martina Staňková, for their valuable assistance with the laboratory work.

Conflicts of Interest Statement The authors have declared no conflicts of interest.

References

1. Kumar M, Nandi A, Yadav RL, Gupta GD, Sharma K. AI-enhanced prediction tools and sensor integration in advanced analytical chemistry techniques. Curr Anal Chem. 2025.
2. Hussain CM, Hussain G, Keçili R. Smart analytical chemistry: integrating green, sustainable, white and AI-driven approaches in modern analysis. Trac-Trends Anal Chem. 2025;191
3. Khan WA, Fallah SAH, Moltajihagh S, Hamdi F, Alipour F, Mollahosseini A, Boczkaj G. Automation and high throughput sample analysis with various platforms in microextraction techniques: a need for ecofriendly, green, and cost-effective sample preparation approaches—a review. Trac-Trends Anal Chem. 2025;189
4. Hosnedlova B, Werle J, Cepova J, Narayanan VHB, Vyslouzilova L, Fernandez C, Parikesit AA, Kepinska M, Klapkova E, Kotaska K, Stepankova O. et al. Electrochemical sensors and biosensors for identification of viruses: a critical review. Crit Rev Anal Chem. 2024.

5. Cuadrado MU, de Castro MDL, Gómez-Nieto MA. Object-oriented techniques for design and development of standard software solutions in automation and data management in analytical chemistry. Trac-Trends Anal Chem. 2006;25(1):66–76.
6. Barabadi H, Ovais M, Shinwari ZK, Saravanan M. Anti-cancer green bionanomaterials: present status and future prospects. Green Chem Lett Rev. 2017;10(4):285–314.
7. Sharma R, Dhillon A, Kumar D. Mentha-stabilized silver nanoparticles for high-performance colorimetric detection of Al(III) in aqueous systems. Sci Rep. 2018;8:13.
8. Sun YG, Xia YN. Shape-controlled synthesis of gold and silver nanoparticles. Science. 2002;298(5601):2176–9.
9. Blazkova I, Nguyen HV, Dostalova S, Kopel P, Stanisavljevic M, Vaculovicova M, Stiborova M, Eckschlager T, Kizek R, Adam V. Apoferritin modified magnetic particles as doxorubicin carriers for anticancer drug delivery. Int J Mol Sci. 2013;14(7):13391–402.
10. de Souza TAJ, Souza LRR, Franchi LP. Silver nanoparticles: an integrated view of green synthesis methods, transformation in the environment, and toxicity. Ecotox Environ Safe. 2019;171:691–700.
11. Hosnedlova B, Kabanov D, Kepinska M, Narayanan VHB, Parikesit AA, Fernandez C, Bjorklund G, Nguyen HV, Farid A, Sochor J, Pholosi A, et al. Effect of biosynthesized silver nanoparticles on bacterial biofilm changes in *S. aureus* and *E. coli*. Nanomaterials. 2022;12(13)
12. Sehnal K, Ly PT, Uhlirova D, Dung ML, Stankova M, Xuan TT, Vsetickova M, Anh NN, Hosnedlová B, Kepinská M, Ruttkay-Nedecky B., et al. Green synthesis of silver nanoparticles using plants originating in tropical areas of Vietnam for biological applications. In: 11th international conference on nanomaterials—research & application (nanocon 2019). 2020. Abs 463–8.
13. Krishnan PD, Banas D, Durai RD, Kabanov D, Hosnedlova B, Kepinska M, Fernandez C, Ruttkay-Nedecky B, Nguyen HV, Farid A, Sochor J, et al. Silver nanomaterials for wound dressing applications. Pharmaceutics. 2020;12(9)
14. Gargulák M, Cízek M, Docekalová M, Kepinska M, Da Silva AF, Surovíková A, Uhlírová D, Stanková M, Ruttkay-Nedecky B, Vavriník A, Sochor J., et al. Theranostic tool for targeting of bacterial cells based on silver nanoparticles. In: 10th anniversary international conference on nanomaterials—research & application (nanocon 2018 (r)). 2019. Abs 370–5.
15. Ruttkay-Nedecky B, Skalickova S, Kepinska M, Cihalova K, Docekalova M, Stankova M, Uhlirova D, Fernandez C, Sochor J, Milnerowicz H, Beklova M, et al. Development of new silver nanoparticles suitable for materials with antimicrobial properties. J Nanosci Nanotechnol. 2019;19(5):2762–9.
16. Chudobova D, Nejdl L, Gumulec J, Krystofova O, Rodrigo MAM, Kynicky J, Ruttkay-Nedecky B, Kopel P, Babula P, Adam V, Kizek R. Complexes of silver(I) ions and silver phosphate nanoparticles with hyaluronic acid and/or chitosan as promising antimicrobial agents for vascular grafts. Int J Mol Sci. 2013;14(7):13592–614.
17. Zheng K, Liu H, Wang Y, Li LP, Li S, Xue J, Huang M. Tumor targeting chemo- and photodynamic therapy packaged in albumin for enhanced anti-tumor efficacy. Int J Nanomedicine. 2020;15:151–67.
18. Farrugia A. Albumin usage in clinical medicine: tradition or therapeutic? Transf Med Rev. 2010;24(1):53–63.
19. Kizek R, Adam V, Hrabeta J, Eckschlager T, Smutny S, Burda JV, Frei E, Stiborova M. Anthracyclines and ellipticines as DNA-damaging anticancer drugs: recent advances. Pharmacol Ther. 2012;133(1):26–39.
20. Gumulec J, Fojtu M, Raudenska M, Sztalmachova M, Skotakova A, Vlachova J, Skalickova S, Nejdl L, Kopel P, Knopfova L, Adam V, et al. Modulation of induced cytotoxicity of doxorubicin by using apoferritin and liposomal cages. Int J Mol Sci. 2014;15(12):22960–77.
21. Ohshima H. Gel electrophoresis of a soft particle. Adv Colloid Interf Sci. 2019;271:16.
22. Mage PL, Csordas AT, Brown T, Klinger D, Eisenstein M, Mitragotri S, Hawker C, Soh HT. Shape-based separation of synthetic microparticles. Nat Mater. 2019;18(1):82.
23. Robertson JD, Rizzello L, Avila-Olias M, Gaitzsch J, Contini C, Magon MS, Renshaw SA, Battaglia G. Purification of nanoparticles by size and shape. Sci Rep. 2016;6:9.

24. Hanauer M, Pierrat S, Zins I, Lotz A, Sonnichsen C. Separation of nanoparticles by gel electrophoresis according to size-and shape. Nano Lett. 2007;7(9):2881–5.
25. Zareba N, Wieclawik K, Kizek R, Hosnedlova B, Kepinska M. The impact of fullerenes as doxorubicin nano-transporters on metallothionein and superoxide dismutase status in mcf-10a cells. Pharmaceutics. 2022;14(1)
26. Skalickova S, Docekalova M, Stankova M, Uhlirova D, Ruttkay-Nedecky B, Fernandez C, Kepinska M, Milnerowicz H, Kizek R. Zinc modified nanotransporter of anticancer drugs for targeted therapy: biophysical analysis. J Nanosci Nanotechnol. 2019;19(5):2483–8.
27. Kepinska M, Kizek R, Milnerowicz H. Metallothionein and superoxide dismutaseantioxidative protein status in fullerene-doxorubicin delivery to mcf-7 human breast cancer cells. Int J Mol Sci. 2018;19(10)
28. Blazkova I, Smerkova K, Blazkova L, Vaculovicova M, Stiborova M, Eckschlager T, Beklova M, Adam V, Kizek R. Doxorubicin interactions with bovine serum albumin revealed by microdialysis with on-line laser-induced fluorescence detection at subpicogram level. Electrophoresis. 2015;36(11–12):1282–8.
29. Skalickova S, Loffelmann M, Gargulak M, Kepinska M, Docekalova M, Uhlirova D, Stankova M, Fernandez C, Milnerowicz H, Ruttkay-Nedecky B, Kizek R. Zinc-modified nanotransporter of doxorubicin for targeted prostate cancer delivery. Nano. 2017;7(12)
30. Heger Z, Kominkova M, Cernei N, Krejcova L, Kopel P, Zitka O, Adam V, Kizek R. Fluorescence resonance energy transfer between green fluorescent protein and doxorubicin enabled by DNA nanotechnology. Electrophoresis. 2014;35(23):3290–301.
31. Kepinska M, Kizek R, Milnerowicz H. Fullerene as a doxorubicin nanotransporter for targeted breast cancer therapy: capillary electrophoresis analysis. Electrophoresis. 2018;39(18):2370–9.
32. Konecna R, Nguyen HV, Stanisavljevic M, Blazkova I, Krizkova S, Vaculovicova M, Stiborova M, Eckschlager T, Zitka O, Adam V, Kizek R. Doxorubicin encapsulation investigated by capillary electrophoresis with laser-induced fluorescence detection. Chromatographia. 2014;77(21–22):1469–76.
33. Krebs F, Zagst H, Stein M, Ratih R, Minkner R, Olabi M, Hartung S, Scheller C, Lapizco-Encinas BH, Sanger-van de Griend C, Garcia CD, et al. Strategies for capillary electrophoresis: method development and validation for pharmaceutical and biological applications-updated and completely revised edition. Electrophoresis. 2023;44(17–18):1279–341.
34. Andleeb F, Hafeezullah, Atiq A, Atiq M. Hemoglobin structure at higher levels of hemoglobin a1c in type 2 diabetes and associated complications. Chin Med J. 2020;133(10):1138–43.
35. Castagnola M, Messana I, Cassiano L, Rabino R, Rossetti DV, Giardina B. The use of capillary electrophoresis for the determination of hemoglobin-variants. Electrophoresis. 1995;16(8):1492–8.
36. Jenkins MA, Hendy J, Smith IL. Evaluation of hemoglobin A(2) quantitation assay and hemoglobin variant screening by capillary electrophoresis. J Capillary Electrophor. 1997;4(3):137–43.
37. Shihabi ZK. Effect of sample composition on electrophoretic migration—application to hemoglobin analysis by capillary electrophoresis and agarose electrophoresis. J Chromatogr A. 2004;1027(1–2):179–84.
38. Shihabi ZK, Hinsdale ME. Simplified hemoglobin chain detection by capillary electrophoresis. Electrophoresis. 2005;26(3):581–5.
39. Shihabi ZK, Hinsdale ME, Daugherty HK. Hemoglobin A(2) quantification by capillary zone electrophoresis. Electrophoresis. 2000;21(4):749–52.
40. Werle J, Dunovska K, Podhajsky J, Cerny M, Cepova J, Parikesit AA, Bjorklud G, Kotaska K, Klapkova E, Prusa R, Werle E. et al. Characterization of hemoglobin variants by capillary electrophoresis, UV-Vis, and FTIR spectroscopy. Electrophoresis. 2025.
41. Strickland SW, Campbell ST, Little RR, Bruns DE, Bazydlo LAL. Recognition of rare hemoglobin variants by hemoglobin A_{1c} measurement procedures. Clin Chim Acta. 2018;476:67–74.
42. Thom CS, Dickson CF, Gell DA, Weiss MJ. Hemoglobin variants: biochemical properties and clinical correlates. Cold Spring Harb Perspect Med. 2013;3(3)

43. Khaled SAA, Ahmed HA, Elbadry MI, NasrEldin E, Hassany SM, Ahmed SA. Hematological, biochemical properties, and clinical correlates of hemoglobin s variant disorder: a new insight into sickle cell trait. J Hematol. 2022;11(3):92–108.
44. Giardine B, Borg J, Higgs DR, Peterson KR, Philipsen S, Maglott D, Singleton BK, Anstee DJ, Basak AN, Clark B, Costa FC, et al. Systematic documentation and analysis of human genetic variation in hemoglobinopathies using the microattribution approach. Nat Genet. 2011;43(4):295–302.
45. Kirk CM, Papadea CN, Lazarchick J. Laboratory recognition of a rare hemoglobinopathy: hemoglobins SS and SG(Philadelphia) associated with α-thalassemia-2. Arch Pathol Lab Med. 1999;123(10):963–6.
46. Kohne E. Compendium of hemoglobinopathies. SEBIA Educational Library; 2019.
47. Chen Z, Shao L, Jiang M, Ma R, Ping H, Wang Z, Ba X, Ma B, Zhou T. Identification of a rare compound heterozygous hemoglobin variant β0-thal [β17(A14) lys > stop, HBB: C.52a > t] and Hb J-Lome [β59(E3) Lys > Asn, HBB: C.180G > C]. Hematology. 2022;27(1):946–50.
48. Cotton F, Lin C, Fontaine B, Gulbis B, Janssens J, Vertongen F. Evaluation of a capillary electrophoresis method for routine determination of hemoglobins A2 and f. Clin Chem. 1999;45(2):237–43.
49. Cardillo M, Katam K, Suravajhala P. Advancements in multi-omics research to address challenges in alzheimer's disease: a systems biology approach utilizing molecular biomarkers and innovative strategies. Front Aging Neuroscience. 2025;17
50. Wang Z, Zhou GC, Cao RC, Zhang GL, Zhang YX, Xiao MY, Liu LB, Zhang XS. Harnessing multi-omics and artificial intelligence: revolutionizing prognosis and treatment in hepatocellular carcinoma. Front Immunol. 2025;16
51. Selby DA, Jakhmola R, Sprang M, Grossmann G, Raki H, Maani N, Pavliuk D, Ewald J, Vollmer S. Visible neural networks for multi-omics integration: a critical review. Front Artif Intell. 2025;8
52. Elfatimi E, Lekbach Y, Prakash S, BenMohamed L. Artificial intelligence and machine learning in the development of vaccines and immunotherapeutics-yesterday, today, and tomorrow. Front Artif Intell. 2025;8
53. Chen WQ, Zhang PZ, Tran TN, Xiao YW, Li SY, Shah VV, Cheng H, Brannan KW, Youker K, Lai L, Fang LH, et al. A visual-omics foundation model to bridge histopathology with spatial transcriptomics. Nat Methods. 2025;22(7)
54. Ma SH, Zeng AGX, Haibe-Kains B, Goldenberg A, Dick JE, Wang B. Moving towards genome-wide data integration for patient stratification with integrate any omics. Nat Mach Intell. 2025;7(1)
55. Vinuesa CG, He YK, Cook MC. A global metric of immune health. Nat Med. 2024;30(9):2411–2.
56. Gohr C, Rodríguez G, Belomestnykh S, Berg-Moelleken D, Chauhan N, Engler JO, Heydebreck LV, Hintz MJ, Kretschmer M, Kruegermeier C, Meinberg J., et al. Artificial intelligence in sustainable development research. Nat Sustain. 2025.
57. Xue BS, Bi XY, Dong ZY, Xu YZ, Liang MH, Fang X, Yuan YZ, Wang RX, Liu SY, Jiao RS, Chen YZ, et al. Deep spectral component filtering as a foundation model for spectral analysis demonstrated in metabolic profiling. Na Mach Intell. 2025;7(5)
58. Yu YH, Li M. Towards highly sensitive deep learning-based end-to-end database search for tandem mass spectrometry. Nat Mach Intell. 2025;7(1)
59. Zhang ZY, Du Y, Shi XJ, Wang K, Qu QJ, Liang Q, Ma XP, He KS, Chi CW, Tang JQ, Liu B, et al. Nir-ii light in clinical oncology: opportunities and challenges. Nat Rev Clin Oncol. 2024;21(6):449–67.
60. Manu VS, Olivieri C, Veglia G. AI-designed NMR spectroscopy RF pulses for fast acquisition at high and ultra-high magnetic fields. Nat Commun. 2023;14(1)
61. Mullowney MW, Duncan KR, Elsayed SS, Garg N, van der Hooft JJJ, Martin NI, Meijer D, Terlouw BR, Biermann F, Blin K, Durairaj J, et al. Artificial intelligence for natural product drug discovery. Nat Rev Drug Discov. 2023;22(11):895–916.

62. Liu G, Catacutan DB, Rathod K, Swanson K, Jin W, Mohammed JC, Chiappino-Pepe A, Syed SA, Fragis M, Rachwalski K, Magolan J, et al. Deep learning-guided discovery of an antibiotic targeting *Acinetobacter baumannii*. Nat Chem Biol. 2023;19(11):1342–50.

MIX
Papier aus verantwortungsvollen Quellen
Paper from responsible sources
FSC® C105338

If you have any concerns about our products,
you can contact us on
ProductSafety@springernature.com

In case Publisher is established outside the EU,
the EU authorized representative is:
Springer Nature Customer Service Center GmbH
Europaplatz 3, 69115 Heidelberg, Germany

Printed by Libri Plureos GmbH
in Hamburg, Germany